ISLAND WARRIORS

I don't know what effect these men will have upon the enemy, but, by God, they frighten me.

The Duke of Wellington

This story shall the good man teach his son,
And Crispin Crispian shall ne'er go by,
From this day to the ending of the world
But we in it shall be remembered—
We few, we happy few, we band of brothers;
For he today that sheds his blood with me
Shall be my brother; be he ne'er so vile,
This day shall gentle his condition;
And gentlemen in England now abed
Shall think themselves accursed they were not here,
And hold their manhoods cheap whiles any speaks
That fought with us upon Saint Crispin's day.

Shakespeare; *Henry V*, Act IV, Scene iii

ISLAND WARRIORS

A MILITARY ODYSSEY AROUND BRITAIN

JOHN SADLER
AND CAPTAIN GRAHAM TRUEMAN

AMBERLEY

First published 2024

Amberley Publishing
The Hill, Stroud
Gloucestershire, GL5 4EP

www.amberley-books.com

British Library Cataloguing in Publication Data.
A catalogue record for this book is available from the British Library.

ISBN 978 1 3981 1479 1 (hardback)
ISBN 978 1 3981 1480 7 (ebook)

1 2 3 4 5 6 7 8 9 10

Typesetting by SJmagic DESIGN SERVICES, India.
Printed in the UK.

Contents

Acknowledgements

Thanks are due to many who have assisted in the writing of this book. Katie Bole at Culloden Battlefield Centre, Inverness; Steve Erskine at Green Howards Museum, Richmond; Bobbi Goldwater at Charge Gallery, Discovery Museum; Ian Martin at KOSB Museum, Berwick Barracks; Diane Stevens at Heugh Battery, Hartlepool; Jules Wooding and Jess Latham at Cumbria's Museum of Military Life, Carlisle; Phil Abramson of MoD Archaeology; Paul Reed; Katherine Neil, Eilidh Young and Freya Samuel from the Highlanders' Museum; Stuart Kennedy of the Royal Scots Dragoon Guards Museum; Vanessa Martin at West Highland Museum; Colonel Nick Lock OBE of the Royal Welch Fusiliers Museum; Matthew Holden from the Soldiers of Gloucestershire Museum; Monica Ramirez Lopez of the Devil's Porridge Museum; Andy Jepson, Stobs Camp, Hawick; Major Paul Knight of the Royal Signals; David Kenyan of Bletchley Park, Milton Keynes; Andrew Whitmarsh of the D-Day Story with Nathalie Boulier of Portsmouth City Council; Dr Daren Bowyer of the Gurkha Museum Trust; Rebecca Blackburn from the Royal Engineers Museum, Chatham; Chris Balmer of Bentwaters Cold War Museum; Ben Hill of the Parachute Regiment Museum with Stephen MacHaye from Norfolk Tank Museum.

I also owe a debt for help and advice over the years to Steve Shannon (formerly of DLI Museum); Peter Hart and Emma O'Brien of the Imperial War Museum; Richard Groocock of the National Archives; Amy Cameron of National Army Museum; staff of Durham County Record Office; the staff of Northumberland County Archives at Woodhorn; staff and trustees of the Fusiliers Museum of

Northumberland, Alnwick; to former colleagues at the North East Centre for Lifelong Learning at the University of Sunderland; staff at the Literary and Philosophical Library, Newcastle; staff at Central Libraries, Newcastle and Gateshead, Clayport Library Durham and Northumberland Libraries at Morpeth, Alnwick, Blyth, Hexham and Cramlington; Lindsey and Colin Durward of Blyth Battery, Blyth, Northumberland; David Fletcher of the Tank Museum Bovington; Rod Mackenzie of the Argyll and Sutherland Highlanders' Museum; Thomas B. Smyth of the Black Watch Museum; Paul Evans of the Royal Artillery Museum; Natalie Worthington from Centre for Juno Beach; Mark Worthington of the Memorial Pegasus; Sir David Kelly; the late Sir Lawrence Pumphrey; Timothy Norton; John Stelling; Mark Pinkney; Sir Paul Nicholson; Major Chris Lawton MBE; Colonel Arthur Charlton; Trevor Sheehan of Defence Photography; Neville Jackson; Colonel Anthony George; John Shepherd; William Hedges; Gavin Edgerley Harris, formerly of the Gurkha Museum; The Gurkha Stories Oral History Project; Stephen Foggden, Captain A. Connelly 2 RGR; Kit Pumphrey; the late Ben Smith; Patrick Mercer; Quentin Matthias; and all of those who have kindly assisted me over the years.

I need to extend special thanks to my friend of nearly two decades and companion on this and many other journeys, Graham Trueman, to Julia Grint for editing, to Beverley Palin for indexing, to my agent Tom Cull, to all at Amberley and finally, as ever, to my wife Ruth for her inestimable patience extended over so many years.

Naturally, I accept full responsibility for all errors. I hope all copyright credits have been correctly attributed but if we have made any mistakes we will undertake to correct them on hearing from anyone affected.

John Sadler

Introduction

> These are the stories our colours have to tell. Not mere gallantry in action – that is a small thing and one inherent to our race. They tell of privations uncomplainingly borne, of difficulties nobly surmounted, of steadfast loyalty to the Crown and of cheerful obedience to orders even when that obedience means certain death.
>
> C. B. Norman: *Battle Honours of the British Army*

If it's a miracle, it's a .577/.450 Martini-Henry miracle. The school Easter holidays, 1964. Nearly at the watershed of leaving primary school and moving on, my pal Doug and I got the bus to Haymarket Cinema in Newcastle (long gone). We saw *Zulu* with Dabulamanzi's impressive impi seen off by Stanley Baker, a young Michael Caine and Nigel Green, pretty much typecast as a burly infantry type with his best don't like it up 'em bayonet ferocity. In fact, the real Colour Sergeant Bourne was quite a slightly built man but who cares. James Booth was gloriously over the top as Henry Hook, dedicated skiver, who delivers a one-man Alamo defending the burning infirmary on the way to his VC. That's what got me going on Tommy Atkins and it's been an undiminished fascination ever since.

War, what is it good for? Well, it forged our several British identities, built the world's biggest empire; beat Napoleon, the Kaiser and Hitler. If all the strands of British were not warrior races, would we just be that nation of shopkeepers the Corsican gentleman called us?

War was Britain's furnace for two thousand years and we are that forging. Rome conquered England but not Scotland and imposed military rule on the north. Saxon raiders, then Vikings and finally

Normans each invaded in turn. England and Scotland spent three hundred years at war with each other – a vicious form of endemic, asymmetric warfare and those scars still linger. Edward III pursued expeditionary warfare against France and established a tradition that has since characterised UK military activity, the projection of force across the globe; most strikingly in my lifetime with the Falklands War of 1982. In 1914 Lord Haldane asked, 'What is the Army for?' Nobody yet has a definitive answer, nor ever will.

All this experience and the many traditions it has fostered are preserved in the aspic of our military museum collections, the broad threads of history and grand strategy, but also that human dimension through a treasure trove of individual stories, and it's these I really aim to tease out. With each site I visit, I am looking at its place in the bigger picture but also focussing on a specific group and personal histories. So, in each instance here, I showcase a single exhibit, one which opens the door to the story I'm telling. This journey around Britain is part of the story. I talked to people everywhere I went (many I knew already) and my travelling companion was Captain Graham Trueman, formerly of 3rd Battalion the Light Infantry. We converse about the places we see, the regions we traverse and the stories we tell. And yes, we certainly enjoy ourselves hugely!

I think I can claim to have a fair bit of form on this front. I've been teaching military history at what was Newcastle University's Centre for Lifelong Learning for twenty-five years. I have managed and curated the North Tyne Heritage Centre, DLI Museum, Bede's World and (a long time ago) the Black Gate, Newcastle. I may claim to be something of an expert on British military swords and firearms with an extensive collection and not infrequently appear as a 'talking head' on terrestrial and digital TV. (End of CV.)

'What's new' and 'why now'? As Britain becomes an island again, possibly a divided one, identities loom large. Recent iconoclasm in the wake of BLM has shown a sharp divide between those who seek to preserve heritage and those who wish to smash it by way of an apology. Recent campaign experiences in Afghanistan and emerging global threats have thrown into stark relief the need to determine the role of the UK's Armed Forces and of its global aspirations in a world which is both unstable and threatening. Trite as this sounds, to ascertain how we move forward we need to both see and understand what went before. Mistakes abound, as inevitably is the case, and what counts is what we learn from these and whether we grasp the many lessons history must teach us. It was Leon Trotsky who said: 'You may not be interested in war, but war is always interested in you…' Vladimir Putin has demonstrated that all too effectively.

Why Museums Matter

Owen Sheers' novel *Resistance* is an alternative history – what might have been had the Germans successfully invaded and if the Home Guard Auxiliers had actually been mobilised. (See Parfield Home Guard Auxiliers, chapter 5.) The heroine gazes on the Hereford 'Mappa Mundi' hidden in a cave and understands what this treasure means. It isn't just potential loot, it is the nation's heritage, its cultural soul the loss of which to a barbarous enemy would be of itself a crushing defeat. We are our heritage, the sum of all our historic parts and this does matter. Our military history, however unwelcome in certain quarters, is a big part of this. Take the county of Durham; there will be hardly a family which did not have relatives who served in the Durham Light Infantry during either or both World Wars. People get this, an attempt to erase the museum and render its collection inaccessible aroused such passion that it materially affected outcomes in local elections. Good.

Everyone should go to Auschwitz. I defy anyone to stand at the entrance to Birkenau, those infamous rail lines and gateway and then deny the Holocaust ever happened. History, properly preserved and interpreted, doesn't just open portals to the past, it offers instruction for the future. There are many rooms in the old Polish cavalry barracks there, but that which has the most profound effect on visitors, as I've witnessed, is a collection of hair harvested from victims of the gas chambers. People are stunned and no, you never get used to it. This is more eloquent than any library collection.

On entering vanquished Paris in 1940, the Germans tore down, as part of their cultural contribution to a Thousand Year Reich, a bronze equestrian stature of General Alexandre Dumas, one of Napoleon's successful cavalry commanders. His crime was to be of mixed race, an affront to Aryan supremacy. Many UK statues remain on some people's hit list, just about everyone my generation was taught to revere. Yes, many of them were probably villains but they're still history. We learn far more from reading books than by burning 'em.

Objects tell stories, ones we can deduce and others which are created when our imagination is set free. I own a couple of 2nd Pattern Fairbairn-Sykes fighting knives, which used to be called 'Commando Daggers' when I was young. War comics abounded with them, being enthusiastically deployed against thuggish Germans and sneering, caricatured Japanese. After the war most were lost or chucked into tool kits as useful bits of kit. The two I've got are pretty standard, no provenance I'm aware of. My 'partner in time' Bev Palin was gifted a 3rd Pattern, previous owner a veteran of the North African campaign with nothing else known. This one tells a story though. The leather scabbard is long gone, desert sands are unkindly abrasive. Instead, it has a replacement made from aluminium, still with Desert Air Force colours visible, never seen another like it. We know the original owner served in the Western Desert, so was he a member of LRDG, Middle East Commandos, SAS or one of the other ad hoc, distinctly informal outfits working against Rommel?

Bev also owns an even more remarkable survivor. An accomplished organist, she picked up a bargain years ago from a bric a brac stall in Lyme Regis. Though unaware of this at the time, the instrument is a chaplain's folding harmonium of Great War vintage. Few remain and, as far as we're aware, this is the only example still being gigged on a regular basis. There's one in Talbot House, Poperinge, and another in the Memorial Museum at Zonnebeke. In 2018, a century after being demobbed, it was played by Bev as a highlight of the Remembrance concert in Durham Cathedral – living history indeed.

A Museum that Never Was...

There is no museum dedicated to the sword makers of Shotley Bridge in County Durham, but there should be. Neither the Union of the Crowns or indeed the Act of Union curbed borderers' tendencies to be quarrelsome. Such inclinations weren't solely confined to the savage north. On 15 November 1712, London was rocked by a 'celebrity' duel between the Tory rake James 4th Duke of Hamilton and the equally rakish Whig, Charles 4th Baron Mohun. Both were disreputable and while Tories bleated about a Whig plot to whittle down their majority (Mohun being the challenger), the affair was actually the culmination of a long-running inheritance wrangle. If it was a plot, it failed, as both combatants died, rushing madly at each other in Hyde Park. It wasn't the kind of skilled swordplay society expected but a drunken brawl fought sober. It's a great one to re-enact, always goes down well. For us anoraks, the blades both used were hollow-ground transitional rapiers and very possibly manufactured at Shotley Bridge.

Sword-making is an ancient art, fully imbued with mystery, from the beautifully cast blades of the Bronze Age (of which our own Ewart Park swords are among the finest), to the dazzling ritual that accompanies the samurai *katana*. In Europe, by far the best were from Solingen in Cleve-Berg, Germany, the 'running wolf' or 'flying fox' emblem being the ultimate kite-mark. Swordsmiths belonged to their own high-status guild and their art was a closely guarded secret, admission was by merit and on excellence, disobeying the guild's rules or betraying trade secrets a criminal offence.

By the late seventeenth century, the hollow-ground blade was state of the art: triangular in section, the long flat planes skilfully ground out to produce a strong, slim blade fusing beauty and function; yet much of this painstaking work was being done by prototype machinery. England couldn't compete; our blades simply weren't of the same quality, we had to rely on imports.

In 1688, Dutch Billy kicked out his Catholic father-in-law James II and, ruling jointly with his wife Mary, took over the whole show. This was good for sword-makers as it meant war with France, war in Ireland and in Scotland. Even when war ended nine years later, England and France were soon at it again with the War of Spanish Succession beginning in 1702. War took up a lot of time in the eighteenth century; soldiers, officers and ordinary civilian gentlemen needed swords; even if we weren't bashing the French, our streets were never safe.

It's possible those German sword-makers who came to settle on Derwentside were fleeing, as they claimed, increasing religious persecution or, perhaps more likely, they were enticed here by the business opportunity – the guilds back home in Solingen were very far from pleased. A shrewd move, nonetheless, the Derwent flows fast for power and soft water is ideal for tempering, gritstone abounds for sharpening and Crowley's ironworks already flourishing. Newcastle, after London, was Britain's second largest port.

There may have been as many as nineteen families, of which the Oleys, Vootz, Mole and Bertrams (who were probably already successful English steelmakers from Blackhall Mill), are perhaps the best known. Each clan brought their own specialism and Adam Oley in particular became famous for the superb quality of his hollow-ground blades A possibly apocryphal story relates how, appearing before a panel testing competing blades, he had his wound around the brim of his hat, so supple was the steel it would flex back perfectly.

These Shotley Bridge sword makers created a significant and highly profitable industrial base on Derwentside and prospered during the endless wars of the era. They made broadswords and back-swords for cavalry, munition quality hangers for the infantry, cutlasses for jack tars and small-swords for gentlemen of fashion. I have a hunting hanger with the distinctive flying fox in my own collection and many more survive as testimony to their workmanship. But after Waterloo, we were done with the French and firearms rendered most swords obsolete. By the 1840s the sword-makers were mostly gone. That's the trouble with history, if it's not just one damn thing after another; what we don't commemorate quickly gets forgotten.

That's rather the point of this digression, aside from the fact I enjoy writing about swords, the past isn't just catalogued and displayed in museums, it's living with us like the mad woman in the attic and we just have to open the door and our minds – I'm not responsible for the consequences.

The God of War

For most of August 2019 I appeared at Bamburgh Castle in the guise of William George Armstrong, First Baron of Cragside (1810–1900), I even grew the sideburns! On several mornings, with a fine screen of haze, luminous after drenching overnight rain, I stood on the gun terrace next to a Napoleonic 32-pounder and marvelled at both the man's scientific genius and his entrepreneurial flair.

Nelson's great gun, cast iron and throwing an iron ball, needed a dozen men to serve it. From the platform it could have hit the tower on Inner Farne, just under a mile away. At half that distance it could easily punch through two feet of oak planking. Yet the basic design hadn't changed much since the time Drake defeated the Armada and suffered from severe limitations. It was the horrors of the Crimean War (1854–1856) that so appalled Armstrong and a thrusting generation of Victorian talent, who realized that smooth-bored cast iron guns could be improved upon.

Armstrong decided to use steel as the core of his new design, with an outer lining of twisted wrought iron cylinders heated onto the steel. This new type of barrel was also rifled, and the gun loaded at the breech, a block or vent which slotted home, secured in place by a tightening screw. This was a tricky and by no means risk-free process and Armstrong experimented by firing across Jesmond Dene in Newcastle! Roundshot, which had been around for about four centuries, was to be superseded by a shell-shaped projectile, thinly coated in lead to hug the rifling and soak up some of the huge recoil.

His first gun was an eighth of the weight of conventional muzzle-loaders, had a much greater range and was far more accurate. Artillery had come of age (a fine example of Armstrong's No. 1 gun can be seen in Newcastle's Discovery Museum). The prototype cost a thousand pounds to manufacture and was presented to the War Office in July 1855. They liked it, though they felt the weapon, a 3-pounder, was too small and they needed a heavier round. Next one up was a 5-pounder, and within a year he was offering a hefty 18-pounder. In November 1859, the investigating committee recommended adoption of the Armstrong gun. At the age of forty-nine, the Newcastle inventor had arrived. All his patents he gave to the nation and the honours flowed in; though in 1863, the army bizarrely reverted to muzzle-loaders and Armstrong had to wait for several more years to see his guns finally and fully entering service.

This was truly a game-changer, the science of gunnery entered a whole new era, the dawn of a modern age and guns from Elswick trailed British armies across the globe. Soon Armstrong was building warships and fitting them with guns of such awesome potency that made the age of fighting sail seem as distant as the trireme. By the time the great inventor died, his guns were weighing in at 110 tons.

All of this led to the era of Industrial War, conflict on a scale never foreseen, where competing technologies ratcheted up levels of human suffering to unimagined heights. This began with the American Civil War (1861–1865) and led to the conflagration of 1914 where Armstrong's factories provided vast quantities of arms and materiel; small arms, machine-guns, quick firing artillery, aircraft, tanks, and vast warships that carried his guns which could throw a shell for twenty miles and more. William, Lord Armstrong was an innovative genius who bestrode the international arms trade like a colossus and whose sprawl of factories at Elswick employed tens of thousands of workers. Making guns is no longer respectable, even if just as necessary. Armstrong was nonetheless the epitome of a truly great industrialist.

As the great man, in his dotage, supervised the wonderful medieval fantasy he was creating or re-creating at Bamburgh, even he could not have foreseen what was to come. By and large, neither did anyone else as the World's empires stumbled into chaos and slaughter. Mankind would pay a very heavy price indeed. But it's still vibrant history, the great citadel of early Northumbrian kings evolves into the ornate fantasy of the man who utterly changed the face of warfare. Aethelfrith ('The Destroyer') would never have grasped the technology, but he'd have got the idea and he'd have loved it.

England's Bloodiest Day

> Towton is a village about ten miles south-west of York. It owes what fame it has to the fact that it was once the scene of a battle. But this was not just any battle. At the battle of Towton more English people were killed than on any other day ever ... while the population of Britain in 1916 was more than forty million, that of England in 1461 was considerably less than four million ... one in every hundred Englishmen died at Towton.
>
> Martin Kettle; *the Guardian*, 25 August 2007.

My fascination with the Wars of the Roses can be traced back to afternoons of black and white TV in the early 1960 with BBC's *Age of Kings* – Shakespeare's History Plays. Subsequently and, still in black and white, interpreted in *Wars of the Roses*. Both landmark series were based on *Henry VI Parts I, II* and *III* and *Richard III*. That Shakespeare could be enjoyable, indeed rivetingly so, came as something of a revelation to one reared on the standard GCE (as it then was), diet of the Bard as purely exam fodder, presented with painstaking dullness. An immediate debt of gratitude is therefore due to the producers and cast.

The town of Ferrybridge in North Yorkshire, lying on the south bank of the River Aire, is somewhere only a Yorkshireman could love. A nondescript remnant of post-industrial and later, post-war utilitarian despondency, it requires a powerful imagination to transpose the landscape and events of early spring 1461 onto such an unpromising canvas. Yet it was here that the Towton campaign began in earnest, both sides disputing the span of the medieval bridge. Its 19th-century replacement is, to be sure, a fine edifice, somewhat diminished by the overwhelming blankness of the more recent motorway crossing a few hundred yards downstream.

Discovering the field of Towton is tricky. As a battlefield tour it involves a fair element of detection; no bad thing perhaps, but quite surprising considering that this was Britain's bloodiest day in a long history of murderous conflict, including both world wars. It seems likely more Englishmen died on that day of major bloodletting than any other, even including 1 July 1916, the disastrous opening day of the Somme offensive.

Admittedly the modern road system, planned with no regard for history, doesn't help. Not until one drives north on the A162 is the route of the Yorkist army clearer. Sherburn-in-Elmet is now by-passed by a loop in the road. The first tantalising evidence is found just beyond the village of Barkston Ash. There, by the side of the road, at the junction of the lane leading west to Saxton, is the stump of a medieval cross, the 'Leper Pot'. It is here, tradition asserts, that Lord Clifford fell. According to Hall he was pierced through the throat by an opportunist archer, having unwisely removed his bevor.

On a muggy morning in late spring, passing through rolling vistas of fields and peaceful hamlets, it seems impossible that such a savage and momentous struggle occurred in this tranquil landscape. Traces are few. Only a ragged cross marks the field, easily missed by the motorist. A rutted pedestrian track leads westwards over fields to the edge of Bloody Meadow with some useful interpretation boards. And that is it. I cannot think of another field of such importance so poorly served. Naseby to be sure is similar and I'm constantly struck by the contrast. Had such ground beenin North America for instance, an entire heritage industry would now be booming. The Towton Battlefield Society, whose members have done and continue to do such sterling work, deserve every praise for their doggedness which, hopefully, will one day result in the construction of a sympathetic visitor centre.

Driving on the A162 from Barkston Ash to Towton, one ascends the rise the Yorkists would have climbed. As you arrive parallel to Dacre's Cross, Castle Hill Wood can be glimpsed away to your left. All this ground now lies under the plough and is privately owned. There is nothing in the village to associate with the battle other than the Hall, a private dwelling. Following the B1217 from the village takes us past the supposed campsite for the Lancastrian vanguard, past the cross, now on your right just off the road and then down to the minor crossroads where a left-hand turn would take you into Saxton, the church and Dacre's tomb there. If you park and begin walking down the bridlepath to your right, you are afforded a good view of Castle Hill Wood and the enticingly mysterious mounds nobody is quite sure of. The best view is after you cross the small stream with farm buildings on your left.

Perhaps the most atmospheric spot of all is the tiny, forgotten St Mary's Chapel which stands alone in a field on the left of the road, south of the crossroads, with the Crooked Billet Inn on your left. This is the sole surviving fragment of the lost village of Lead, associated with the Tyas family. A 13th-century foundation, restored prior to World War Two and simple to the point of austerity, it fits the landscape as though cast down by divine providence – probably it was attached to the now vanished manor-house. This triangle of undulating farmland enclosed within three roads, A162 in the east, B1217 in the west and the minor road westwards from Sherburn-in-Elmet which links up with the latter, encloses the site of Towton – Palm Sunday Field, England's bloodiest day.

Between the great Yorkshire Rivers of Wharfe and Aire, the land rises in a sweep of that magnesian limestone belt which extends across North Yorkshire, furnishing the stone that built York Minster. This tract is cut by lateral gorges through which race the Humber's tributaries, Ure, Nidd, Wharfe, Aire, Went and Don. Once, the land was dressed by a dense covering of ancient elms, giving the historical name Elmet, the *Sylva Elmetae* of Bede.

When the Elizabethan antiquary John Leland visited the field, some four score years after the fight, locals advised that many of the fallen had been interred in Saxton churchyard:

> They lay afore in five pittes, yet appering half a mile of by north in Saxton feldes. The subsequent Act of Attainder referred to the battle as having occurred on Sonday called comynly Palme Sonday, the xxix day of Marche ... in a feld bitwene the Townes of Shirbourne in Elmett and Tadcastr ... called Saxtonfeld and Tawtonfeeld, in the Shire of York.

There is thus a general consensus that the action took place between the two hamlets of Saxton and Towton. The ground is elevated, though not to any great extent, a mere 150 feet or so, but sufficiently to form a distinct space. Eastwards, the land slopes away to merge into the flatness of the Vale of York. To the west it declines sharply, falling into a defile above the Cock Beck. From Saxton, line of approach for the Yorkists, the ascent is clear and abrupt, from Towton and thus for the Lancastrians, gentler. The level ground atop the tabletop is bisected by a small ravine which declines from North Acres in the east to a sharper trough in the west, where it debouches into the Vale of the Cock.

This area is known as Towton Dale and that steep slope down to the water became the infamous Bloody Meadow, scene of the most concentrated carnage. It is possible the low ground by the river was altogether wetter and given to flooding, though there is no direct evidence to support this. These precipitous banks only level out further south by the Crooked Billet and St Mary's. Two Gentry families, Tyas and Skargill, are linked to Lead and it may be, as Andrew Boardman suggests, that a portion of the Neville affinity encamped here. He offers some interesting insight on a possible heraldic provenance of the name 'Crooked Billet', though earlier authors, particularly Dr Leadman, associate the name more with the arms of Lord Fauconberg. The place and the delightful chapel, which was certainly in situ in 1461, have a singularly melancholic, enigmatic air. Visitors, or at least those with a romantic tendency, are left with a sensation that the stones have a story to tell!

The angle of descent to the east at the extremity of North Acres is generally shallower, but by the excellent vantage point of the Hawthorn tree, dips more radically. From here one can obtain a clear view of the field, the cockpit, the arena of Towton where the two sides would slog it out in the whorls of bloodied slush for, if the chroniclers are to be believed, a full ten hours.

'In this valley on one savage day, a greater proportion of our population died than in a whole year of the First World War. It was by far the bloodiest battle in our history, yet Towton is largely forgotten and untaught' (A. A. Gill). Writing in the *Sunday Times* magazine in the wet summer of 2008, the late journalist and author A. A. Gill visited the field of Towton: 'It would be impossible to walk here and not feel the dread underfoot – the echo of desperate events vibrating just behind the hearing.' One has merely to visit the mute testimony of the war cemeteries which mark the line of the Western Front to feel a tangible sense of loss. The German burials at Langemarck just north of Ypres are particularly gaunt and austere; a nation's suffering written in stone. Towton has no such commemoration, merely the cross, unannounced, and generally unnoticed. This seems bizarrely inequitable for the losses of Palm Sunday Field, with only Marston Moor in 1644 coming close (in both senses, as the two sites are close).

During the rather better summer of 1996, builders undertaking new works accidentally uncovered the site of a mass grave, now The Towton Mass Grave Project. The remains of 43 dead were recovered from the pit, a mere 6m x 2m and 50 cm deep. Here was the Holy Grail of Towton scholars, that which had eluded generations of searchers, grave evidence, mortal traces of those who had died in the carnage. Rumour of another even more tantalising fragment, an intact muster roll, have circulated, spoken of in hushed tones, as precious to some as the Dead Sea Scrolls.

As Gill points out, Towton, for all its magnitude and drama, is a poor relation to other more renowned fields such as Hastings, Bosworth, or Culloden. If anyone retains any shred of schooldays history, it is the tale of Harold and the arrow in 1066. Partisans and detractors of that most enigmatic of Yorkists and controversial of monarchs, Richard III, have kept the memory of Bosworth alive and still happily contentious. Culloden, a field which does indeed inspire dread, especially when wind and driven sleet lash the hallowed ground, has a magnificent new visitor centre, entirely befitting.

But Palm Sunday Field stays bare, despite the dedicated efforts of the Towton Battlefield Society. Part of the difficulty lies in the nature and ownership of the ground, which does not readily lend itself to 'development' as a heritage site. Many might prefer it stayed that way. Yet Towton is *huge* – the biggest, bloodiest, and arguably most dramatic battle, where numbers of combatants and casualties exceed the combined total of the three other fields mentioned.

Apart from the Ricardian element, which enjoys almost Tolkienesque status and is frequently pursued with as much grasp of historical reality, the Wars of the Roses – a term unknown to participants – has been largely consigned to academics and anoraks. Castle Museum York does have a good display from the period, but the Towton field itself, the living exhibit, is largely unmarked.

The Troubles

I was privileged to have as my companion on this journey my old friend Graham Trueman who served in 3rd Battalion Light Infantry 1980–1988, trained at Sandhurst, reached the rank of captain, and saw active service during several tours in Northern Ireland (Operation *Banner*). I've been obsessed with the study of war for all of my life, but I never served the colours. I think it was Dr Johnson who remarked that every man wished he had. Doing the job for real gives you an enhanced perspective, a developed eye for ground and for potential threat. Graham and I have had many conversations over the years and undertaken a good number of tours. Such experience, analysis and insight are priceless.

'Repent ye crapheads, God is airborne/ We are not lazy/ We do not skive/ That's how us Paras/ Stay alive.' Introduced by this graffiti memorably inscribed on the wall of a urinal in Northern Ireland, Graham summarises his experiences thus:

'These reflections are based on my recollections from rural counter insurgency ('COIN') operations in the early 1980s. Units were deployed to Northern Ireland on two-year or six-month tours. The longer tour was regarded very much as a standard posting, just as a unit may be posted to Catterick for a couple of years. Generally, one rifle company would be on internal security ('IS') duties and the rest of the unit would undertake standard training. A six-month stint was classed as an emergency tour. The whole unit was involved in IS operations and the tempo was fairly high.

Pre-deployment training was divided into two types, training for rural operations and training for urban operations. There was a degree of overlap. Training was battalion run, with occasional specialist input from Northern Ireland Training and Advisory Teams ('NITAT'). Thetford Training Area (now called Stanford Training Area) was used for field training. Ranges at Lydd and Hythe for specialist weapons training, where the main objective was to shoot the moving terrorist with a gun, as opposed to the woman with a pram. Shooting up the pram was more fun. Most nerves were calmed on being told there was more violence in Los Angeles in a day than in the Province in a 6-month period. A useful mantra for junior officers is enshrined in the 'Platoon Commander's Aide Memoire'. [See https://zulu-alpha.co.za/wp-content/uploads/2013/03/tampt2.1.pdf.]

My Battalion took over the Fermanagh border area from one of the Parachute Battalions. Rumour had it that the Paras were frustrated at being the only battalion in the regiment not to deploy to the South Atlantic and had vented their frustration on the locals. So going out on a familiarisation patrol sporting a Para beret, to hide the fact a battalion changeover was imminent, did make me rather think I was heading out to a party I hadn't been invited to!

Some rifle company headquarters were situated in Ulster Defence Regiment (UDR) facilities. These were rather like TA centres, with multi-gyms and bars. Rifle platoons operated from Victorian Royal Ulster Constabulary (RUC) police stations, fortified with high fences, watch towers, CCTV masts and anti-mortar nets. Although these were active police stations, local defence was an Army responsibility.

Watchtowers were manned 24 hours a day and equipped with general purpose machine guns (GPMGs).

Accommodation was crowded, mostly in Portakabins. Some had helipads inside the perimeter, some relied on nearby fields, always a concern as this limited the choice of landing zone. Some platoon locations had section-manned permanent vehicle checkpoints (PVCPs). These were a bit rough and scary and looked temporary rather than permanent. The platoon based in the RUC station at Belleek had a PVCP at the head of Lough Erne, 3 or 4 miles away by road, depending on whether you went by road in the Province, or took a short cut through the Republic.

Two specific incidents I remember very clearly. Firstly, the RUC playing chicken in armoured Ford Cortinas in a carpark at a Lough Erne visitor centre when they were showing me around the area, followed by a water pistol battle using the Cortinas as cover. Secondly, the alleged contact at one platoon base in an RUC station, which was separated from the border by a river. Apparently, someone took a shot at them from the Republic and soldiers were then queuing up to have a shot back. It transpired the whole thing was probably sparked by an ND (negligent discharge) which was then covered up by claiming there had been a contact and returning fire. A new mythology!

My strategic and operational awareness, as a platoon commander, was probably not what it should have been. The Troubles had been rumbling on since my childhood. My understanding of the situation was that the Protestants and Catholics did not like each other, that neither liked us and we didn't like them. A bit basic on reflection, but I thought little more about it. I was never concerned about which battalion or unit was where. I had a rough idea about my battalion deployments, and an exact idea of other platoon locations in my company.

One platoon was co-located with Company HQ, the other two in police stations closer to the border. There was daily liaison with the RUC. We were in the Province in support of the RUC. Technically, we provided military aid to the civil power (MACP). More often than not a member of the RUC came out on patrol with us. In larger scale operations, we might put a cordon in place while they conducted a search. Often, if we were doing vehicle checks, it was the RUC constable who would actually stop the vehicles and question the driver. Relations with the police at this level were good. They were armed with WWII-era US M1 .30 calibre carbines and .357 Magnum Ruger revolvers.

Operationally we spent our time patrolling, mounting cordons, doing vehicle checks and base security. I cannot actually recall being given an overall 'mission', or even a set of orders. My understanding was that our presence would deter terrorist activity. Dominating my area of responsibility was the key. I was largely left alone to get on with it. Direction came from the int cell in each location. These were jointly manned by RUC and a battalion Intelligence Section corporal. The information from the cell basically steered who we watched, where we went and when. Critically, patrol history, displayed graphically, prevented the creation of patterns of activity.

Tactically, patterns were key. Or rather the avoidance of patterns. In simple terms, if we always patrolled at a certain time, or followed the same routes, the terrorist would either be able to avoid us and operate without interference or carry out attacks on the military. We referred to our sorties as patrols – 'Eagle' Patrols, always a buzz. [Eagle patrols were conducted by air using AAC and RAF helicopters, *Apocalypse Now* stuff with open doors and automatic rifles swung on slings.] We watched border crossing points, cordons, established a general presence on the ground and looked out for illegal crossing points. Farmers' daughter patrols were often difficult to forget. Generally, patrols were of eight men, divided into 'bricks' of four. It is worth noting that four-man and eight-man patrols rather than platoon-sized patrols were introduced in Northern Ireland by Lieutenant-Colonel Keith Spacie, CO 3 Para, later Major-General, as a new battalion tactic in 1974. A year later this patrol technique was adopted by the Army as a whole.

I say patterns were key, but we weren't totally governed by them. Our role was to get the job done and do it properly, awareness of the risks of falling into familiar patterns of behaviour was just one aspect, albeit a very important one of our tactical deployment on the ground. Patterns weren't just important for us soldiers either, or the RUC. Civilians were players too, whether they wished to be or not. If our blokes patronised a sweet shop, the proprietors might find themselves targeted for fraternisation.

We went to farmhouses while on patrols. Where these were run by the Protestant community, the lucky section that went indoors as opposed to guarding a perimeter would enjoy a magnificent cream tea, much coveted, though such hospitality led the providers into further risk of targeting. The army were the new kids on the block, parachuted into a long-running and tense sectarian row, always seething, sometimes bursting into flame. No such thing as neutrality.

One evening some of the RUC at our base asked me if I would like to go to the pub for a drink. I was quite chuffed to be asked, but also concerned that any pub they went to would be regarded as a safe pub, and therefore by default not a safe pub because of patterns (again). I insisted on taking a 9mm Browning pistol, stuffed down the back of my jeans, after all, they were all packing revolvers, so why shouldn't I? It was all quite exhilarating. Not surprisingly everybody seemed to know who and what I was, but what I was worried about most was the thought of a loaded Browning falling down my trousers. It did not. And I had a pleasant evening.

A friend who was on a two-year tour and based in Ballykinler invited two of us to a punk rock disco in his Officers' Mess. To my great surprise, we got permission to go and the MTO (Motor Transport Officer) even let us borrow a Ford Escort Estate to get there. We borrowed leather jackets, found chains and clip-on earrings, discovered that Ingrams Shaving Lather would spike our hair and we were ready to go. And we went well armed, the footwell had a pair of Browning pistols and a couple of Sterling sub-machine guns. The RUC lent us a police radio. God knows what would have happened had we been stopped at an Army or RUC checkpoint, never mind a PIRA one. As it happened nobody stopped us, I think everybody was at the disco!

The popular British press, mostly *The Sun*, did its best to keep us in the minds of the British public. Christmas presents and cards from strangers were in abundance one year. By the early '80s public awareness of the situation on the ground in the Province seemed to be low. It was just something which was rumbling along in the background. From a soldier's point of view, reality was somewhat different. Being on operations for six months is quite different from day-to-day living. It sounds obvious, but it needs saying.

All the more so when you are on operations a mere 200 miles from home, and when in many ways, when you walked (patrolled) down a street, you could be at home. The shops were the same, the font on road signs was the same, the weather and vegetation were the same, cars were the same. And although they looked the same, the people were not all the same. Whereas on a street in Morpeth if you look directly at someone, you generally get a response, be it a smile or a nod, even a good morning, in Ulster quite often there was just nothing. Blank.

Rural Ulster is seriously rural. Intelligence suggested we watch a small single-storey dwelling at the end of a lane, about 300 yards from a road at the end of a track. Cattle lived in one half of the building. The only activity was an old lady carrying milk in a bucket to the main road where she poured it into a churn, to be collected by the dairy lorry. Probably the only time I ever saw a British peasant, French meaning of the word.

Some of the old timers were interesting. Our task one day was to put a cordon around an unauthorised border crossing, a single tarmac track, used by locals as a shortcut, farmers to smuggle cattle so they could claim EEC grants for the same cattle on both sides of the border, and gunmen, up to no good. The Irish Army put a cordon in on the southern side of the border. Engineers dug up the road and huge lengths of railway line were cemented in. One spectator turned out to be an old IRA member, as opposed to PIRA (Provisional Irish Republican Army). He recounted tales of derring-do in the 1920s and 30s and had absolutely no time for PIRA. Mercenary scum, in his opinion, using politics as a cover for crime. He was quite charming!

I wouldn't want anyone to think I am belittling what happened over there with these rather irreverent tales. Hours and hours were spent patrolling or on stag (guard duty) and lots of people were injured or killed, and right was not all on one side. Thankfully now it over, mostly. It was rather nice to go back a few years ago and see that checkpoints had been removed and watchtowers demolished and actually to think that I contributed, just a little, to the relative peace that is now enjoyed by all communities.'

1

Northeast England and Border

> He who wishes peace should prepare for war; he who desires victory should carefully train his soldiers; he who wants favourable results should fight relying on skill, not on chance.
>
> Vegetius (*c.* AD 390)

> My son was killed while laughing at some jest. I would I knew
> What it was, and it might serve me in a time when jests are few.
>
> Kipling: *Epitaphs of the War*

Northumberland and the English side of the border have a very long exposure to war. Bronze and Iron Age hillforts abound. Emperor Hadrian, in a move which has proved an inestimable blessing to the local heritage tourism economy, built his astonishing wall with its chain of forts and militarized hinterland. Vikings beat the place up, Scots and English squabbled bloodily for centuries, vast coastal fortresses, lordly castles, manorial towers and yeomen's bastles[1] proliferated at a rate that would embarrass North Korea. Hartlepool was the only place on the UK mainland to see serious action in the Great War; Northumbrian beaches and dales were studded with coastal defences and pillboxes arrayed in rather mystifying stop lines.

This was the Threap (or Waste) Land. From 1296, when Edward I ('Longshanks') came up north to chastise his reluctantly rebellious Scottish 'subject' John Baliol and got started by trashing Berwick upon Tweed pretty massively, the border lands became a war zone. Until 1603, when James VI of Scotland neatly and bloodlessly reversed the process. It is for Graham and me, the Homeland and we've spent much of our lives traversing its rolling hills and secret valleys, not to mention rehearsing some rather more modern tactics.

By the ruined shepherd's bothy in North Northumberland, as desolate and wild as it gets – but country that has been defended since the Bronze Age:

John: How does an infantryman look at ground?

Graham: Not like you.

John: Go on then...

Graham: He looks at the ground from a tactical perspective, opportunities, threats etc, where might an attacker be lurking, how does he, the infantryman, seize and control the ground, how does he construct defences, who's on either flank or can the bastards sneak around behind you. Terrain is the main tool of his trade, he needs to ensure he dominates the ground and that he can hold is securely, beyond that it's just scenery.

John: You proper soldiers have been taking that view of this chunk of real estate, empty as it now appears for the last 3,000 years, we've got a Bronze/Iron Age fort with a WWII Home Guard impromptu firing range across the deep valley of the burn and a pillbox skulking between the ditches of the old fort.

Graham: Stop lines?

John: For sure, not that they'd ever have stopped anyone.

Not surprisingly, this martial heritage has left its mark not just upon the landscape but on the people too with some cracking museums, only a few of which I've space to profile here. This is not intended to demean those I've missed out such as the lovely Royal Northumberland Fusiliers Museum in Alnwick Castle, or the eclectic Northeast Land Sea and Air Museum, Washington, and glorious Blyth Battery.

Northumberland Hussars Museum, Charge Gallery, Discovery Museum, Newcastle upon Tyne

On 16 August 1819, British cavalry turned their sabres against protestors in St Peter's Fields Manchester and put numerous of them ruthlessly to the sword. The Peterloo Massacre outraged public opinion and yet gentlemen of property felt that revolutionary trends were undermining the fabric of society, threatening to shake the established, God-given order of things. Something had to be done. Raising local mounted forces, drawn from the tenantry and affinities of local magnates, landed and industrial, was considered the best means whereby those of substance might 'render effectual aid to the civil power in case of disturbance'. The Duke of Northumberland, a man of very great estate, had called a meeting of county magistrates in October of that year. Within two months, the Northumberland and Newcastle Volunteer Corps of Cavalry was raised. For the most part, it comprised 'gentlemen of great respectability, such as merchants, brokers, and tradesmen, together with officials of the collieries, and tenantry and retainers of the landed proprietors'.[2]

For obvious reasons, the Yeomanry were regarded as tools of oppression and these well-groomed class-warriors were soon dubbed 'The Noodles'; 'Blue bummed bumblers/ Cock-tail tumblers/ Fireside soldiers/ Dorn't gan te waar'.[3] Oddly, in mining areas of Northumberland, colliers volunteered. The lure of working with horses outweighed any disaffection. Indeed, the Ashington troop was soon amongst the smartest. There was stiff competition, gentlemen need to dress up. With their smart, tailored uniforms, they took on a dandified air. Besides, if this was training for war, it smacked of romance and derring-do, of stiff parades, 'of jewelled hilts/ for daggers in plaid socks; of smart salutes/ And care of arms; and leave; and pay arrears/ Esprit de corps; and hints for young recruits'.[4]

Reviews were glittering affairs, everything -shone:

> The regiment was reviewed on the Town Moor by the Duke of Northumberland, who wore the full uniform of a Lord Lieutenant, complete with the insignia of the Order of the Garter. The weather being favourable there was a numerous assemblage of the beau-monde on the ground to witness the spectacle; among others was Her Grace the Duchess of Northumberland, in a splendid equipage drawn by six beautiful blood horses.[5]

Highlight of the social calendar was a grand ball in Newcastle's Assembly rooms, held every year through the 19th century from 1828. In that year: 'In all, 343 ladies and gentlemen, who included nearly the whole of the gentry of the neighbourhood.'[6] It was interrupted only once by an uncivil outbreak of the new plague, King Cholera, whose ravages discourteously took no heed of civil society.

Service in the militia had traditionally been far less glamorous, decidedly unpopular. Limited forms of conscription in the eighteenth century during the Seven Years War (1756–1763) had led to riots, one of which at Hexham in 1761 ended in bloodshed with about forty deaths as panicky North Yorkshire militia opened fire on a hostile crowd.[7] Those fears over possible French aggression in the late 1850s had led to the raising of Rifle Volunteers, altogether more fashionable than unwilling and workaday militia but, unlike their mounted contemporaries in the Yeomanry, unwaged. Inevitable rivalries arose:

> The Noodles tho a' gentlemen/ Respected to be sure/ They nivor like the rifle çurs/ Fixed ramrods of the Moor; the riflemen black-legged us a/ Undermined wor daily pay/ But smash, aw'll fight him for a quart/ Then for war lads, clear the way'.
>
> Joe Wilson: *The Noodle an' Rifleman's Dispute*[8]

South Africa changed all that. Those remarkably effective Boer riflemen inflicted a series of costly and humiliating defeats on imperial forces. To beat these hardy farmers-cum-guerrillas in their own country would require more fast-moving light cavalry than the Army had available. So, the *Noodles* went to war. Many rushed to volunteer, gentlemen of means and estate subscribed generously so that 355 local horsemen formed three full squadrons of Imperial Yeomanry.[9] This was triple the number originally hoped for. Now there was real fighting in a very real war and the part-time soldiers performed creditably.

Haldane's reforms of 1908 abolished old distinctions between Rifle Volunteers and Yeomanry. The *Noodles* were now the Northumberland Hussar Yeomanry. These measures, though long overdue, were not entirely popular. Rates of pay came down and county gentlemen, who had been used to doing things very much in their own way, suddenly found themselves altogether more accountable to that great Whitehall behemoth: the War Office. All the new Territorial units from Northumberland, Durham and North Yorkshire were to form part of the newly created Northumbrian Division, commanded by General R .S. S. Baden-Powell, hero of Mafeking and founder of the Scouting Movement. Territorial divisions were intended for homeland defence, though many in the ranks chose to volunteer for overseas service.

Even as dark clouds began to mass over Europe, this Edwardian twilight seemed set to endure: tradition garnished with the laurels of empire, boosted by centuries of imperial expansion, fuelled by unparalleled industrial might. Annual camps stretched through sixteen languorous days of summer. Riding and manoeuvre, saddle soap and champagne.

Uniforms elegantly tailored to show the rider to best advantage, to impress ladies with coquettish smiles and parasols, in dazzling white: 'A feature of these annual camps was the regimental sports, which were made the occasion not only of an attractive display of horsemanship, in which keen competition was witnessed, but of social gatherings in which friends joined officers and men in a pleasurable outing.'[10] A shining twilight before the apocalypse, a world that writers such as Ford Madox Ford would capture in prose, perhaps never better than in his *Parade's End*.

It was during this epoch, when the Queen Empress ruled nearly a quarter of the earth's inhabitants, that the Royal Military Tournament got underway, a chivalric tribute to the Spartan fighting qualities of British soldiers and sailors. Officers duelled with sabres and from horseback hacked imaginary foes represented by melons in a ritual dubbed 'cleaving the Turk's head'. Political correctness was making an early appearance: manly displays or martial prowess were to be lauded, but 'without encouraging the display of those brutal and degrading passions, which induce a couple of vagabonds, with a dislike for work, to batter each other with their fists for a wager, till one or the other can no longer stand upright.'[11]

That brewing conflict would be a very different type of engagement. It was to be industrial warfare on a scale and intensity never experienced before. Technology had vastly increased the killing power of both artillery and small arms. The new quick firers[12] would move the guns to prime slot as arbiters on the field. Tommy could fire twelve to fifteen aimed rounds per minute from his Short Magazine Lee Enfield (SMLE).[13] Steamships could transport him over oceans in a fraction of the time sail would have taken.

Railways could move him, his horses, equipment, rations, fodder, and ammunition over great distances on land. Tinned food ensured he could be fed and sustained year-round in trenches. Machine-guns could spew out 500-600 rounds per minute and kill at 2,000 yards. The war itself would witness exponential development in the role of aircraft and in the introduction of new, even more terrible weapons. Tanks, poison gas and flamethrowers would add fresh dimensions of horror. It would prove a very unlovely war.

When, in that August of 1914, war was declared, the Noodles were commanded by Lord Ridley of Blagdon, though he was already a sick man (he died in 1916). At this point 'A' Squadron was in billets on Gosforth Racecourse, 'B' Squadron somewhat distant at Castle Eden in East Durham and 'C' Squadron in south-east Northumberland. All three were soon concentrated at the racecourse before, on 11 September, being ordered down to Hampshire under Major (then Lieutenant-Colonel) Cookson of Meldon Hall (formerly of the Lifeguards) in mid-Northumberland. They were destined for Belgium as divisional cavalry.

They were going to war. Those halcyon days of festive pomp, of war imagined, would become a distant memory. On 5 October 1914, the regiment sailed on the *Minneapolis* from Southampton for an uneventful night passage to Zeebrugge.

> The morrow broke cold and wet as we steamed slowly into harbour... It was late in the afternoon before we set off down a long typical Belgian road toward Bruges. Our reception ... was ecstatic. At every hamlet along that poplar-lined stretch of pave the inhabitants would raise a cheer for 'les Anglais' while little urchins would clamour for buttons and badges ... pretty girls would almost drag us from our saddles to kiss us and to shake our hands.[14]

Private Daglish from Morpeth, when a young lady waving her handkerchief shouted *Vivent les Anglais*, replied in thick Northumbrian, *Vary canny, hoo's yorsel* (I'm very well, how are you?)[15]

Captain Grant, who narrated the experience of the Noodles during the 1st Battle of Ypres, had previously served in South Africa. The Hussars were the first Territorial unit to be shipped out. By 08.00 hours on 8 October the yeomanry were in the saddle, riding as the eyes and ears of 7th Division. They hoped to fight Uhlans, 'but to our disappointment encountered none. Instead, we discovered how unsuitable pave roads are for cavalry, and that mounted men do only less footslogging than infantry when a division is moving *en masse*.'[16]

Ostend proved less congenial. The streets were chaotic as stunned refugees struggled to get clear of the Teutonic horde: 'Fearsome rumours as to the fate of Antwerp flew from mouth to mouth ... at this period of our soldiering we were credulous and drank them in.'[17] It was now they realised that their task was not to assist the Belgian Army as much as to cover its withdrawal, 'to shield that heroic remnant from annihilation'. They entrained for Ghent where their earlier rapturous welcome was replayed. The Noodles deployed on outpost duty and for the first time came under enemy fire, happily ineffectual:

> I was picturing to myself Saturday night at home and thinking how little the boys there were dreaming of what we were doing that night, when

suddenly a succession of reports sounded in the air. If must confess I could not determine whether they were rifle shots or not. Just then a shadow loomed up before me, and with an effort I spluttered out 'Halt! Who goes there?' I had my finger on the trigger and was ready for him. I felt, I must confess, much relieved when immediately there came the whispered assurance 'friend'. It was an infantryman, like myself, on outpost duty, and he enquired if I had heard anything lately. 'Yes, I replied, 'I think it must have been the rumbling of transport wagons on the cobbled road.' 'No, mate,' was his rejoinder,' it was 15 rounds rapid.' The battle was drawing nearer.[18]

Isaac Armorer Patterson was a native of Hexham. By 1914 he had already been serving as a Territorial for a decade. He was engaged to Daisy Barrasford, also of Hexham who, when her fiancé went to war, became a nurse, based in Birmingham. The couple saw little of each other through four years of war, though they corresponded regularly. Photographs show Armorer's tunic pockets bulging with bundles of Daisy's letters, as many as thirty at a time. On 22 August 1918, he was posted missing in action, and it wasn't until October, as war's end approached, that Daisy, who had kept writing, learned her faith had been vindicated. Armorer was a POW, he had been wounded in the leg and captured.[19]

The action in which Patterson (a subaltern) was wounded took place around 08.30 hours by a railway loop, just east of Happy Valley, and a mile north of Bray-sur-Somme. The Hussars had charged German defences, MG posts linked by infantry trenches, and Armorer's horse was shot from under him.

Wounded be damned was what I said when I got hit. My next thought was to get into a small drain three yards from me. I made it to the drain and lay down full length. It was about nine inches deep and twelve wide, but one can make oneself very small when only twenty yards from a German machine gun. During the first five minutes, I had hidden my sword. I stuck it into the side of the drain and covered the hilt over.

What I saw when I looked out was not a pleasant sight; a German officer with his revolver drawn coming towards me. He looked at me and passed on to see if there were any more fellows lying about. Then he made signs for me to get up, but I could not. He helped me up and called over one of his men to help me back to their gun line. I thought – 'I am both wounded and a prisoner, two things I never dreamt of.'

The Germans stripped off Patterson's puttees to expose the damaged area of flesh. 'I had a look at my wound, a hole straight through the shin bone and out at the back of the knee, a little to the right.'[20] Soon, he was joined by more wounded captives from his regiment, Private Moore and Corporal Dance. The injured officer was despoiled of his kit, revolver, ammunition, glasses and the badge of the cavalryman, his spurs.

> I had an awful time trying to walk back to the dressing station. It was a small hell. I had the German guard on one side and Cpl Dance on the other and used both as crutches Happily, the party ran into a Red Cross orderly who improvised a makeshift stretcher from branches and groundsheet and, with a comrade, made up a stretcher party. They carried me about a mile to the dressing station at Bouvray Farm. We went through a lot of heavy shelling. During one of our halts, I saw one of our planes bring down a balloon and get clear of the German aircraft.[21]
>
> We finally reached the ADS, located in an old dugout; I'd had a good dose of my flask by this time and gave the bearers a drink.

Next, he was conveyed by motor ambulance along the straight Roman Road from Albert to Bapaume, through Mericourt. Then the operating theatre: 'The first thing I saw was a human leg lying under the operating table, it had been taken off above the knee. My own had swollen terribly and did not look at all well. I thought it was all broken to pieces, judging from the appearance. Now, I had my leg thoroughly examined and put into proper splints. The hospital was fairly clean and not badly looked after. My knee was about the size of a football and my leg about three times normal size. I was on the serious surgical ward and most of the German patients had legs or arms off.'[22]

His next transfer was to a base hospital well behind the Front. It was only a week since his capture and this unit proved 'one of the dirtiest places I have ever seen.' Here he met an Irish NCO called Buckley. The latter was senior man amongst the local British POWs held in the nearby working camp. 'He told me how they were being treated and how they'd been treated since March and through the rest of the summer. He told me that since March, around 20 per cent of the prisoners had died.'[23]

On 29 August, 'I was put on a stretcher and loaded onto a Red Cross train, which was really quite clean ... on 30 August we arrived at Valenciennes where I was taken to a POW hospital. On the ward there were eleven British officers all more or less wounded. The food here was very poor, thank goodness there were some Frenchmen in the hospital who would sell their biscuits for ½ Mark each. 6 September I had the splints taken off and was in a terrible funk, I thought my leg was going to fall off!'[24]

Just over a fortnight later he was moved to Antwerp and St George's Barracks, which was being used as a POW hospital. 'Food was a bit better here than Valenciennes. There was also a canteen where you could buy brandy, cigarettes, and caramel. It was possible to have both a bath and haircut and on Wednesdays we ate boiled meat and potatoes. I got a newspaper from our Belgian helpers, and it showed the great advances our armies were making. I knew almost every village that was being taken; there was great excitement as first one place, then another fell.'[25] 25 October:

> We left Antwerp for Karlsruhe. The Belgians provided us with a half-chicken each, two eggs, four apples and a bottle of John Crabbe's whisky (pre-war stuff). Had it not been for these gifts we would have been very badly off. We passed through Aachen, Cologne, Mainz, Heidelberg, and Frankfurt. We arrived at around three in the afternoon on 27 October and were put up in the Hotel Europa. Next day we were marched to a temporary camp then, on 30 October made ready to move again. Before leaving we heard that Austria had signed an armistice. Our party comprised 58 officers and we had a second class carriage right through. We passed back through Hamburg and Frankfurt to Pillau [now Baltiysk] arriving on 02 November.[26]
>
> We were marched to a small fishing village of about a hundred inhabitants, it was as near to the sea as was possible without being in the sea. The camp used to be a German hospital, it was a new camp, no canteen, no parcels, in fact nothing, hardly any rations. We were thoroughly searched and made to strip off. During the searching I found a German trying to pinch a packet of tea from an officer at which I spoke up and caused a bit of a stir. On 10 November, I went to Holy Communion, a thing I never dreamt of [Armorer was chapel], the padre was a South African with one arm.[27]

His single entry for 11 November simply says, 'Thank God.'[28]

Armistice meant an end to hostilities, but the POWs, as ever, were more immediately concerned with rations or rather, their lack. Their daily allowance was meagre in the extreme; Germany was near destitute: '8.00 am – coffee and bread allowance for the day (275g)/ 1.00 pm – Potatoes and cabbage/ 4.00 pm – coffee/ 7.00 pm – soup (not good). If anybody wants to take their weight down, I strongly advise the above rations. There is not sufficient to feed a new-born baby. We often talk about the dog's supper at home and would give anything for it. This place as a whole is simply bloody! We were in wooden huts, twenty-four to a hut, each hut divided into two compartments, twelve in each and a stove in each compartment. The beds were the best part of the camp, and we make the most of it, going to bed about 9.30 pm and getting up around midday.'[29]

By 5 December, there were signs their incarceration was nearing its end. Danish Red Cross officials and parcels were expected. The representative duly arrived. The promised manna from heaven did not. By early afternoon, rumours were buzzing that they would soon be repatriated, sailing from Danzig within days.

> This news was received with great applause. On 5 December I was invited out by a Captain Robertson. This dinner was the first substantial meal I'd had since I was made prisoner. We had sausages, a tine of pork and beans, boiled potatoes, sultana pudding and a cup of cocoa, a few English cigarettes, and a pipe of Players Medium; bliss.[30] On December 8 we received orders to entrain; we paraded at 4.00 pm and arrived in Danzig at 7.00 am. The first thing we saw when we got off the train was the white ensign flying on HMS *Coventry*. As soon as we stepped on the gangplank, we knew we were prisoners no more. We were free; it was hard to believe. The band played 'Rule Britannia'.

He finally returned home on 14 December 1918.[31] Daisy and Armorer were married on 15 January 1919. As they left Hexham by train for their honeymoon in Cumberland, the Abbey bells rang out in celebration.

For a long while the Hussars Museum was crammed into the ancient Holy Jesus Hospital in Newcastle, before being moved into what is now the Discovery Museum in the 1990s, nicely fitted into the Old Coop building on Blandford Street. A few years ago, the place had a major makeover as the Charge Gallery, now also commemorating

13/18 and 15/19 Hussars linking to their successor unit the Light Dragoons (part of Royal Armoured Corps). Bobbi Goldwater is the curator and one of the team who crafted the makeover – a tribute to just what can be achieved with limited resources but a great deal of talent and commitment.

Bobbi has several 'favourite' artefacts. One is a Tarleton 'Emsdorf' helmet from 1760 worn by Private J. Williams of 15th Light Dragoons. Another is a remarkable survivor, a soldier's pelisse (a short fur-trimmed and fitted jacket), from the Crimean War era; few OR tunics of this sort survive. Private George Allen of 19th Hussars went off to serve in the Sudan (1880–1884) with his own gold-plated spoon and fork; and an astonishing letter survives from the Christmas Truce of 1914. That the NH took part is known but rather glossed over in Grant's history; such personal accounts which have endured provide a very tangible and poignant link.

My own favourite is a brass-barrelled Maxim machine gun, a proper belter with provenance. The gun was taken by Boers following the defeat of Dr Sir Leander Starr Jameson's ill-fated foray in 1896 (the Jameson Raid) but recaptured by British 5th Imperial Yeomanry at Brukspruit on 19 September 1900. Amongst the troopers who recovered the Maxim was William Scott, son of a wealthy yeoman, Henry Hall Scott from Alnwick. The farmer had pledged £1,000 and a son towards the raising of this force. The son did very well, winning a DCM and Mention-in-Dispatches. William became a noted trainer and jockey until his career ended abruptly with a fatal fall at Kelso Races in March 1925.

At Kelso Races:

John: I love horses. Did I tell you about 'Butcha' Hornby?

Graham: Yes, but don't let that deter you...

John: Well, first light on a late August morning in 1914 near a village named Casteau, the occasional flick of a tail or plume of hot breath rising as the horses waited on the road to Mons. A clatter of hooves on the road alerted mount and rider alike to the approach of another group of horsemen, and the troops of the 4th (Royal Irish) Dragoon Guards readied themselves.

Captain Charles Beck Hornby, commanding the two reserve troops of around thirty men in each, was a career soldier, and the son of a Colonel. Harrow-educated and commissioned at 18 in 1901, he had risen through the ranks serving in the Second Boer War. Like many of the British Expeditionary Force in 1914, at 31 he brought experience with him.

The reconnoitring German cavalrymen, Cuirassiers of the 9th Cavalry Division, sensed danger, perhaps the ill-timed stamp of a hoof or a twig breaking under the boot of the dismounted British waiting in ambush. The four lancers began to retreat to the relative safety of a larger group of Uhlans but were pursued by Hornby's two reserve squads who had remained mounted, the 1st with swords drawn, followed by the 4th.

Hornby's troop charged. Armed with the 1908 pattern cavalry blade, they found themselves pitted against lances. Developed in the aftermath of Balaklava, where the curved blades of previous issues had been designed for the slash rather than the thrust and had proved less than ideal faced with heavy Russian winter uniforms, the P1908 instead had a straight, needle-like blade which would sink straight through its human target. The long blade, with its bowl hilt to protect the hand and pistol grip for stability, was perfectly designed to counter the Uhlan's lance-parrying and exploit the lance's length to get inside the German guard. Although not primarily designed as a slashing weapon, the sharpened front edge is equally lethal. Hornby is said to have personally despatched three of the German party.

Thus was the British sword first blooded in the Great War, the first German casualty of a mechanised war forever defined by firearms and artillery downed by a weapon that would have been recognisable to the Ancient World. Hornby's sword remains in the collection of the Yorkshire Regiment, in remarkably good condition despite his apology to the armourer for having left blood on the blade before sheathing it.

Supported by a troop of Jägers, the Uhlans turned and fought. In the following mêlée, with 4th Squad providing covering fire to the 1st, Drummer Edward Thomas is reputed to have fired the first British shots of the war. They suffered no casualties except for one horse, euthanised and handed over to a Belgian butcher, taking three German prisoners – described by a doctor as 'Ploughboys'. Contrary to the terrifying war machine of the propaganda, the conscripts were young and had had minimal training.

Hornby was awarded the DSO for his role in the skirmish but was seriously wounded some weeks later. Promoted to Major, he went on to be mentioned in despatches twice and was awarded the Croix de

Guerre in 1917. Despite retiring from the army in 1922, he served as a Captain in the RAC during the Second World War. He died in his native Hexham, Northumberland in 1949. Good story.

Graham: Typical bloody cavalry.

King's Own Scottish Borderers Museum, Berwick upon Tweed

One of those who marched up the dusty pave to Mons was J. B. W. Pennyman from Ormesby Hall near Middlesbrough.[32] He served as machine-gun officer (MGO) in 2nd Battalion KOSB, from the start of hostilities until being wounded on the Aisne in October 1914. The Battalion embarked from Dublin on 13 August 1914: 'Left Dublin on the Bibby Liner *Gloucestershire* – expecting a hostile demonstration from the citizens but none occurred. 14 August: We passed quite close to the Cornish coast. I'm sure there was a general, though not expressed, idea that for each man this might be his last sight of England. We were packed very tight, and the ship and each regiment messed in the saloon. At Havre, teams of London dockers were already in situ.'[33]

17 August: 'Left camp at 04.00 hours, entrained at 06.00. We all, officers, and men, performed our ablutions and shaved at the station. Breakfast was at 07.00 hours, and we left two hours later, still having no idea of our destination. We were in a passenger train and reasonably comfortable, arriving at Rouen about midday. We were cheered by the French who were begging souvenirs; half the men gave away their badges. Every man we saw in the fields first pretended to twist his moustache up to his eyes and then to cut his throat to show by what means the Kaiser should die! About 22.00 or 23.00 hours we passed through Amiens.'[34]

22 August: 'We crossed into Belgium today. The inhabitants had heard we were starving, and all classes loaded us with fruit (mostly unripe), cigarettes, tobacco, matches and wine. It was an awful job getting the battalion along through these towns and I shall never forget it ... we might have been marching through the slums of Glasgow, except for the language of the people. At about 17.00 hours we arrived at the Mons-Conde Canal, where we took up our posts for the night. I put my guns in a little sandbag fort which I built on the lock abutment with the guns trained straight down the canal.'[35]

23 August: 'We were not attacked early on, so I looked around for a better location for my guns. A line of houses stood on the opposite

side of the canal, and it seemed the only place I could get any field of fire would be from the second floor of the highest house. I knew we should have to leave on the arrival of the first shell.' The section hacked loopholes through the walls and piled sandbags so the guns could fully cover the lock and its approaches. By late morning things were beginning to hot up.[36]

A, B and C companies were lining the meadows in front facing half-right. We could see the Boche deploying around 900 yards in front. They were beginning to filter into woods across the canal in front of our position and were engaged by A and C companies. A group of Boche, possibly an MG section, were spotted in front and we laid out four of them with our first traverse. We then took on any Germans we saw in the open and did considerable damage. This was our first experience of killing people. It was rather horrible but satisfying.[37]

Our infantry was heavily engaged and taking casualties. D company was attempting to winkle the Germans out of the wood. Suddenly I saw the front edge of the wood lined with Germans and surmised that they were going to try and rush D company, so I concentrated the full fire from both guns on the fringes of the wood and the Germans retreated.' Their eyrie was now a prime target: 'Gilmartin asked me whether our sandbags were bullet proof and, as he spoke; one bullet just came through and dropped on the floor. It was really red hot. Soon the fire on us became so hot indeed that bullets started coming through the walls.'[38]

It was time to move. 'I went along the line seeking a good fire position; shells were by now pounding the houses. The colonel was walking calmly up and down sucking on his old pipe and not caring a damn. By 18.00 hours we had orders to withdraw, an orderly retirement covered by the YLI. The swarm of refugees tumbling back included our hosts of the night before from the lock-keepers house. This was now blown to pieces.'[39]

Mons, the first British battle of the war, had cost the BEF some 1,600 casualties, though the Germans had received a very bloody nose and sustained much greater losses. The brunt of the fighting had fallen on Smith-Dorrien's 2nd Corps. Haig had come up on his right during the day. With his Gallic allies on his right pulling back, French ordered a retreat.[40]

24 August: 'We retreated to Wasmes, took up a good defensive site overlooking the KOSB position. The country in front was nothing but

a collection of coal mines with enormous heaps of slag. There was a terrific artillery duel all over. Again, we retired, constantly under heavy shell fire. Our limber lost a wheel in a most exposed position, and we had to lift it free and shoulder some of the contents.' The MG section was continually obliged to halt and take up a rearguard position, exhausting work in high summer heat. Though covered by a British howitzer battery, the German guns reached out unending talons of fire. Shortage of transport was becoming acute.[41]

'I went back into the village and was told there wasn't a cart available. A wretched refugee woman came along leading a pony and cart. There was nothing for it but to commandeer this. It was most heart-breaking, but it had to be done. The pony was the most miserable creature I ever saw, so we christened him 'EEP, EEP Ouray', which is what all the country folk cheered when they saw us.' [42] As the division continued its retreat along weary miles over which they'd earlier advanced, they drew on towards Le Cateau. It was harvest time. Heat lay heavy as a thick blanket, caking men and animals in sweat and dust. Le Cateau was in fact a larger battle than Mons. Smith-Dorrien wished to deliver what he termed 'a stopping blow' to halt the Germans and allow his corps to fall back unmolested by close pursuit. Although the BEF saw off a number of German attacks and then continued its withdrawal in good order, the day was dearly bought; some 7,812 men and 38 guns were lost.[43]

26 August: 'At 16.30 hours we relieved another regiment in some trenches ... well dug in and well-sited, but only very short, holding a platoon at most, and very far apart. About dawn the CO and adjutant galloped round the trenches, said we had to stay on and not retire under any circumstances. I had an excellent field of fire, about 800 yards, and we thought we would do some damage before we were all put out of action.' To their right front stood Le Cateau church and a crossroads. The poplar lined Cateau–Cambrai road ran before them.

'At about 19.00 hours we saw enormous masses of German infantry deploying around three miles ahead and realised we were in for a big thing. There was much heavy shelling but little damage; the Boche advanced to within around 2,000 yards. Troops to our front, being without orders, began to retire but the colonel just stood there on his great horse till they were all back in their trenches and then he trotted back. By some miracle he wasn't hit.'[44]

Towards evening, the KOSB position was becoming steadily more exposed. Enfilade fire was snapping over their heads. 'My [machine] guns had not fired all day, but as other troops were retiring as did we. I pointed out a place to the rear which we could use as an RV and, at the word "go" we just ran like wild rabbits bolting out of a burrow, men in my section each carrying two or three ammunition boxes.'

Matters swiftly descended into confusion. Men became separated and valuable ammo was lost. Initially, only one of the precious guns could be located. The other had to be retrieved. This was providential as loss of the prized Vickers would have been symbolically catastrophic. The army, following Wellington, refused to give up its guns. An artillery limber sufficed for ready transport and the retreat continued.

As darkness began to fall, Pennyman rounded up a platoon-sized group of stragglers and marched on. 'We lined a potato field and prepared to give any German Uhlans a warm reception. None came and we were relieved by our own cavalry. The battalion had suffered casualties and men captured. The Colonel was wounded and captured in the retreat. I was knocked flat by a shell, though seemingly unhurt. A few days afterward, however, my attention was called to a neat little bullet hole in my Glengarry strings.' A very near miss indeed.[45]

27 August: 'Our retreat continued; the men were very tired. I noticed some horrible looking carcases in the mud by the roadside. They looked so dirty and beastly that nobody had touched them but, on closer examination, I found them to be perfectly good British ration meat. So, we hacked some flesh off and went on till we saw a chance of cooking them.' By the village of Beaurevoir, the KOSBs harvested potatoes and later cooked a grand stew in a vast witch's cauldron that an old lady kindly provided for the task. 'All said, this was the best meal we'd had since leaving Dublin!'[46] As the British marched out of the village, there was more cheering, rudely cut short by incoming salvoes of German shells. Next stop was the township of Ham, but as fears of prowling Uhlans circulated, the exhausted borderers kept marching. On this time towards Muille, over the unyielding pave with summer's full heat still blazing. 'Here the brigadier was regrouping, we had marched 35 miles from le Cateau, and the men were beat to such a degree it was more like herding sheep than leading soldiers.' [47]

28 August: 'The RV was now to be at Noyon, another 20 miles away; lots of stragglers were coming in, some marching, a lucky few in motor lorries. In that week we'd marched over a hundred miles, fought three battles and two scraps' [skirmishes]. Next day, 29 August passed as a much-needed rest, followed by a moonlit march in the cool of night. More hard marches followed between 29 August–1 September

and the state of the men's feet giving rise to worry: 'It is a curious fact that all the reservists' boots were too small for them, either their feet swell in Civvy Street, or they swelled instead under the strain of constant marching.'[48]

1 September: another rearguard action. 'The enemy had no artillery or else it was very soon put out of action, and we gather what little artillery we had with us did tremendous execution amongst the enemy's infantry. After about two hours firing, we had apparently got the better of the Germans in front of us and it was rather annoying to have to retire. After fighting all day, we fell back to Silly le Loy where I was billeted in a very nice farm. Dined that night at a table and ate with knives and forks; plenty vin du pays and most exquisite brandy and liqueur rum.'[49]

Though the Schlieffen Plan appeared to be working, there were serious cracks. Von Moltke's timidity and the swelling crisis on the Eastern Front led to a thinning of the concept. With the right flank weakened, there was opportunity for a significant Allied riposte. Joffre had not lost his head and was far closer to the fighting than his opponent. His driver was in fact a star of early Formula One. Dazzling panache at the wheel enabled his commander to keep in constant touch with his front-line officers.[50] Joffre's blow, the Battle of the Marne, achieved no breakthrough. Whilst the Kaiser's armies did not crumble, the confidence of their commander did. The order came for the Germans to retreat – *the miracle of the Marne.*

Now the BEF were swinging west of Paris. 'We were hoping for a siege, at least we'd get a rest!' By 4 September they had reached Coulommes, 'ate apples, drank cider, had a rest day, and moved off at 23.00 hours, passed the field of Crecy.' 6 September: 'Our retreat was at an end, we marched back to Villeneuve, some Germans appeared in the early hours, but they were falling back now, and we captured some Uhlan stragglers.'

Next day: 'Marched to Boissy … saw a strange RAMC major having dinner with us and thought we had a new doctor [the battalion MO had been wounded in the retreat], then we found out he had been put under arrest for stopping on the line of march to buy something!

8 September: 'We engaged the German rearguard north-east of Doue. We were faced by German horse artillery and two cavalry regiments plus one heavy gun. The KOSB were ¾ mile in front of the British gun line and the Boche gun was plumping very large HE shells all amongst the battalion. Miraculously we sustained no casualties.'[51]

Despite this intense weight of fire, the order for the infantry to advance came through: 'We had to cross the brow of a hill – about five hundred yards of perfectly open ground of which they already had the range to a tee. The battalion doubled across this in two lines, went down through a thick wood and reformed in a railway cutting at the bottom. Nothing could have bettered the German artillery's efforts and I'm sure they thought they'd wiped us out. As a matter of fact, we had twelve casualties, killed, and wounded. When we got through the wood, we saw the River Marne in front of us. We chased their horse artillery off with MG fire, then, full of fury, the MG section charged on, looking for more Boche. We passed staff cars full of senior officers looking on till we found a wood we were told was full of Boche. Like pheasant shooters we waited for the quarry to emerge, but none did!'[52]

Thursday 10 September, the pursuit continued. 'We marched all day through unsavoury German remains. I found an abandoned MG limber. It was a most beautiful little conveyance, and all the fittings were A1. 11 September: I was in charge of a small group of German POWs. They were extraordinarily docile and well-behaved ... all the prisoners I saw were decent looking young men, quite good class and well-nourished. Their equipment and uniforms were excellent. I saw no sign of any atrocities nor heard of any, though of course these were widely reported.'[53]

Within two days, the KOSB were approaching the River Aisne. Here, the fluid front would harden as the Germans dug in along the favourable high ground beyond the river. The war of manoeuvre was nearly over. Stalemate was looming, even if it was not in the least obvious to the combatants. As the Borderers approached the village of Sermoise, they were held up by plunging, long-range fire. Only at dusk did they move forward into the village: 'People were very plucky, gave us coffee and anything they had to eat. Our orders were to get to the river – this proved difficult in the late summer dark, the only bridge was under fire. The battalion was very tired we'd had only an hour's sleep in the last twenty-four. In the confused night the MG section got lost and we ended up back in the village where the West Kents doctor told us we could now get over the bridge.'[54]

After a confused and difficult night, Pennyman and his MG section made for the bridge at first light. The river was no more than fifty yards across, the bridge had been blown but 'The RE had made a raft, but it had a very nasty habit of sinking, and when we got there, we found three drowned men ... oddly the Boche were not contending the crossing.' Once over, the Borderers faced a difficult task. 'The village of Missy was about ½ mile ahead, the riverbank was thickly wooded, a belt of trees thirty yards wide sloping down to the water's edge. Beyond, say ½ mile of parkland rising to wooded hills full of Boche. The West Kents were just ahead of the wood; but we couldn't advance and retreating meant swimming for it. By 08.00 hours the battalion was fully across and deployed in the trees. The Boche advanced into the copse some 700 yards ahead but our MG fire sent hem running back as quick as they could.'[55]

Enemy fire soon intensified. The Germans were trying to infiltrate the woods and a single platoon was detailed to occupy the ground. Murderous small arms fire forced them back. 'The river was alive with bullets. We lost a good many men. I wanted either to advance or retire, but the West Kents colonel ordered us to hold on. We swept the enemy ground with MG fire. One gun jammed, the first time this had happened. There was so much noise, verbal communication was almost impossible. A bullet went into the ground very close to me as I was working the other gun. I thought it might be a sniper who had caught sight of me, so I moved three or four yards to our right. Next thing I remember was a sensation like a blow from a cricket ball in the chest. It knocked me clean down and I remember shouting as I fell and bleeding profusely at the mouth.'[56]

Evacuation was difficult and protracted: 'We travelled in spring-less lorries to the railhead at Ouichy. The large, covered entrance to the station was entirely full of wounded. Mostly pretty bad: I and four other officers were put in the ticket office where we lay all night.[57] The old ticket man's daughter provided for us next morning. When she saw I couldn't get up to drink, she bent down and fed me like a trained nurse. Later our stretchers were put into the familiar carriages "Hommes 40; Chevaux 8". Spring hooks had been fitted to hold stretchers ... a stretcher is a very cold, hard thing to lie on for the best part of a week.'[58] Along the Aisne there was stalemate, and the lines began to harden. The war of manoeuvre was very nearly over. But Pennyman survived.

Daniel Logan Laidlaw VC,[59] dubbed 'the Piper of Loos', was born in Berwickshire into a military family. It was an older brother who taught young Dan the pipes and persuaded him to transfer from the DLI into the KOSB. A reservist by 1914, he re-enlisted and reached the rank of acting corporal in 15 KOSB, taking part in the battle of Loos during September 1915. It was Laidlaw – whose legendary courage, calmly walking along the parapet, was reminiscent of Caesar's 10th Legion standard bearer during the tricky landing off the south coast of England in 55 BC – who inspired his comrades to go over the top and keep moving forwards. He kept playing until he was hit. His citation reads:

> For most conspicuous bravery prior to an assault on German trenches near Loos and Hill 70 on 25th September 1915. During the worst of the bombardment, when the attack was about to commence, Piper Laidlaw, seeing that his company was somewhat shaken from the effects of gas, with absolute coolness and disregard of danger, mounted the parapet, marched up and down and played the company out of the trench. The effect of his splendid example was immediate, and the company dashed out to the assault; Piper Laidlaw continued playing his pipes till he was wounded.[60]

I was fortunate enough on the 90th anniversary of his feat to be piped across the same ground by his great-grandson – a highly evocative experience. Happily, the Germans weren't firing but the *vin blanc* was flowing.

Laidler received his VC from King George V himself at Buckingham Palace in early 1916 and later gained a Croix de Guerre. Twice promoted, he survived the war and was demobbed in April 1919. Look at a picture of Laidlaw, every inch an NCO, formidable bristling moustache, no-nonsense face and granite eyes. He couldn't be anything other than a borderer. If we want to imagine what kind of face snarled from beneath a steel bonnet, his will surely fit.

Now the KOSB museum is located in the marvellous Vanbrugh Barracks, abutting Berwick's Elizabethan Walls, two masterpieces of military engineering cheek by jowl. The galleries are gloriously old-fashioned, 'quirky' I think is the right word. Ian Martin, the curator, is rightly proud of the regiment, its history and collections, including a super archive that I have plundered extensively over the years. I've more than a few favourites here, but the one I cherish most in a WWII-era Thompson SMG which is inscribed 'donated by the IRA 1985'.

This is the type of museum I remember from my long-departed youth, a collection maintained by veterans and ordered by trustees, all or most of whom served. Attractive and informative as such galleries are, they have a distinctly limited shelf-life as the gap between those who served and their immediate families widens, and the ties begin to fray. Modern museum goers often have no direct link to the old regiment, traditions vanish into the haze and computer games prove more compelling. I find this infinitely sad, if inevitable.

At the DLI Museum, the annual military vehicle fest over August Bank Holiday. Dozens of military vehicles of all types and hue, whole platoons of re-enactors unleashed in their full kit and happy to be amongst their own tribe: Brits, Germans, Russians, even some East German Border Guards with a cracking Gaz jeep. Every type of small arm, bayonet and fighting knife. We arrived in Graham's splendid Fox armoured car:

John: Canny bacon sandwiches these...

Graham: That's your third.

John: They're free courtesy of the field kitchen and Durham County Council, what's not to like?

Graham: Bloody good event this.

John: They're not all sad fantasists like me, a surprising number have, like you, actually done this for real.

Graham: It is the tribalism. OK, this is to war rather less than voyeurism is to sex but it's about belonging, being among your own kind and able to speak freely and not feel like a berk asking anorak questions.

John: No shortage of anoraks for sure.

Heugh Battery, Hartlepool

The small port of Hartlepool on the coast of East Durham seems an unlikely setting for a Great War battle. Nonetheless, the Headland and town would earn the unenviable distinction of being the scene of a dramatic and costly fight on the cold, mist-shrouded morning of 16 December 1914. Fear of seaborne attack had preceded the Kaiser's Navy by more than a generation, though Victorian planners were more concerned with closer neighbours just over the Channel.

Hostile French intentions, however imaginary, led to a rash of coastal fortifications being thrown up and two were to be erected by Hartlepool. One in Fairy Cove and another by the lighthouse on the ancient Headland. This town within a town juts out into the cold reaches of the North Sea, its fitful moods and constant winds driving wild-capped breakers raging against the embattled point.

'The great guns roared, the fire flew/ It was a grand display/ The seagulls screamed an' flapped their wings/ And flew nor' far away/ The greet roond-shot went plish for plash/ Into the tortured deep/ They made the crabs and lobsters hop/ An' the fish could get nee sleep.'[61]

Heugh Battery received a new lease of life in 1908 when the introduction of an improved Vickers 6–inch Mark VII gun greatly added to earlier, far less telling firepower. The Victorian pile was refurbished two years later, and a pair of the new cannon installed. In 1907, the smaller Lighthouse Battery, lying a short distance south on the very tip of the Headland, was also equipped with one of these newer guns. Service in South Africa followed when the part-time gunners were mobilised against the Boers. During the early years of the century, coastal defences were augmented by a Coastal Patrol comprising several light cruisers and destroyers. By the time shots were fired in December 1914, work was in hand to further strengthen defences with the construction of an Examination Battery complete with searchlights. This was not ready when the German cruisers struck.[62]

Wary of encountering the Royal Navy in a major fight, the German Home Fleet had in 1914 adopted a more peripheral strategy involving U-boats, mine-laying and coastal raiding. In November, Great Yarmouth had been shelled, a mere pinprick, but this emboldened Admiral von Ingenohl to devise a more ambitious plan for an attack on Hartlepool, Scarborough, and Whitby.

These exposed east coast ports were sufficiently remote from any major concentration of British warships and the raiders would comprise only a portion of the Kaiser's squadron, the remainder lurking in wait to ambush any rash pursuers. On the night of 15 December, the German cruisers made steam with Admiral von Hipper leading three powerful ships, *Seydlitz*, *Moltke* and *Blucher*, against Hartlepool. The first two were newer, armed with mighty 11-inch and 5.9-inch guns, formidably armoured. *Blucher* was both smaller and more venerable, though her 8.2 inch guns could still wreak significant havoc.[63]

Like latter-day corsairs, these sleek, dark raiders cruised parallel to the unsuspecting coast, sliding past lights from coastal hamlets and the flare of heat from a locomotive's boiler. Fishing smacks and merchantmen were disregarded, surprise more valuable. Hartlepool and the mouth of the Tees soon lay before their guns, still silent beneath the dark cloak of winter's night. A quartet of fast destroyers sallied out to take on the invaders, the still night air rent with the crash of great guns reverberating over the waves. Three English vessels, seeing they were hopelessly outgunned, soon turned to flee behind trailing smoke. One rushed on heedless into the teeth of withering fire that brought her, wallowing, to a dead halt. In this one-sided fight, a pair of British torpedoes passed close by two of the raiders but without inflicting any damage.

Dawn was now creeping over the headland and at 07.30 hrs a small detachment of 18th (Pals) Battalion Durham Light Infantry deployed behind a sand-bagged perimeter set up between the Heugh and Lighthouse Batteries. The infantry had turned out from their billets in the Borough Hall, hobnails striking on cobbles, an early morning rendition of *Tipperary* echoing through the murk. One of those who marched was Charles H. Moss, a native of Seaham who would serve throughout the war. Something of a radical, he was not unduly blinded by patriotic fervour.

> I was with a specially selected detachment of the Durham Pals. After six weeks of elementary infantry training, we were suddenly sent to West Hartlepool to take up coastal defence duties. I was billeted with my company in the Central Estate School where we slept on the bare boards of the classroom. It was between 5 o'clock and six o'clock on the morning of 16th December when we were all separately woken by the orderly corporal. We were quietly ordered to 'Fall In' in the schoolyard at once 'with no lights and no noise'. We had no idea why this was being done but it was obvious that 'the wind was up' about something. When we got 'Fell In' we were issued with old-fashioned dirty white haversacks and several khaki bandoliers of .303 ammunition. It was said the haversacks were South African War issue.[64]

In charge of the 'Saturday-Night' gunners that day was Lieutenant-Colonel Lancelot Robson, a lifelong artillery enthusiast who had commanded the guns since 1908. He had moved his HQ from Fairy Cove into Heugh Battery and had shifted the signalling station into the lighthouse. He was aware of the risk of attack, alerted by Admiralty signal the previous evening: 'A special sharp lookout was needed to be kept along all East Coast at dawn tomorrow.'[65] Charles Moss and his comrades were now in the front line: 'The main rumour was that the

Germans were going to invade England on this part of the coast, and we were to provide the first line of defence. It was a cold and frosty morning with rather a thick mist out to sea, with not a ship of any sort to be seen.'[66]

Undisturbed by the brave if doomed British attempt at interception, the raiders peeled off to deliver their separate attacks. *Seydlitz* intended to steam north and engage Cemetery Battery whilst *Moltke* would pound Heugh. The smaller *Blucher* was to lag southwards targeting harbour installations and shipping in the port. Thick woollen skeins of winter fog drifted along the coast. The enemy warships glided like wraiths, mistaken in the gloom for British battlecruisers.

When Lieutenant-Colonel Robson finally caught sight of a grey-hulled monster sliding through the swell, he realised this did not resemble any British vessel. In the battery, Captain Trechmann, despite his Germanic sounding name a British officer, made the key assumption that these were indeed the enemy. Swiftly, gunners Tyson and Williamson were recalled from guard duty and assigned action stations at No. 1 gun. This is the point where men might rightly feel a tremor of anxiety, but one of the Durham men gleefully observed, 'Lads, I have been waiting for this for the last eighteen bloody years, let's give 'em the works!'[67]

Lieutenant-Colonel Robson was still trying to clarify the new ships' identities by telephone to coastguards, but the ominous thunder of the naval fight offshore dispelled any doubts. Robson had done well to abandon Fairy Cove and take station within Heugh Battery. Inside his cramped eyrie, all was bustle as rangefinders tracked the enemy ships. The first two would be engaged by Heugh, while at the lighthouse, Trechmann's cousin Dickie was shadowing the last of the three with his single gun.

For a long moment it seemed as though everyone was holding his breath, then, with the roar of an approaching express train, huge shells from *Seydlitz* blasted the terrace houses immediately to the rear of the lighthouse. The calm domestic scene of the Headland was instantly transformed into nightmare. Masonry and tiles tumbled in clouds of dust that seemed to envelop the battery. Robson's communications were amongst the first casualties. He found he couldn't reach Lighthouse Battery by telephone, nor even communicate with his own gunners. Nonetheless, his men were well trained, and this opening German salvo was their cue to return fire. British shells, far smaller than those from the raiders' massive guns, arced out to sea. More enemy rounds crashed and glissaded off the fort's glacis, careering crazily overhead, an inferno of sound.

For a moment, the infantry in their sandbagged emplacement believed the firing was all from British guns: 'When I saw some flashes of fire about a mile out to sea, followed by the thunder of big guns and shriek of shells close over our heads, our fellows started cheering because the first salvo seemed to be fired out to sea, which made us think it was our navy in action. We soon found out how wrong we were when we heard shells exploding in the town!'[68] The Durhams were ordered from their flimsy redoubt out into the dunes where they deployed in open order: 'I was very conscious of the heat and stench of the shells as they roared close over the top of us and also how deathly cold I had gone, the excitement must have had something to do with it.'[69]

Temporarily safe behind the rear wall of the lighthouse, gunners Houston and Spence detailed as stretcher bearers crouched as shells fell around them. The machine gun was gone, crew dead or dying, their puny sandbagged redoubt split and scattered. Both men ran towards their comrades, one was plainly dead another two badly hit. As they knelt by the casualties, the next salvo engulfed them too. An officer and other DLI men now rushed up to render whatever assistance they could. One soldier, deep in shock, persisted in marching up and down the roadway like a clockwork automaton.[70]

Dickie Trechmann had to resort to a megaphone to bellow orders as shells fell all around and his own gun barked in response. *Blucher* was 4,200 yards distant, but smoke and mist obscured her. Their first two rounds fell short. Their aim had to be adjusted after each round, so thick was the blanket of fog and cordite. With the third round they scored a direct hit, shearing off part of the fore-bridge. A fourth round misfired; standard drill required the round to be left in the breech for ten minutes to avoid accidental detonation, fatal for the crew. Sergeant Douthwaite ordered the rest out while he risked extracting the shell and dumping it in water. The gun stayed in action.[71]

Blucher's assigned roles were firstly to attack harbour installations and secondly to cover her two sister ships, either from the attentions of British torpedoes or fire from the shore. When Lighthouse Battery's shell crashed into her, it destroyed subsidiary armament and killed nine crewmen.[72] Undeterred, the raider began to close towards the shore. An agonising delay ensued for the Durham men whilst their blocked breech was cleared, then more of their rounds struck one of *Blucher's* gun turrets. It was unclear if the British gunners had damaged the warship's steering gear, but she appeared to come close to running aground before turning and switching fire against British cruisers now sliding out of the harbour, inflicting considerable damage on HMS *Patrol*.[73]

Robson's guns at Heugh Battery experienced similar problems with misfires as the electrical detonation system malfunctioned alarmingly. Moving to percussion firing, the guns blazed at the raiders, focusing their fire on *Seydlitz* in the lead. Vibration from the guns was making the work of the rangefinders almost impossible and Robson, whilst he knew his position was extremely exposed, believed it to be his duty to remain. Consequently, he and his assistant Captain Walsh kept station with commendable sang-froid as battle raged around them. Robson had time to contemplate the considerable gulf between exercises and the real thing and decided if he should survive to see another, he'd arrange for the battery commander to be knocked out in the opening stages.

In Lighthouse Battery, the gun was constantly plagued by misfires, though this did not prevent the loaders toiling below from sending up a constant supply of shells and explosive. These now began to build up on the surface, potentially lethal as German rounds continued to fall. Captain Jack Farmer, struggling to clear the gun, was only alerted to the danger when a German shell bounced directly in front and skidded overhead.

The gun was cleared, and another eight rounds fired. Half of these battered their target before *Blucher* steamed northwards, effectively out of range. Gunners at Heugh were still duelling with *Seydlitz*'s massive broadside, at least one enemy round came perilously close to No. 1 gun, a mere six feet in front of Gun Captain Mallin, but again it went over the glacis and exploded in a field behind. Mallin and his crew were spared, the grazing donkey was not so fortunate. Another round slammed home terrifyingly close to the ammunition lockers. Providentially it was a dud.[74]

Despite the high drama, the fight had its near farcical moments as the Durhams watched:

> At the height of the bombardment some fishermen were beaching their boat, one of them got wounded. Sergeant Heal and Corporal Brewerton ran down to the shore and carried the wounded man to a safer place. One remarkable sight was to see a train come puffing along the embankment

> from old to West Hartlepool when the bombardment started. The train was in full view of the sea. When it got about halfway on its journey, the driver stopped the train. He and the fireman ran down the land side of the embankment while all the passengers also jumped or rolled down to comparative safety. There the train stood throughout the rest of the bombardment, a perfect target for the German gunners yet it did not receive one direct hit. When the shelling finished the driver and fireman got back onto the train and away it went.[75]

As *Seydlitz* steamed clear, Trechmann ordered his guns to direct their fire against *Moltke*. He also ordered the gunners to switch to high explosive (HE) and aim at the ship's super-structure, as the belt armour on the Germans' hulls rendered them safe from the 6-inch rounds. After a few fell short, the British shells now began to take effect, sweating gunners cheering 'Go on shove 'em through! Shove 'em through! Give the buggers hell!' Such enthusiasm, understandable if unprofessional, earned a swift rebuke from their officer: 'Do stop your dancing about and save your breath for your gun corrections!'[76]

Infantry were still fanned out in the dunes but without any enemy in sight: 'There was a road which led to Sunderland, quite close to where I was and I could see packs of terrified dogs running along with heads down and tails between their legs, all hell bent in getting away from the awful destruction and pandemonium.'[77] By this point, the gunners realised they were no longer being fired at. The enemy ships had switched their murderous attention to the docks and to the unprotected settlement behind.

Blucher was blasting the shoreline immediately in front of the battery, plastering the rocks with smoke shells, attempting to blind the guns. Trechmann now decided to give her his full attention and the men swapped back to AP rounds. Though the shells did little damage, *Blucher* began to steam clear as the other two raiders swung around to loose more rounds towards the old Cemetery Battery site. *As Blucher* slid into sheltering fog, the other two cruisers, still under fire from Lighthouse Battery, also retreated.

Dickie Trechmann ordered his men to cease firing. The enemy was now some 9,500 yards distant. It was 08.53 hrs. After the fury of battle, silence descended and sweating, smoke-grimed gunners realised they had survived. These Saturday Night soldiers had endured their ordeal and stood by their guns. They had taken on the might of the Kaiser's Navy and seen them off. As they paused and looked back over the town, they began to understand the cost. The infantry saw it too, close up: 'As we marched back to our billets, we saw some of the terrific damage that the shelling had done, whole streets of houses had been wrecked. One street near the school had their front walls completely blown out, exposing the contents of the rooms as though they had suffered an earthquake.'[78]

Hartlepool was ablaze. A flaming gasometer lit the morning sky like a torch held aloft for the dead. There were many dead and many more injured, over a hundred in all, dazed survivors walking shattered streets, ruined houses bulging like split sacks, the calm day-to-day of their ordinary lives suddenly snatched away. The war had reached Britain.

Diane Stephens curates the Heugh Battery Museum. The old fort welded to the uncompromising spit of the Headland, a place that seems disconnected from the Port of Hartlepool, preserved it seems in another time. Lighthouse Battery has gone, but the elegant terraces burst by German guns on the day were rebuilt. The Heugh is primarily maintained and run by volunteers, many of whom were ex-gunners who have done marvellous work in restoring an eclectic range of artillery pieces. The guns still point out to sea and Diane sees the Battery as a living part of the town's history, a source of pride to the community where young and old can connect with their heritage and draw encouragement and inspiration. A place not just of commemoration but one which happily fulfils a less martial role, bringing visitors and revenue to the Headland.

It is this physical connection Diane treasures, the way the Battery moulds to the topography with the smooth sweep of the glacis and fit of the defences, almost like a prehistoric *dun*. One object she particularly treasures is a skeletonized Mark III Lee Enfield, a battlefield relic from the Somme, wooden furniture long rotted away, just the bare bones; what stories could it tell?

The Green Howards Museum, Richmond

Steve Erskine, like so many curators of regional military museums, is a great enthusiast and has always been most helpful during my numerous research visits. It's a great museum, slap in the middle of the lovely market town of Richmond and squeezed into a brilliant conversion of the former Trinity Church. One of my personal favourites a few years ago was a special exhibition on personal protection, which featured notorious bushranger Ned Kelly's homemade armour, a Poundland Robocop. One of the best of many fine exhibits is a military coat worn by Major General Samuel Townsend (1732–1794), who served with the regiment from 1760 to 1769. He saw active service at the siege of Belle Isle in 1761. Military uniform mirrors civilian fashions. In the 1780s the style of this coat would have been typical formal attire of the fashionable 18th-century man. The coat is single-breasted with three hook fastenings at the chest. It is then cut to sweep away to the hips. The coat is heavily tailored to exaggerate the hourglass shape of the wearer. Narrow sleeves were very fashionable, and the fabric was cut to reduce excess material around the elbow. Large pockets, which span almost the width of each tail, were cut into the lining.

Stanley Elton Hollis was born on Teesside in 1912, so he was in his thirties when he landed on Gold Beach with 6th Battalion Green Howards. He began his career as a navigational officer in the Merchant Navy but that ended with a serious bout of fever. Before the war he worked locally as a lorry driver and joined the TA, 4th Battalion, transferring to the 6th on mobilisation. He had an active war, very in fact, promoted to sergeant at Dunkirk. He then served in North Africa and Sicily, being wounded during the savage fight for Primosole Bridge. Just as well, he'd never be more severely tested than at Gold Beach on D-Day.

His battalion had trained by Inverary on Scotland's rugged north-west coast. Reveille on 6 June for the Green Howards was around 02.30 with a decent breakfast for those who had the stomach. They would transfer from their transport *Empire Lance* to landing craft, which they loaded up with kit and ammo beforehand. They were about six miles out, a grey, choppy and gusty morning, the cold hour before action on a sour belly. Stan was passed a box of condoms by his CO Major Ronnie Lofthouse and told to give them out. Very likely sex wasn't much on their minds. Hollis' earthy response was – 'Sir, are we going to fight 'em or f**k 'em?'[79] In fact, the contraceptives were intended as waterproof tompions to keep their rifles dry.

Getting down via the nets into the landing craft wasn't easy, with the violent pitching of both vessels and the weight of kit. If the motion of the ship was bad, this was much worse and the laden bobbing craft had to cruise around in circles until like two families of ducks they set off in line abreast, 'A' Company on the left and 'B' on the right. As they chugged steadily if not speedily (around 5 kph) towards the hostile shore, which took a good hour or so, the whole Allied bombardment provided a deafening chorus, the thump of those big naval guns, crack of 25-pounders, bright salvoes of streaming rockets, rounds spurting sand from the beach. Hollis profoundly hoped this might do for the mines.[80]

Stan identified a German strongpoint dead ahead as they cruised in to land (in fact it was a railway shelter), and grabbing a Lewis gun he rattled off a full pan of ammo (the marks were there for years apparently).[81] The gun was stripped of its water cooling jacket and as he hefted it clear of the bracket, forgetting it would be red hot by now, he badly blistered his hand.

They plunged into the surf. Just ahead of Hollis another old sweat, Sergeant Hill, slipped in a shell hole and went under dragged down by his kit; the craft's propellers chopped him to bits. As they got to the shore, their beach-master went down. Mortar rounds were coming in at a depressingly high rate, though they weren't getting any small arms fire. The plan called for Stan to lead mortar-men and Bren gunners from each platoon to charge ahead and set up at the high water mark, providing smoke and covering fire to get the rest through the belt of mines ahead. On their right a tank brewed up and the turret hatch bowled along the sand, a lethal projectile, but no-one was hurt.[82]

One enemy plane swooped down overhead, but they got away with that, too. Up the beach and onto a low ridge of dunes festooned with thick wire entanglements. Birds were sitting apparently unconcerned on the coils. One joker suggested they had no choice as there wasn't any room left in the sky. Ahead now was a dense belt of mines. 'D' company were first through after their assault engineers had gapped, Stan and the others followed the reassuring lines of white tape. Beyond the minefield lay Meuvaines Ridge and Mont Fleury Batteries.

Once through the hedge beyond the belt of mines, the Green Howards were fully exposed to the attention of the German defenders dug in on the higher ground and a storm of fire descended as they struggled forward; it turned out Private Mullally the joker had offered his last quip. Inch by fire-swept inch they crawled forward. Major Lofthouse had spotted the pillbox that was doing most of the damage. Hollis saw it, too, and stormed forward, his Sten chattering. He made it and lobbed a grenade in, killing two defenders and persuading the rest to give up – 'they were quite willing to forget all about the war.'[83]

He barged ahead, up a shallow communications trench aiming for a larger bunker the inhabitants of which went into the bag, 'about eighteen or twenty'. A pretty decent haul and it turned out these were the fire control team for the battery up ahead. As the battalion advanced up the hill, Stan looked back just for an instant and saw, from his higher position, the whole vast armada laid out – 'it gave us a great feeling of confidence.' It was only 09.30 and they could see enemy bolting from their positions, but not that far, falling back behind a sheltering wall and firing. Hollis saw one German crazily loping along the top of the wall. Swapping his Sten for an Enfield rifle, he brought the fellow down first shot but was lightly wounded in the face just after, 'not a lot of damage – a lot of blood, it looked a lot worse than it was.'[84]

On they went into the village of Crepon. With Lieutenant Patrick now dead, Stan was commanding 16 Platoon and the Major ordered the company to check/clear the several farmhouses lining the approach road. Stan broke into one of the silent steadings; it seemed deserted except for one terrified boy, perhaps ten or eleven. The effect of seeing this ferocious, blood-garnished veteran bursting in must have been terrifying. As he came out to check the rear, an enemy round smacked off the back yard wall, fragments whizzing. Aside from a pair of excited local canines, he could just about make out an enemy gun.

After reporting back to Lofthouse, Hollis detailed half a squad to start shooting up the hedge in front of the enemy. This drew a sharp response and most of the group were killed straight off. Another approach was needed and the major ordered Hollis to get hold of a PIAT[85] with a brace of Brens and creep forward through an extensive rhubarb patch which gave out onto a line of trees. They got forward without drawing fire, and Stan blasted off a round from the unwieldy PIAT. It missed and the German gun shot up the farmhouse in retaliation; stone timber, plaster and dust erupted. Stan crawled back to report, but intensive bursts of fire from both sides told him his two Bren-gunners were in trouble.

Stan grabbed a Bren himself and started crawling back, sheltered briefly behind what was left of one of the walls, then in a lull charged forward spraying the distant hedge. Under cover of this dramatic diversion, the other two blokes made it back unscathed. A tank arrived and moved ahead of what was now a general advance, 'C' Company on the left, 'A' and 'B' on the right, Stan's platoon was left of HQ, sheltering behind the tank. It was raining. The tracks chewed through mud and then stopped; small arms fire all around them. More men were going down.

Now, Stan could see a lively pair of Germans firing down the narrow lane, one shooting, one covering/re-loading; the Tommies were very exposed and getting shot up. Stan decided he'd see to this duo and felt in his pouch for a grenade. While he'd seen to it earlier that everyone else had plenty ammo and bombs, he had forgotten his own supply![86] Hastily 'borrowing' a grenade, he chucked it. 'I used to throw them like a cricket ball.' He dashed out under the cover, he hoped, of the blast. Second mistake, he'd forgotten to pull the pin, but the Germans were still heads down waiting for the blast, so he was able to do for them both.

His platoon was down to just over a dozen, but now they'd reached the enemy-held hedge, there was another just beyond, so the sunken ditch between meant Stan and co could crawl in relative safety, albeit in wet discomfort, between the two. As they slopped through the mire, it was obvious they'd not been seen, so effectively came on the flank of those Germans still holding out. Bringing up whatever automatic weapons they still had, Hollis got them to spray the far hedge and beyond. Enemy fire immediately slackened and then stopped.[87] By now more armour had appeared and Stan and Sergeant Major Moffat from 'B' Company were able to get a badly wounded officer, Captain Young, hoisted onto the hull and casevaced.

Stan Hollis won the VC for his actions, the only one to be awarded on D-Day. His citation reads:

> In Normandy on 06 June 1944 Company Sergeant-Major Hollis went with his company commander to investigate two German pillboxes which had been by-passed as the company moved inland from the beaches. Hollis instantly rushed straight at the pillbox, firing his Sten gun into the first pillbox, He jumped on top of the pillbox, re-charged his magazine, threw a grenade in through the door and fired his Sten gun into it, killing two Germans and taking the remainder prisoners.
>
> [Later the same day] C.S.M. Hollis pushed right forward to engage the [field] gun with a PIAT from a house at 50 yards range... He later

found that two of his men had stayed behind in the house... In full view of the enemy who were continually firing at him, he went forward alone to distract their attention from the other men. Under cover of his diversion, the two men were able to get back.

Wherever the fighting was heaviest [he] appeared, displaying the utmost gallantry... It was largely through his heroism and resource that the Company's objectives were gained, and casualties were not heavier... He saved the lives of many of his men.[88]

The Durham Light Infantry Museum (DLI), Durham

The Somme and its bitter autumn of endless cloying mud. Though further fighting was not formally called off until 18 November, general exhaustion and vile conditions meant the final gasps were sporadic. For the DLI battalions comprising 151 Brigade, there was to be one final test, bitterer and bloodier than those which had gone before. This was to be an assault upon the Butte de Warlancourt:

> Gird trench ran east and west from the Albert-Bapaume road towards Gueudecourt. On the left lay the Butte de Warlancourt, a mound or tumulus some forty feet high, reported to be an ancient burial-place similar to those found on Salisbury Plain. This, in September, when the Battalion first entered the Somme fighting, stood out from the surrounding country a green, conical-shaped hill. Of little or no strategic importance, except that it provided observation of all ground toward High Wood, Martinpuich and east of that village. It had been so battered by the daily shelling that all signs of vegetation had now disappeared, and it stood a shapeless, pockmarked mass of chalk. Beyond the Gird Trench lay a stretch of undulating country with Bapaume clearly visible in the distance, and midway almost hidden in a small valley, Le Barque. The remainder of the attack frontage held no special feature except a considerable amount of dead ground to the rear of the objective.[89]

The Butte, a squat carbuncle, lay in the path of any further British advance. The Durhams of 151 Brigade (6th, 8th and 9th Battalions DLI), were to attack at 09.10 on 5 November. Their brigadier was Roland Bradford who had won his VC in desperate fighting around le Sars during the previous month. Roland, at 25, the youngest ever Brigadier-General in the British Army, and his brother George were to be the only siblings who both won the supreme accolade. Neither survived the war. Roland would die at Cambrai in November 1917, his brother in the course of the great Zeebrugge Raid on St George's Day 1918.

At this late phase in the Somme battle, conditions were dreadful; sodden, heaving ground churned into a noxious, cloying soup, raked and scarred by endless shelling. Yet, despite Roland Bradford's deep misgivings, the attack was to proceed. The brigade would advance each battalion with three companies up; four waves, thirty paces between each and in columns of platoons. Their combined task was to secure the Butte itself, the quarry in front of it, sections of Gird and Gird Support trenches. Conditions were worse than ever, pounding rain matched the familiar, seemingly endless tattoo of guns. Wild wind buffeted the struggling Tommies, wading through deep muck, weighed down by heavy pack and wet wool.

Lance-Corporal Harry Cruddace 1/6th Battalion DLI was amongst those waiting for the order as dank, unfriendly dawn crept near:

> Every man then looking to the loading of his rifle and the fixing of his bayonet as the Zero Hour was almost upon us. Serious men gazed eagerly into one another's faces and some muttered thoughts of God and their loved ones at home. Chums clasped hands and said 'Cheerio' not knowing what the day had in store from them. Officers were in eager conversation with the senior NCOs regarding the readiness of the men. The enemy kept up a ceaseless bombardment of our trenches in conjunction with the merciless rain and cold.[90]

As ill-luck would have it, the Durhams were to assault the German trenches at a point when the garrison was being relieved, so the defenders would be far more numerous than anticipated.

> At last, the Zero Hour arrived, and the officers' whistles sounded the advance. Immediately the first wave mounted the trench and made off in the direction of the enemy trenches. They were met by terrific and annihilating fire and crumpled up like snow in summer. The second wave was, by this time, on its way. I was in that wave and placed my gun sections in single file to make a lesser target. The enemy barrage was doing enormous damage and our fighting strength was fast diminishing.[91]

Dragged down by the sinking slime, the numbed Tommies inched forward into enemy fire:

> Out of my two sections of fourteen men there were two of us left – a No. 1 on the gun by the name of Private Allen and myself. I pushed on with one gun and a quantity of ammunition to about 30 yards from the German trench and took up a position in a shell hole. We opened fire on the opposing troops who formed an excellent target. In taking up my position, in the excitement I placed myself on the right side of the gun instead of on the left, which was fortunate for me. After firing one or two magazines, the enemy found us with a machine gun and succeeded in wounding my No.1 in four places down his left side.[92]

The hurricane of enemy fire cut swathes through 6th Battalion and the 8th were similarly winnowed, together with the Australians on their flank. Only 9th Battalion clawed out a precarious foothold in the German lines, a series of posts or mini strongholds being carved from Gird Trench, the Butte and Quarry areas.

Bradford's men found their hard-won gains very difficult to hold. Despite the skill of his deployment, the Germans were intent on ejecting the surviving Durhams. Artillery rounds and the angry buzz of snipers blew and snapped overhead: 'The Germans in the dugout on the northeast edge of the Butte brought a machine gun into position and were worrying us from behind. Many gallant efforts were made throughout the day to capture this dugout without success. All our parties who tried to rush it were destroyed by the German machine-gun fire and the large numbers of snipers in Warlencourt.'[93] As morning dragged towards noon, sporadic counterattacks began to nibble at the margins. These were seen off. During the early afternoon, more determined efforts with bombing parties were also beaten back.

Though the Durhams held the Butte, that surviving enemy dugout was a constant menace. Successive attacks caused mounting casualties and the dense curtain of artillery shells ploughing the wastes of No-Man's-Land prevented any attempt to bring over reinforcements. Mid-afternoon and the pressure was relentless:

> We have been driven out of Gird Front Line and I believe my posts there are captured. I have tried to get back, but the enemy is in considerable force and is still counter attacking. It is taking me all my time to hold Butte Alley… About 6 p.m. the Germans made a determined

> counterattack preceded by a terrific bombardment and we were able to get to close quarters. A tough struggle ensued but our men who had now been reinforced by the Reserve Company and who showed the traditional superiority of the British in hand-to-hand fighting, succeeded in driving out the enemy.[94]

Darkness brought no relief from an increasing tempo of German assaults, outflanking the Durhams rapidly shrinking salient. As 5 November drew to a close, the bloodied and exhausted survivors were being forced back towards their own lines:

> The Germans still holding out in the Butte dugout came out and advanced over the Butte. The enemy advanced throwing bombs. A party of about twenty Germans worked round our left flank and attacked the Quarry in the rear. The enemy was in great strength. This attack was perfectly organised and was pushed in with great energy and determination. The enemy advanced beyond Butte Alley and evidently had intended to capture Maxwell Trench. Our men resisted heroically but after a desperate stand were driven back to Maxwell trench, so that by 1 a.m. on the 6th we were in the same position as on the morning of the 5th prior to the assault.[95]

For the Durhams, this Calvary was over; many men had been lost and no gains had been made.

From 2006–2009, I was a heritage officer for the County Council, based at DLI Museum where I worked with the brilliant and dedicated curator Steve Shannon and his team. There can hardly be a family in Durham that didn't have some connection with the regimental museum, one of the most famous in Britain's Order of Battle. Although amalgamated in 1968, the name lives on with today's Rifles (2007).

Generally regarded as the first purpose-built military museum in the UK when it opened in 1966, the year in which regimental colours were formally laid up in Durham cathedral, it may have been the case that the building was intended initially to become the home for what would later become the sprawling Beamish, Museum of the North. It was an innovative, modernist structure, set by a small lake, constructed on former industrial ground. It was a wonderful space, and a decent chunk of my boyhood was spent there. It helped nurture the seed that would lead me to becoming a military historian. Even during my tenure, we were fighting a rear-guard action against a council anxious to close the place and in all probability, sell off the land to developers. The subsequent closure happily sparked a backlash and may have influenced the prevailing junta's demolition in the 2020 local elections. Plans for a re-opening are afoot.

What has to be the most totemic object (for me anyway) is an Ordnance QF 2-pounder anti-tank gun of WWII provenance. This is the famous 'Wakenshaw' gun (see below), painted in desert colours and bearing the numerous scars of its final battle. Of itself, the gun isn't in any way remarkable. Manufactured by Vickers Armstong, it entered service in 1936 and despite a number of serious shortcomings, stayed in the field as late as the Arab-Israeli war in 1948. It's a well-constructed, light, and robust weapon, 40 mm calibre and normally operated by a three-man crew. It was fitted to a range of armoured cars and tanks. Initially effective enough, it was quickly outclassed by up-armoured panzers and performed weakly in the Western desert until replaced by the heftier 6-pounder. Nonetheless, around 12,000 were built.

The next war, Western Desert 1942; after the Fall of Tobruk and the humiliation of the 'Gazala Gallop', it was on to the 'Msus Stakes'. Here, it was the turn of 9 DLI (as part of 151 Brigade), to feel the weight of Rommel's onslaught. By 02.00 hours on 27 June 1942 the ring was tightening, and three hours later the assault began in earnest. Early in the action there occurred an incident which soon passed into legend and is remembered by the DLI and its successors today:

> Quite early in the action a German tracked vehicle towed a light gun to within close range of one of the Battalion's two-pounder anti-tank guns, sited on a forward slope in front of the infantry positions. The Battalion gun crew opened fire and put it out of action with a round through the engine. The Germans replied with another mobile gun and either killed or wounded the men manning the 2-pounder. They then moved forward towards the damaged tractor and tried to get the light gun in action against the infantry.[96]

Private Adam Wakenshaw had already lost one arm, yet he crawled back to the damaged two-pounder and fired a further five rounds. The German vehicle burst into flames and the gun was damaged. Another German round struck the position, killing the single other survivor and wounding Wakenshaw again. Despite these dreadful injuries, he somehow crawled back to the weapon and managed to insert a sixth and final shell into the chamber just before a direct hit killed him. His body was found that evening. For such astonishing gallantry he was awarded a posthumous VC. In that day's fighting, often hand-to-hand, the 9th Battalion suffered grievous loss, with virtually all three forward companies being destroyed. Only the HQ elements remained to finally withdraw.

Palace Green – England's death march. 'I beseech you in the bowels of Christ think it possible you may be mistaken.' So, Oliver Cromwell, not quite yet Lord Protector, implored the Scottish Parliament to abandon their shaky alliance with Charles II. He failed to persuade. The campaign which followed, desultory at the outset, ended with Cromwell's decisive victory at Dunbar on 3 September 1650. Many regarded this as his finest hour. What ensued was not.

That September, approximately five thousand men began a forced march from the battlefield of Dunbar to Durham, destined for the Southern ports. It took them seven days, without food or medical care and with little water. They were now property; the chattels of a ruthless regime determined to eradicate any possibility of further threat. Hundreds died or were summarily executed on this English via dolorosa.

Those who survived long enough to reach Durham found no respite, only disease and despair. Exhausted, starving and dreadfully weakened, perhaps another 1,700 died there, most probably from fever and dysentery. For those who survived, hard labour awaited them. They faced forced exile as virtual slaves in a harsh new world across the Atlantic. And what were the prospects for their families left behind to fend for themselves? It could reasonably be defined as a war crime.

Transportation to a life of indentured servitude featured throughout the 17th century. Its use after the Monmouth rebellion in 1685 gave Rafael Sabatini the major plot line for his 1922 novel *Captain Blood*, memorably filmed a few years later with Errol Flynn in the title role. Some accounts suggest the full tally of Scottish prisoners after the battle was in the region of ten thousand. Almost half of these were non-combatants, camp-followers, tradesmen and the like, who were released without sanction. The uniformed captives, around five thousand (a more exact number cannot be given), were deemed too great a threat and had to be neutralised. Militarily, this was sound, as 1651 would demonstrate. Charles II was well able to raise enough men for his doomed march to Worcester.

The men of Dunbar embarked on a journey into servitude. The long convoy (five to six miles in length) was initially shepherded the 20 miles (32 km) to Berwick upon Tweed guarded by a single troop of 25 cavalry/dragoons; or so the record maintains. It seems impossible that a single troop, even mounted, could control so large a number. We know most of the captives were quite young, 18–25 with some even younger. Cromwell saw a commercial opportunity here. Transportation as an indentured servant had long been a means of capitalising on the American colonies' need for semi-skilled and skilled labour.

Their initial ordeal ended on 11 September when they were marched over Framwellgate Bridge into Durham and the sanctuary of the great Norman cathedral. They had already spent a night in a church, St Nicholas in Newcastle, where their disordered bellies had resulted in such fouling that the burgesses had been obliged to pay for a major clean-up. By now, many were so weakened that disease spread easily. Of the three and a half thousand counted through the cathedral's doors nearly half, seventeen hundred, died within a short space of time. Their remains were buried in pits dug on the city's Palace Green, then open ground as the name suggests. Within a decade the graves were lost when the area was developed and built over, as if hiding a guilty secret, one that didn't come to light for over 360 years.

The surviving Scots presented their English captors with a significant problem. Holding such a large number of prisoners would be costly. However, letting them go could prove very dangerous. One week after the battle, the Council of State, England's governing junta, decided to turn the problem over to the powerful Committee of Safety. They informed veteran parliamentarian Sir Arthur Hesilrige, Governor of Newcastle, that he could dispose of as many of the Scots as he deemed fit to the coal mines and other industries. Armed with that authority Hesilrige consigned forty men to work as 'indentured servants' (effectively forced labour) in the salt works at Shields.

He then sold another 40 off as general labourers and set up a trade in linen, twelve of his prisoners becoming weavers. He may have been making use of existing skills. Dental analysis carried out on one of the recently rediscovered bodies showed damage to the teeth consistent with regularly using them to saw thread ends. Hesilrige was clearly a strong believer in private enterprise and was not above using his position to build up his personal wealth and then flaunting it.

Alongside these developments, the Council of State received several applications from entrepreneurs in the American colonies hungry for cheap labour. On 16 September, negotiations began. The petitioners, John Becx and Joshua Foote, conferred with their partners, the ominously named 'Undertakers of the Iron Works' and three days later, Hesilrige was directed to transport a hundred and fifty prisoners of war to New England. The brokers insisted they should only receive strong, healthy specimens, best quality.

These chosen men were sent down to London by ship where they stayed in limbo while Parliament made sure they could cause no further trouble in the Americas. It was not until 11 November that the Council finally issued sailing orders to the master of the *Unity*, Augustine Walker. He weighed anchor right away. The passage from London to Boston normally took six weeks and was likely very disagreeable. Becx and Foote acquired these Scottish assets purely as a commercial venture. They paid five pounds a head and could anticipate selling each man's contract for five or six times that amount in the New World.

In November 2013, during construction of a new café for Durham University's Palace Green Library on the city's UNESCO World Heritage Site, human remains were uncovered by university archaeologists. The jumbled skeletons of what would prove to be twenty-eight individuals were subsequently excavated from two burial pits. It was the start of five years of meticulous investigation. A team of experts from Archaeological Services, Durham University – the University's commercial archaeology consultancy unit – and academics from Durham's Archaeology and Earth Sciences departments worked together to excavate and analyse the bones.

From the outset, the Durham team acknowledged the possibility that these might be some of the Scots soldiers of 1650. There has long been folklore about these men and what they did in the Cathedral where they were held. Indeed, one local woman who visited regularly as a child has difficulty remembering when she first heard the tale, so ingrained a part of her childhood had it been. The driver of the digger which unearthed the remains was also quick to suggest they must be of the Scots prisoners.

In May 2018, the 28 men were reburied in Elvet Hill Road Cemetery in Durham, less than a mile from the spot where they were discovered. There was huge interest in the event, particularly in Scotland where newspapers had covered the Durham discovery in some detail from day one. Handfuls of Scottish earth were thrown onto the coffins and great care was taken to reflect the traditions of worship of these 17th-century Presbyterians. The service was put together by Durham Cathedral, representatives from the Church of Scotland and the Scottish Episcopal Church. Metrical Psalms from the 1650 Scottish Psalter and a Bible reading from the 1611 King James Version were included – an expression of the wish of all involved to honour the traditions of the dead.

These men do not have a dedicated a museum per se, but a fascinating exhibition was staged on Palace Green in 2018 that profiled the work and findings of the archaeologists. It was a shame at the time and still is that some form of permanent exhibit was n0t possible. This is a dark moment in British history which deserves to have an enduring spotlight.[97]

At Palace Green café:

John: Part of the problem with history, aside from just being one bloody thing after another is that some events disappear without trace. Who'd have said this serene place was the site of a war crime?

Graham: It's not always that we forget, sometimes we don't want to remember, and society goes in for collective amnesia. Once the red mist dissipates, folk just prefer to forget and those who were there don't want to talk about it, partly because only those who were there would understand, and they don't need to be reminded.

John: This will be a story to tell your grandchildren.

Graham: And mightily bored they'll be...

Richmond Castle and the 'Conchies'

After conscription was introduced in 1916, conscientious objectors ('conchies') had a rough ride during the conflict itself, when perceived shirking provoked contempt. Only after 1918 as the true cost began to sink in, would attitudes shift:

You'd leave me cold though this our arguing
Endured till dawn, you lack the essential thing.
Healthy and leisured, pious, gentle, learned,
'You feel forbidden to fight' and thus unconcerned
With that vast horror which defeat would bring.

Hence, tho' you touched of David's lyre the string,
Or wrote with quill plucked from an angel's wing,
Unless you 'gave your body to be burned',
You'd leave me cold.

'You feel it wrong to nurse.' While others fling
Red life in the scale, what is your offering?
Levite! In you the milk of pity has turned!
For if sore wounded, and with eyes that yearned,
I lay in the Jericho Road a-perishing,
You'd leave me cold!

H.M.W: 'To a Conscientious Objector' (30 December 1916)

Richmond is a very pleasant town. As well as the Green Howards museum it possesses a very fine Norman castle with a long and colourful history, but I'm going to focus here on its role as a gaol for those who, during the First World War, refused to take up arms.

James Sadler, my grandfather, was a bricklayer, born in Byker in 1890 into a nonconformist liberal tradition, he and his three brothers had links to the socialist movement. all were members of the Peace Pledge Union and latterly Independent Labour Party. James was an absolutist and refused all forms of service. He was incarcerated in Richmond Castle and family tradition asserts he was taken out, tied to a firing post and blindfolded, having been given a final chance to enlist. He refused and the execution did not proceed. He went on hunger strike and was force fed, he became so weak he was discharged from prison and sent home in the belief he was already dying. He did survive, but his health never fully recovered. He died aged 70 in 1960.

Among these 'absolutists' was Wearside man Norman Gaudie, a 28-year-old accounts clerk on the North-Eastern Railway and a footballer at Sunderland AFC. The son of a Quaker, Gaudie was a committed pacifist and held an absolute objection to any form of military service, including non-combatant roles that would aid the war effort.

In March 1916, Gaudie made a claim for total military exemption at a tribunal in East Boldon. This was refused, however, and he was exempted from combatant service only. Having subsequently refused to comply with a notice to report to the army, Gaudie was arrested and appeared before Jarrow Police Court on 20 April 1916 charged with being absent under the Military Service Act. Pleading not guilty, Gaudie was fined 40s, which he said he did not intend to pay, and he was remanded in custody to await a military escort. Gaudie was then taken to Newcastle Barracks where he was subsequently posted to the 2nd Northern Non-Combatant Corps based at Richmond Castle, North Yorkshire.

On his arrival, Gaudie refused all military orders and was part of a group of 15 other absolutists who were forcibly stripped, dressed in khaki uniform, and locked into the guardroom cells. Soon to be known as the 'Richmond Sixteen', the group included Quakers, Methodists, Jehovah's Witnesses, and socialists, and would become a source of contention for both civil and military authorities. News of their treatment quickly reached the public, and the subject was raised in the Houses of Parliament.

While few initially felt much sympathy for the plight of COs, the brutal manner of their incarceration began to attract significant criticism. Over time, some grew to respect the steadfast beliefs of the men and their determination to stick to their principles under such duress. On 13 May 1916, the *Northern Echo* published the following article describing the treatment of the men at Richmond Castle:

> He heard on Saturday of a case of a man Yorkshire who was taken to Richmond, and because he resisted, he was frog-marched through the town and came back in a bleeding condition. This was not the kind of treatment to put a stop to this problem. These men might be unwise, stupid, foolish, but that was utterly un-English treatment. He knew of many cases, on the other hand, of extraordinary kindness shown by the military. Bullying should be sternly dealt with by the War Office. He wanted, too, to stop the large number of new cases that would arise unless something further were done.

For Lord Kitchener, already under huge pressure following the crippling losses sustained during the first two years of war, the situation was becoming increasingly damaging. Determined to take decisive action, he ordered the men to be removed from the castle and shipped to France. Once there, they would be on 'active service' and could be sentenced to death for any refusal of a direct military order. On 29 May 1916, Gaudie and the other absolutists were forcibly removed from their cells and secretly transported to Southampton, where they were joined by another group of Objectors. The group was then shipped to France aboard the torpedo ship *St Tudno* and landed at Le Havre the following day. After spending the night at Cinder City Camp, Gaudie and his group were transported north to Boulogne, where they were eventually housed at Henriville Camp. There they were told that another group of COs had already been executed by firing squad; a lie.

Despite the threat of execution, the men maintained their stance and continued to ignore military orders. Eventually, they were given 24 hours to make a final decision and were allowed free time in Boulogne to discuss the matter. They were told if they accepted service in the NCC, their previous non-compliance would be forgotten. The following day, Gaudie and his group announced their intention to continue steadfast in their absolute objection to any form of war service. It was a decision that further antagonized the military hierarchy.

On 6 June 1916, the group were sent to the docks in Boulogne and ordered to unload cargo from a recently arrived ship. Refusing, the men were then individually asked if they were prepared to move the load, fifteen of the sixteen said they were not. They were then taken to the nearby field punishment barracks and locked inside cramped and unsanitary cells. One of the men, Charles Herbert Senior, later wrote of the ordeal:

> [We] were put into a dark cell with atrocious toilet facilities. It was down a passage and the only air that entered this cell had to get down this passage and through inch spaces between 8-to-10-inch board that formed the front of the cell, and no outlet. It reminded me of the Black Hole of Calcutta I had read about when I was a boy. The size was such that seven could lie close together on one side, five on the other side with our feet meeting in the middle.

On 13 June 1916, the fifteen Richmond men appeared before a Field General Court-Martial charged with 'Disobeying, in such a manner as to show a willful defiance of authority, a lawful command given personally by his superior officer.' It was a charge that carried the maximum penalty of death. Each man was then read the full details of the charge he faced: 'At Boulogne, on 06 June, 1916, when, as one of a working party at the docks, he was personally ordered by Sergeant —, the non-commissioned officer in charge of the party to commence work, he did not do so, saying "My religious convictions won't allow me to do it" or words to that effect.'

During his Court-Martial, Gaudie explained the reasoning behind his refusal to undertake any form of military service with perfect clarity: 'My motive for refusing is because my "religious convictions" prevent me from taking any part in the military system whatsoever and I am therefore bound to disobey any military orders in loyalty to those convictions, which are based on the spirit and teaching of Christ.'

Inevitably, each man was found guilty as charged and on 24 June 1916 the group was sentenced in a very public show of military law. The sentence was undoubtedly intended as a deterrent to others. Forced to stand in front of hundreds of assembled troops, the men were handed the maximum penalty – death by firing squad. There was to be a dramatic twist. After a substantial pause, the men were subsequently informed the sentence had been commuted to 10 years penal servitude by the Prime Minister, Herbert Asquith MP.

It was a victory of sorts for the men and one that had been made possible by a sequence of events that had begun some weeks earlier. Shocked by the plight of the Richmond Sixteen, the MP for York, Arthur Rowntree, had begun a campaign for their release and had already succeeded in increasing public sympathy for their cause. Furthermore, news of their secret transit to France had also leaked out when a member of the group had thrown a note from the train window while it passed through London. For the government and prime minister, it was a political nightmare.

Arguably the most important factor in deciding the fate of the group, however, came on 5 June 1916 when the Royal Navy armoured cruiser, HMS *Hampshire*, hit a mine west of the Orkney Islands and sank with the loss of all but a few onboard. Among the dead was Lord Kitchener, who was travelling to the Russian port of Archangel for diplomatic talks. With Kitchener gone, the prime minister had little stomach for the political storm that would be caused should the Richmond Sixteen be executed. Nonetheless, difficult times still lay ahead for the group as they began the journey home to begin their sentence. Writing in his 1922 book *The Courage that Brings Peace*, Gaudie said:

> From the Boulogne guard room, they were sent to Rouen and confined in a prison foul and disgusting beyond words, whence they were taken down to the quay to be shipped back to England. As they waited there a crowd gathered, for behind them was a hospital ship and some French officers in high command were inspecting it. The officers left but the crowd lingered for a man had begun to address them. He praised the generals who would shrink from no sacrifice to drive the enemy out of France and the brave men who had come out of England to their help, and who now lay wounded and perhaps dying in the ship.
>
> Then he pointed to the 'despicable cowards' – who had come indeed, but in the face of the enemy had refused to fight. They should have been shot, but instead were on their way home. The crowd grew angry, dangerous, and jostled and pushed until they were almost thrown in the water. There was only one of the group who spoke French. He said to a small boy near him, 'We are trying to follow the Good Saviour.' The boy's face suddenly changed and an old man near him asked what had been said. The words were repeated and repeated again and behold slowly and silently the crowd moved away.

Despite the Great War ending in November 1918, Britain's conchies were to remain behind bars, prompting a number to go on hunger strike in protest at their continued incarceration. In 1919, the last of the COs were finally released, having served as much as three years in unsanitary and draconian labour camps and civil prisons around the country. After their release, the men became social outcasts in a country that had seen almost one million of its men killed. Many more suffered long-term physical and psychological effects – including Norman Gaudie, who was unable to resume his pre-war career in football. After returning to the north-east to become a director at a well-known sugar manufacturer, Gaudie died in Boldon in 1955, aged 67. He remained a committed pacifist and steadfast in his beliefs to the end.

English Heritage that manages the site has added an excellent and poignant exhibition about the Richmond 16.

2

Scotland

No drums they wished, whose thought were tied
To girls and jobs and mother,
Who rose and drilled and killed and died
Because they saw no other,
Who died without the hero's throb,
And if they trembled, hit it,
Who did not fancy much their job
But thought it best and did it.

Michael Thwaites: *Epitaph on a New Army* (November 1939)

Scotland has a long military heritage, from Calgacus defying Agricola and the might of Rome's vast Empire and Robert Bruce humiliating Edward II and the might of England, to generations of Scottish soldiers who forged the British Empire and held it (and sometimes did rather well out of it). Inevitably, this is a contentious point, but I've spent an awful lot of time north of the border and I can say most Scots are proud of their martial past and what their sons achieved in world wars – rightly so.

During the summer of 2022, Bev and I spent some time portraying David I and Queen Margaret for the education/entertainment of visitors, and very popular we proved too, I easily lost count of the number of 'selfies' we posed for. I never miss an opportunity to visit the Scottish National War Memorial, located within Crown Square. It is magnificent and the Roll of Honour contains the names of 147,000 men and women who lost their lives during the Great War, with 50,000 more from the Second. One in five Scots who saw service from 1914–1918 failed to return.[1] The work was commissioned in 1919, led by the architect Sir Robert Lorimer and was formally opened by the then Prince of Wales on 14 July 1927. It is a space of deep beauty, respect and tranquillity, a most fitting tribute to a nation's loss.

Royal Scots Dragoon Guards, Edinburgh Castle

Ensign Charles Ewart (1769–1846) became a hero of the Battle of Waterloo on 18 June 1815, which finally decided the outcome of the French Revolutionary and Napoleonic Wars. It also decided who would be top dog in Europe and by proxy the world for the next century – Britain and her vast, expanding mercantile empire. Waterloo fascinated the Victorians and those who distinguished themselves on the day were lionized. Ewart was from the Scottish West March near Moffat and joined the Royal Scots Dragoon Guards enlisting at Kilmarnock, aged twenty. He served in several campaigns and rose to become an NCO. He was tall, well over six feet, heavily built, phenomenally strong and a master swordsman.

> Charles Ewart was about 6 feet high, firmly knit, with a countenance indicative of high mental and physical energies. In the retreat of the British through Holland after the disastrous Battle of Nijmegen, Ewart one day heard the wailing sounds of a child coming from close to the roadside. He dismounted and found a woman and child lying in the snow. The mother was dead, but the infant, still clinging to life, was in the act of suckling from the breast of its lifeless parent. Ewart rescued the baby and on reaching the encampment for the evening, met with his colonel who offered to defray the expenses of a wet nurse. Great difficulties were encountered in this; however, he was successful and also fortunate in discovering the father of the child, a sergeant in the 60th regiment. Some years later the father, now a sergeant-major and father of a healthy son, fortuitously met with Ewart. He tried to give Ewart a sum of money, which upon being firmly rejected, was replaced with the gift of a silver watch, and this was finally accepted as a memento of the events.[2]

His finest hour would be on the Ridge of Mont St. Jean when his regiment made their famous if ultimately very costly charge against

D'Erlon's Corps as Boney's great hammer blow, directed at the centre of the Allied line, began to waver:

> At Waterloo, Ewart was involved in a hand-to-hand combat with an officer, whom he was about to cut down, when a young ensign of the Greys offered to instead take the Frenchman to the rear as a prisoner. No sooner had he agreed to the request when he heard the report of a pistol and upon turning saw the ensign falling from his horse, and the officer in the act of replacing the weapon with which he had dispatched the life of his preserver. Thus enraged, Ewart cut down the officer, deaf to his pleadings for mercy.
>
> Dashing forward he now found himself close to the standard-bearer of one of the French 45th Regiment du Ligne. A short and deadly conflict ensued and as the staff had stuck fast in the ground, he was able to lay hold of it without further trouble. Looking round he saw a lancer single him out, gallop forward and hurl his spear at his breast. He had just enough strength to ward off the blow, so that the lance only grazed his side; then raising himself up in his stirrups, he brought his opponent to the ground with one cut of his sword. When riding away with the Eagle he experienced another narrow escape, for a wounded Frenchman, who he had taken for dead, raised himself up on one elbow and fired at him as he passed. The ball fortunately missed him, and he was able to take his prize to the rear.[3]

Scott, in the year after the battle, invited Ewart to speak at a commemorative dinner in Leith but the ensign (as he now was) replied that 'he would rather fight the battle of Waterloo over again, than face so large an assemblage.'[4] Just over two hundred years later I was lecturing in Newcastle Central Library on Waterloo and one of the attendees brought in some memorabilia of an ancestor who had fought. He had served with the Royal Scots Greys, and I was able to heft his original 1796 pattern heavy cavalry sabre, the Holy Grail of Napoleonic sword buffs, with full provenance. He asked me if it was valuable and I'm proud to say I replied honestly that yes indeed, it was!

Stuart Kennedy curates the Royal Scots Dragoon Guards Museum, inside Edinburgh Castle. As mentioned earlier, we were there during the summer posing as King David I and St Margaret (his mother); 'Margaret' is 30 years my junior, but we can allow for artistic licence. The place was heaving, to the extent we feared structural overload. Many, indeed, most of the visitors were from overseas, at long last released from the Covid curse.

What Stuart wants is to ensure the museum is something the wider regimental family can both identify with and be proud of. I get this completely. This was part of my mission when I worked at DLI Museum and you have to realise this need changes the farther in time you get from the date of amalgamation. At the same time, the place has to offer a good all-round visitor experience. Naturally, Ewart's eagle is a prize exhibit, inevitably and quite rightly, as its capture was a legendary feat of arms.

Stuart has also organised an exhibition of e-blueys dating from the time of the 1991 Gulf War – Operation *Granby*. These are letters written by serving Armed Forces members, stationed in live theatres, to their families back home in the UK. They're downloaded and printed at various locations in Britain, then delivered to intended recipients by Royal Mail.

This is something different and in my view quite brilliant. Any of us who have had family serving in the Armed Forces remember the e-bluey, letters from the front, personal and, certainly for me, deeply poignant, something which will resonate deeply with all who served and their families at home. I'm not sure what Ewart would have thought of it.

Black Watch Museum, Perth

Duncan Campbell was one of those who took part in the disastrous attack ordered by General James Abercrombie against the defences of Fort Carillon in July 1758. This was one of the many actions fought in the Seven Years War (1756–1763) in the vast arena of the North American wilderness. Ten days after the fight, Duncan Campbell, serving with the 42nd, succumbed to his wounds. Some three years before, Campbell is said to have provided shelter to one who turned out to have been his brother's murderer.

This was a fine choice for a Highland gentleman; to deny the call of vengeance according to the obligations of the host. A humane man, he spared the assassin and gave him refuge. A little while after, his dead brother appeared to him in a dream in which he promised they would meet again at a place called Ticonderoga, which at that point and, unsurprisingly, Campbell had never heard of. On the very afternoon of the day on which the fateful engagement had been played out in New York State, a ghostly panoply of fighting men appeared in the clouds over Inveraray Castle, Clan Campbell was thus given notice it would soon mourn many of its lost sons.

The name Ticonderoga derives from the Iroquois expression *tekontaroken*, 'at the junction of two waterways', an exotic-sounding but sensible name, as the place stands astride the strategically significant narrow passage of Lake Champlain, where a short traverse cuts through to the northern flank of Lake George. Thus, the fort dominated the busy trade highway between English-held Lake Hudson and the French-controlled St. Lawrence River. Such was the importance of its location that four major actions were fought there over two decades.

When the French began to build in 1755, they named their new work Fort Carillon, the choice inspired by the musical resonance of tumbling falls nearby. Construction was spread over two years and the ramparts were not completed until 1757. By then, local rivalry between France and England had flared as a wider consequence of the Seven Years War, Pitt the Elder's expansionist triumph that a later populist leader, Winston Churchill, branded as 'The First World War'. It was from there the French launched a successful campaign against Fort William Henry, an episode familiar to all readers of James Fenimore Cooper's *Last of the Mohicans*.

By the following year, the British in North America had been substantially reinforced whilst the French under Louis-Joseph de Montcalm were forced back onto the defensive. Two campaigns were planned: General Amherst would lead an attempt on the great French bastion of Louisburg, whilst Abercrombie, as C-in-C, would assault Fort Carillon. His forces were considerable, comprising some 7,000 British redcoats and 9,000 colonial militia, well supported by a formidable artillery train. His regular battalions included the 27th, 44th, 46th, 55th and the 42nd Highland Regiment (The Black Watch). For the difficult and dangerous business of scouting and skirmishing he could count upon over a thousand light infantry. Col. Gage, commanding the 80th foot, was a devoted student of that celebrated exponent of irregular warfare Richard Rogers, whose Rangers had already made a name for themselves as masters of commando-style operations.

Against this formidable deployment Montcalm could muster no more than 3,600 defenders all told, made up of seven battalions of French regulars supported by marines, French colonials and native allies. He knew the blow must fall upon the bastion of Ticonderoga and thus looked to his defences. The site enjoyed a superb tactical position, pushed out on a spur that jutted like a salient against the course of the river, the rushing waters providing additional cover on three sides. The fourth flank was partly screened by a bog and covered by a timber outwork, a substantial palisade nine feet high, the timbers soundly buttressed, and the ground before sown with stakes and other obstacles. The fort itself was a classic star-shaped construction with four corner strengths or bastions, two freestanding works, or ravelins, guarding the most exposed flanks and the whole built upon a swell of higher ground. It was altogether a most commanding position.

Any successful assault could only be envisaged after the major logistical effort needed to transport men, their equipment and supplies, the great guns, ammunition and quantities of powder had been competently undertaken. The officer responsible was Brigadier Viscount George Augustus Howe, one of several remarkable brothers, all of whom were destined to leave their mark on North America. Howe, colonel of the 55th had made a particular study of the irregular pattern of warfare that so suited the terrain and had spent time with Rogers. The many lessons learnt he put to good effect, stripping his fellow officers of much of their more luxurious baggage, lightening the men's load, adapting their uniform, weapons and drill. He went so far as to be seen washing his own linen in rivers, a chore normally undertaken for officers by the regimental women whom he'd sent back. This was all a novelty!

In the warm flush of summer, the end of June, Abercrombie's task force was mustering on the banks of Lake George, the ruins of Fort William Henry a potent reminder. Thanks to Howe's efficiency, some 800 barges or bateaux, plus 90 whaleboats had been laboriously hauled overland. Each of the boats would carry 22 troops with all their gear and 30 days' supply of foodstuffs. On 5 July with commendable discipline and élan the regiments embarked, and the great flotilla of heavily laden craft proceeded sedately down the length of Lake George. It was indeed an awesome sight, the might of the burgeoning British Empire on the move, a rich tableau of scarlet and gold, the deep blue of the colonials, the latent power of the great guns lashed to rafts, the silken standards lifting in a gentle breeze. Around them the sylvan glories of an untouched wilderness; tall stands of pine and fir, the light glancing off the calm, burnished waters of the lake, hills rising on every side.

By dawn the following day, 6 July, the long passage of Lake George had been accomplished without mishap and the battalions were approaching the narrows; ahead of them was the expanse of Lake Champlain and at its headland reared the great mass of Fort Ticonderoga, an unconquered citadel. Montcalm, however outnumbered, was still vigilant and the shoreline was held by a body of skirmishers; fire from the boats drove them back beneath the shelter of the trees and the attackers disembarked without further opposition.

Once ashore, the men were detailed into marching columns, but this terrain was not the level arena of Flanders, it was a dense web of tangled undergrowth, fallen trees and confusing slopes. Abercrombie's guides proved less than competent, and the British were soon in difficulties. In the confusion, the right, commanded by Howe, became involved in a running fight with the French, taking casualties, amongst whom was their gallant commander, one whose loss would be keenly felt. Major-General Stewart writing of the action some decades after, commented:

> The march was continued in the same order [July 7], but the ground not having been previously examined, and the guides proving extremely ignorant, the columns came in contact, and were thrown in confusion. A detachment of the enemy, which got bewildered in the wood, fell in with the right column, at the head of which was Lord Howe. A smart skirmish ensued, in which the enemy were driven back and scattered with considerable loss. This petty advantage was dearly purchased by the death of Lord Howe.[5]

Abercrombie, deprived of his subordinate's wise counsel, now took a fateful and disastrous decision. He could have sensibly deployed his colonials to interdict communications between the beleaguered fort and possible reinforcement from Fort St. Frederic (Crown Point). He could then have begun the laborious process of dragging the guns from their barges to the eminence of Mount Defiance and from there commenced a bombardment, which, given the weight of shot he disposed, must surely have soon battered the fort into submission. He chose instead to rely on the distinctly unsound advice of a junior officer of engineers, who opined that the impromptu works covering the approach were flimsy and could be rushed with the bayonet. With drums beating and banners unfurled, the regular battalions of foot moved into the attack with purpose and precision, but the defences were far stronger than had been suggested, the line of the breastwork was held by Montcalm's veterans.

A withering hail of fire greeting the leading ranks who were flensed away as their comrades behind struggled to fight through the lines of obstacles – this was grim, close-quarter work, which presaged the horror of the trenches; the men could make no headway: 'Masses of infuriated men could not go forward and would not go back; straining for an enemy they could not reach, and firing on an enemy they could not see, they were caught in the entanglement of felled trees. Shooting, yelling, and cursing, they were assailed all the while with bullets, which killed them by scores, stretched them on the ground, or hung them on jagged branches, in strange attitudes of death.'[6]

Thus far, the 42nd and the 55th had been kept in reserve, but as the assault faltered they were now sent in. Formed nearly two decades before, the 42nd (Lord John Murray's Highlanders – the Black Watch) saw themselves as an elite, drawn mainly from the ranks of Clan Campbell, the soldiers were men of quality and some standing. Their élan was ferocious, and they fell upon the defences with gusto. The battle now reached its savage denouement as the Highlanders fought and clawed their way forward. Captain John Campbell and a handful of his company became the only British forces to gain a foothold in the enemy position, but unsupported and disdaining retreat, they were cut down to a man. As the celebrated historian Francis Parkman wrote:

> Then the highland soldiers of the 42nd could endure no longer. Impatient of their position in the rear, they rushed forward, hewed their way through the obstacles with their broadswords. Since no ladders had been provided, they made strenuous efforts to carry the breastworks, partly by mounting on each other's shoulders and partly by fixing their feet in holes, which they had excavated with their swords and bayonets in the face of the work. The defenders were so well prepared that the instant an assailant reached the top, that instant he was thrown or shot down.[7]

Losses were frightful. The 42nd, which had suffered most grievously, had lost eight officers, nine NCOs and 297 other ranks killed, 17 officers, ten NCOs and 306 men hurt – a total loss of 647; the regiment was reduced to a mere shell. The attack was a disastrous failure, though in recognition of its sacrifice the 42nd was awarded by George II with the designation 'Royal' – never was such a distinction better merited.

One of those who served during National Service was the acclaimed Scottish actor Iain Cuthbertson (1930–2009). I loved his portrayal of the manipulative Border warden Ker of Cessford in the 1960s TV series *The Borderers*, wonderfully Machiavellian. A graduate of Aberdeen University, he had originally thought of a career in the Foreign Office. In the regiment he acted as an advocate for unruly squaddies. One was up for being firstly, late on parade and secondly, assaulting his senior officer. Cuthbertson got him off the more serious charge: 'Thanks awfully fur pretendin' ah didnae dae it sur.' Not sure if this affected his subsequent life choice in becoming an actor.

It was possibly Marshal Wade who created the Black Watch. During the 1720s after the failure of the '15 the Hanoverian government authorised that great military engineer to raise six 'watch' companies as militia. He recruited three of these from Whig Clan Campbell and the other three from Fraser of Lovat, Munro and Grant. Officially embodied at Aberfeldy in 1739, this new unit had a wide-ranging brief to keep order in the highlands – an unenviable assignment. In Gaelic they were known as *Am Freiceadan Dubh* – Black or Dark Watch. This may have derived from their dark-coloured plaids or been a slur against their pro-government role, betraying their highland comrades. Possibly, it referred to their role in stamping out blackmail and extortion, rife in the lawless glens. However it came about, the name stuck. It was at Fontenoy in 1745, that the Black Watch showed their mettle and became a battle-hardened elite. Their exploits since are legion.

They are now 3rd Battalion, Royal Regiment of Scotland (3 Scots). Until the Childers Reforms they achieved fame as 42nd (Royal Highland) Regiment of Foot (The Black Watch). In 1881 they were amalgamated with 73rd (Perthshire) Regiment of Foot. Known as the Black Watch (Royal Highlanders), they became The Black Watch (Royal Highland Regiment) in 1931, a title which endured until 2006, the senior Highland Regiment. Today, the extensive collections are fittingly housed in Balhousie Castle Perth. Dating back to the 12th century (though mostly Victorian Gothic), it's spot on, a proper tribute to a rightly famous formation.

Culloden Battlefield Centre, Inverness

Long, long ago, as a history-hungry boy of eleven, I watched a BBC drama-documentary on the battle of Culloden[8] based on John Prebble's magisterial and (by the restrained standards of the day) visceral account of the battle, the first I'd read where you could smell the powder residue and hear the screams. I had read the book that summer and the programme was aired just before the Christmas holidays. The next year, 1965, as part of our family's annual pilgrimage to the highlands – a Sprite Musketeer caravan dragged at a very stately pace by our grey Austin Cambridge saloon – we went to the battlefield.

At this time most of the ground was irritatingly covered in Forestry Commission trees, gulag-friendly planting, which obscured just about everything; yet, in a low, drifting mist, wholly unintentionally they cloaked the site in grim melancholy, a highland dirge in conifers. My imagination took off, as boy's imaginings are wont to do. The stubby headstones of clan graves, the dark frisson of the Well of the Dead and Old Leanach bothy, completed the effect. They still do, despite meticulous re-rendering of the ground and first-class interpretation, Culloden never surrenders its hold. Sixty years on and I'm still there.

It was dank and dreach on 16 April 1746 when a ragged army deployed over unforgiving ground, frequent showers of stinging sleet driven by a keen-edged wind biting their exhausted faces. The Jacobites seem to have hoped or prayed that the government army would not attack that day, but by first light the long lines of scarlet-coated foot, the squadrons of dragoons, guns and limbers were on the march. By 10.00 a.m. Jacobite scouts reported the royal army just four miles distant. A hurried council of war was held. Lord George argued that the men were in no state to fight, the ground was wrong, they were outnumbered and outgunned, but the prince was adamant that the army must stand. There could be no further retreat. By 11 o'clock the armies were in sight of each other across two miles of open moorland, a dismal, dreary day, the grey skies still heavy with intermittent squalls.

Despite their fear and exhaustion, the clans still found sufficient stamina to argue over who should have the right. As a placatory gesture, the prince awarded the honour to Murray's Athollmen, who drew up flanked by the Camerons, Frasers, Clan Chattan, Farquharson's, MacLachlans, Macleans, John Roy Stewart, and Chisholm's, and Clan Donald, deployed in bad grace. Clanranald, Keppoch, Glencoe and Glengarry on the left; the second line comprised, from the right, Lord Lewis Gordon, Glenbucket, Perth, the Royal Scots and Irish Picquets. The remnants of the cavalry were drawn up behind.

Murray commanded the right, Drummond the centre, Perth the left. The Jacobite guns were positioned in batteries of four on each flank and in the centre of the line. The gunners were volunteers who barely knew their trade – in contrast to Cumberland's dedicated professionals. The rebels were positioned so that the flanks rested upon, on the right, the walls of the Culwhiniac enclosure and on the left those of Culloden Park. Beyond the Culchunaig and Culwhiniac enclosures lay the course of the Nairn and a bare 500 yards beyond the Jacobite front the government troops were deploying.[9]

Earlier the duke had ordered the shift from column into two lines, each of six battalions with three in reserve. On the left, his position rested on Leanach farm and the secure barrier of the river. He had hoped to have a morass on his right, but finding that he had advanced beyond it, ordered up two of the reserve battalions to take the extreme right of both the first and second lines. To consolidate his position on the left, he ordered Wolfe's regiment to move slightly ahead of the front line and redeploy at right angles to it *en potence*, their backs to the gaggle of buildings at Leanach. The redcoats could pour enfilading fire into any immediate attack upon the line.

Brevet-Colonel Belford had supervised hauling the guns on the rough march from Nairn. He had sixteen 3-pounders, sited in pairs between the front-line battalions. The trusty Coehorns were split into two batteries of three and positioned ahead of the flanks. Lord Albemarle was to command the first line, General Huske the second, Hawley was given the cavalry, with Kingston's horses covering the right. More dragoons crowded the extreme left and guarded the flank of the second line, which from right to left comprised Battereau's, Howard's, Fleming's, Conway's, Bligh's and Sempill's, with Blakeney's in reserve.[10]

The swift-footed Campbells, for whom this was a clan affair, had begun to infiltrate the Culwhiniac enclosure, the obvious possibilities of which Murray had not overlooked. He had also requested that the regiment on the right should tear down the walls of the Culloden enclosure to give them more space to manoeuvre. Other officers suggested that the walls be lined with marksmen. Both ideas were dismissed out of hand by the prince, who, despite his insistence on fighting on Drummossie Moor, had failed to reconnoitre the ground to his front, which was extremely soft and waterlogged, heavy going even for Highlanders, with the wind and rain beating in their faces.

Shortly before one o'clock, Cumberland treated his troops to a short, soldierly oration. He advised his redcoats to remember that they fought for King, Country, and ancient liberties. Looking across the expanse of that dismal moor, few on the rebel side, with the obvious exceptions of Charles and O'Sullivan, can have had any illusions as to the likely course of the coming battle. Viewing the long lines of scarlet and gold, the forest of gleaming bayonets, the black muzzles of the guns, many must have quailed, but they did not flee. Lord George is said to have muttered, 'We are putting an end to a bad affair.' At one in the afternoon (by James Wolfe's watch), the rebel guns began a desultory cannonade and the royal gunners responded in kind.

The last battle to be fought on British soil had begun.

It is said that the rebel gunners were provoked by the elegant figure of Lord Bury, who had cantered to within 100 yards to mark their dispositions. If his lordship was the intended target, the artillerymen were way off, as the ball passed clean over his head and over the heads of the first-line battalions before pulping some unfortunate in the rear. Another ball struck a few yards away from the duke, missing his ample figure but killing another two almost directly in front.[11]

Belford's gunners totally outclassed their scratch opponents, most of whom were soon dead or in flight. With the brief unequal duel ended, the Royal Artillery concentrated their fire on the clan regiments, the roundshot tearing great gaps, strewing the heather with bloodied plaid and mangled limbs. This was not warfare as Highlanders understood it. It was not in their nature to stand helpless and exposed while being shot down like deer. The carnage was dreadful. Mingled with the cries of the injured and dying was the order to close up, repeated as every ball found its mark. Whilst the number of actual casualties caused by the government guns was likely quite few, the psychological effect was considerable. Highlanders had never been trained or expected to endure the nightmare of being pounded by artillery.

Charles had taken up position on a slight eminence north of Culchunaig, and behind the centre of the second line – an unsuitable vantage point. His view was restricted, his mood indecisive. Lord George and the chiefs pleaded for the order to charge, to be allowed to close with the enemy, sweep across that dreadful killing ground before the regiments were utterly decimated by the relentless and galling fire. The prince was guarded by the scant remains of the horse, and several were killed around him as the gunners found the range. At last, Charles gave the order, sending his young aide Lachlan MacLachlan towards Perth on the left, but the youth was struck down before he could deliver the message. Lord George had earlier sent Ker of Graden to Charles, begging that those of the left should echelon forward to close the greater gap on that wing – the opposing lines funnelled towards the right.[12]

Cumberland had taken station towards the front of his second line ahead of Howard's. From here he could see exactly how the battle was developing, at this point satisfactorily. Belford's sweating gunners were winning the day for him. Around 13.30 hours. a sudden squall flung wind, rain, and sleet in the rebels' faces; Murray could barely control the fury of his Athollmen, 'so impatient they were like to break their ranks'. With a defiant roar the Macintoshes and Clan Chattan burst from the centre, followed by the Athollmen and the Camerons into the stinging sleet.

As the Highlanders burst over the moor, the first rank of the government battalions dropped to one knee, the remaining ranks remained standing, firing by platoons, dense rolling volleys dropping the clansmen as Belford's men began loading grape. The wet ground and the pelting rain made the going hard. Those in the centre crowded onto the old drove road, where the surface was firmer, but this channelled them to the right, where they blundered into the struggling regiments on that wing. And that fire never slackened; eighteen officers of Clan Chattan and a score from the Athol Brigade fell before they even saw the government troops, so thick was the smoke of battle. Archaeology has indicated that the walls of the enclosure were only knee height, so no impediment to Highlanders, but that the government mortars, contrary to previous opinion, delivered a continual 'stonk'.[13]

As Clan Chattan closed with Barrell's and Munro's, the fury of the onslaught pushed the 'Old Tangeerers' back, but they held, flickering bayonets clanged and scraped, Price's and Cholmondeley's blasted the attackers' flanks whilst Wolfe's men raked the Athollmen. Over a thousand Highlanders were hacking at the redcoat lines with Barrell's and Munro's bearing the brunt. Officers and men fell beneath vengeful broadswords, but the cry of 'Give them the bayonet' still sounded and the veteran infantry lunged and parried, striking to the right, driving the needle-sharp points through flesh and bone, their white gaiters spattered with blood.

With his front line under such enormous pressure, Cumberland ordered up Sempill's and Bligh's from the second; unbloodied, these battalions fired and fell on with the bayonet. Lochiel had already been carried off, both ankles smashed by grape; Murray charging at the head of the Athollmen had his horse shot from under him and received several gashes in the savage mêlée. Seeing that his men could make no headway, he dashed back to entreat the second line to advance – pleas that were greeted with a marked lack of enthusiasm.

On the left, and with more ground to cover, Clan Donald never reached Pulteney's and the Royals, whose rippling volleys punched great gaps in the charging lines. Still sulking from their relegation to this wing, hampered by the uncertain ground and mercilessly pummelled by shot and canister, the advance past the sheltering walls of the Culloden enclosure opened their flank and for once, the government dragoons did not miss the opportunity. Keppoch and many of his officers were killed and the clan regiments would have been completely cut to pieces if Stapleton had not led forward his Royal Scots and Irish to give covering fire.[14]

Murray had previously noted with alarm that the Campbell militia were encircling the Jacobite right by breaking down the rough masonry of the enclosures and clearing a path for Hawley's dragoons. To counter this, Lord George ordered some of Lord Lewis's Gordons men to face about and guard the flank. This still did not prevent some Campbells from lining the northern walls of the Culwhiniac enclosure and blazing away at the Athollmen now streaming back in disorder.

To add to the rebels' growing difficulties, five squadrons of dragoons had been lurking along the track that ran up from the Nairn past Culwhiniac farm; the obliging militia heaved a mass of stones out of their way and the horse cantered towards Culchunaig and the Jacobite rear. Alerted to this fresh hazard, Murray dispatched Gordon of Avochie to hold the Campbells with FitzJames's horse in support. Another posse of Campbells struck at the retreating Camerons who still had sufficient fight in them for a vigorous riposte.

Hawley, with ten squadrons, was finally and firmly established behind the Jacobites, but as his men advanced along the track, sunken through generations of use, Elcho's Lifeguards and FitzJames's horse bravely barred the way. A spirited charge could have annihilated the handful of rebel horse, but Hawley's men had difficulty in manoeuvring, impeded by the lane, and both sides merely swapped pistol shots. This would not be to the general's liking. The rebel cavalry, pathetically few as they were, did their comrades good service buying vital minutes whilst the broken regiments fell back. The prince, the wreck of his army about him, may have attempted to rally the MacDonalds on the left but was persuaded by Sheridan and Hay of Restalrig to quit the field – he had had his try at generalship.

Lord George Murray was left to salvage what he could, rallying Ogilvy's and some of the Angus men. He, together with Stapleton's battalions, formed a screen, keeping the threatening dragoons at bay whilst the defeated Highlanders either splashed over the Nairn towards Badenoch or fled along the road to Inverness. Over a thousand lay dead on that terrible field, piled in heaps before the survivors of Barrell's and Munro's. Barrell's alone had lost 17 killed and 108 wounded out of a roll call of 438. In total, Cumberland had lost around 50 dead and many more injured. He had, however, utterly crushed the rising and secured his father's crown.[15]

Beyond the battlefield the fields and lanes were choked with rebel wounded, enthusiastically sabred by dragoons who, drunk with slaughter, cantered down the Inverness road cutting down any they encountered, including numbers of luckless civilians. A host of captives were taken, including Balmerino and Kilmarnock. Strathallan had preferred to cheat the headsman by a suicidal charge into the midst of Cobhams. More prisoners were dragged from refuge in the cottages and barns; most got no further than the nearest convenient wall. The Royal Scots and Irish, being regular French units, were allowed honourable surrender. Brigadier Stapleton died of his wounds.

Indefatigable Murray, together with Drummond and Perth, had managed to keep some 2,000 men together and under arms. A rendezvous was arranged at Ruthven, but their prince did not appear. Instead, he sent a note which contained some helpful advice: 'Let every man seek his own safety the best way he can.' For some, including Murray, this was the beginning of a life in exile, for others a career 'skulking' in the heather; for the prince, a short interlude of wandering, to be followed by a wasted life of acrimony, despair, and dipsomania.

For the Highlands and the clans, that spring and summer were to be an ordeal by fire, slaughter and terror, their dress and weapons proscribed, their steadings burnt, their byres empty. Barely a generation later, absentee lairds and the more profitable farming of sheep finished what Cumberland's men had begun. The 'pacification' of the clans which followed the debacle at Culloden – and that term must, as a savage euphemism, rank alongside such loathsome modernisms as 'ethnic cleansing' – destroyed the ancient life of the glens. This and the Clearances, almost equally traumatic, which followed barely a generation after, completed the destruction.

I rate Culloden Battlefield Centre as the best in the UK. The old, admirable but rather inadequate centre, which did in fact encroach onto the field, was replaced in the first decade of this century by a state-of-the-art building, standing clear of the actual ground. Superb dioramas, interactive maps, loads of kit and a brilliantly evocative exhibition of Jacobite arms, 'The Swords and the Sorrows', creates an impression both comprehensive and indelible. I first encountered the *Outlander* phenomenon in the well-stocked bookshop there. I confess I had not heard of Diana Gabaldon until I saw a huge stack of her hefty output.

Outdoors, on the ground, the positions of both armies are clearly marked, and you can walk (or run if you've a mind), the line of the Jacobite charge – the moment in time when a whole way of life came to an abrupt end. History is rarely so obligingly precise. Culloden Battlefield Centre is what Towton could be, it brings the history to life in a wholly accessible manner, the ground is beautifully interpreted; overall, a first-class visitor experience.

In the visitor centre café:

John: I'd not really heard of *Outlander* till I saw how many books there are on display here, a lot more than there are of mine, it's so unfair!

Graham: Diana Gabbeldon's stuff is fantastically popular, it's History-Lite' of course but I enjoyed it, I do confess me, and the TV series is good as well. It's not history as such but it's good fun and great locations.

John: I get it now and she's done wonders for the Scottish heritage industry. When I was writing about Culloden ten years ago or more and I spent a couple of days on the ground here, I counted a couple of dozen bus tours at least coming in every day, even if it was July, and most were from the US. When I first started coming here as a lad, admittedly that's half a century ago, you could generally count on having the place pretty much to yourself, now it's full of visitors.

Graham: Like you, I was stunned by the BBC drama-documentary from the early '60s, black and white but as gritty as it gets, budget was about a fiver. Didn't the same director go on to make *The War Game*?

John: He did and that was *so* gritty the Beeb decided they didn't dare show it, just in case anybody understood what the concept of nuclear war actually implied. It wasn't on screen till the '80s if I'm right, alongside a more anodyne contemporary version, *Threads* I think that was called.

Graham: *Culloden* kind of had the same effect in relation to the highlands, from John Prebble's book of course, and he certainly doesn't spare anyone's sensibilities when he describes the clear-up in the glens after the battle, atrocities stacking up like dominoes, nasty business by any standards and largely indiscriminate. Small wonder Scots call Cumberland 'The Butcher'.

John: The whole summer of 1746 was a serious stain on the reputation of the British Army, but it's a myth to think this was wholly English v Scots. It was more of a dynastic civil war. Cumberland had nearly as many Scottish soldiers on his side and the bulk of the ethnic cleansing afterward was joyfully conducted by Scottish officers and not all lowlanders. It was, in effect, racial. The lowlanders saw themselves as a different race, ashamed of their savage, plaid-wearing upland contemporaries whom they viewed as illiterate popish trash, *the mi-run mor nan gall*, the 'great ill-will of the Lowlander'. What is astonishing is the way, within a decade or so, the UK Government had found a way to utilise the martial fury of the clans in its own service and how instantly they took to it.

Graham: You could say the rest was history.

Fort George, Ardesier – The Thin Red Line

Merrily danced the Quaker's wife,
And merrily danced the Quaker;
But the Quaker's lad went off to the wars,
A sad surprise for the Quaker![16]

Graham and I have forayed north across the line many times, both of us feel a deep affinity with the Western Highlands, Graham was in fact educated there. Some years ago, with Adam Barr, Graham's stepson and a gifted photographer, we motored via Bay of Largs and vast Dumbarton Rock to Glen Coe, our objective to take photos for the book I was then writing on the infamous massacre of 1692. For all of us, the Glen has a deep hold, If any event in British history deserves a decent museum and visitor centre it's what happened here on the night of 13 February that year, but there is depressingly very little, aside from the much later monument, hidden away just behind the village. We've seen Culloden more than a few times, and if we're there we have to go to Ardersier Point and the great fortress.

Fort George is unique, 'awesome' is a much-overworked term, but it is indeed so. In one respect the great fort, *An Gearastan* ('Garrison' in Gaelic) proved to be an expensive white elephant. At £200,000 in the money of the day, it cost twice its original budget (just to show calamitous overspend on public sector projects isn't a modern phenomenon). It was the ultimate deterrent to a lost cause, Jacobitism, in reality dead in the water since Culloden.

The place was not completed until 1769, designed and partly built by John, Robert and James Adam. An architectural gem, it is sited on Ardesier Point overlooking the regal sweep of the Moray Firth, served by its own harbour to permit resupply by the Navy.[17] Its predecessor had been located centrally in Inverness (now the pleasing gothic revival courts complex). A medieval motte, there was a makeover by Cromwell and again in 1727, but it was slighted by the Jacobites prior to their *Gotterdammerung* on Drummossie Moor.

This is top-end Vauban design, with projecting bastions, redoubts, casemates and sober but imposing barrack piles, an Adams take on Vanbrugh and now also home to the Queen's Own Highlanders Museum. So far as I'm aware it's the only real fortress in the UK that pays full homage to the soaring genius of Sebastien Le Prestre Vauban (1633–1707). Europe, especially Flanders and the old Barrier Lands are of course, studded with his mighty constructions, chewed through by Marlborough in the War of Spanish Succession (1702–1711). The siege of Lille during autumn 1708 showed just how much punishment these citadels could withstand.

For me certainly, among the huge wealth of objects this vast enclosure holds, one of the most dramatic – and so resonant in terms of siege warfare – is a 19th-century 13-inch mortar, its smooth-bored stubby barrel fixed at a 45-degree angle. It fired a 200 lb (91 kg) shell, with a range of 2,900 yards (2,700 m), using 9 lb (4.1 kg) of black powder. It has echoes of Sebastopol or Vicksburg, not to mention the siege of Fort William Henry as depicted in *Last of the Mohicans*.

Fort George is not far either from Culloden battlefield as mentioned, or indeed the remarkable Clava Stones.[18]. I keep saying this about locations in this book but it's true, Fort George is utterly remarkable. It reminds me of some vast forgotten outpost, like the decaying pile in *Desert of the Tartars*.[19] It was built for a war that never happened, effectively redundant by the time it was commissioned. Yet it endures, brooding over a flat and largely empty landscape, an odd juxtaposition of serving military garrison, museum, tourist attraction and good vantage for viewing dolphins out in the grey waters of the Moray Firth.

Apparently, the MoD plans to withdraw by 2030; then what? The upkeep must be enormous, and I can't see that revenue from visitors could possibly hope to match outgoings, but I know of nowhere else which so strongly and so poignantly signals the Scottish Highlander's transition from despised rebel to loyal soldier. Before the '45, the bulk of Scots regiments had been drawn almost exclusively from the Lowlands, where hatred of the Gael ran deep. Aside from, as noted earlier, those independent watch companies raised by Wade later in the 18th century, Highlanders had been viewed with about as much sympathy and understanding as, say, Pathans or Zulus a century later.

It was not long, however, before Whitehall realised that in the Crown's turbulent northern subjects whose traditional dress and weapons were proscribed under the Disarming Act of 1747, they possessed a reservoir of available manpower, accustomed by tradition to the use of arms, brave beyond compare and, above all, expendable. Surely it was preferable that these uncouth clansmen be encouraged to take the king's shilling and fight and die for him abroad rather than be left idle in their own dark glens, where conspiracy and rebellion had been seen to flourish. Pitt the Elder, in a celebrated speech to the House in 1766, cannily praised the martial vigour of the Highlanders and stressed the advantages of recruitment.

The unequivocal loyalty the clansmen demonstrated towards their chiefs and tacksmen translated quite easily into the regimental system, especially when units were raised locally. The Duchess of Gordon is said to have boosted her husband's recruiting drive by offering a kiss to every red-blooded Highlander contemplating taking the king's shilling. The 100th and 92nd, immortalised as the Gordon Highlanders, brought nothing but honour to the name. The natural valour of the Highlanders, once safely channelled, brought respectability, paid for in blood. The hit-and-run nature of warfare in the American wilderness suited the Highland temperament, whilst their contemporaries in other regiments, such as the KOSB (25th), won fame at Minden in 1759 when six foot regiments marched against the French army, their disciplined volleys beating horse, foot, and guns alike. Of the eleven regiments raised in Scotland for service in the Seven Years War (1756–63) only one battalion was from the Lowlands.

I'm very grateful to Eilidh Young and Freya Samuel from the Highlanders' Museum who shared with me their deep love of the collections in their care. The setting is unique and nowhere quite has the atmosphere of the elegantly sober barrack piles at Fort George. Both stress the importance of this history of the highland regiments within the great redoubt intended to overawe and intimidate.

We can't leave Fort George without mentioning Wade's Roads. Field Marshal George Wade (1673–1748) was an Irish soldier and military engineer famous for building things; he wasn't distinguished as a field commander or tactician, but he knew how to build. Confronted by any crisis, he generally set to digging. Wrong-footed by agile George Murray before Carlisle in 1745, he took his frustration out on the landscape and began work on his famous military road (now the B6318). This is a huge boon for visitors to Hadrian's rather earlier construction, but sadly, long swathes of the Wall ended up as hardcore! At one point the aged Marshal even had his own verse added to the National Anthem, thanking him for the deliverance. Sadly, it's been dropped.

Once the dust had settled from the Jacobite attempts of 1715 and 1719, the Whig administration reflected on a solution to the ongoing problem of restless highland swashbucklers in an untamed and largely uncharted wilderness dumped on the fringes of a state emerging into the Age of Reason. You can't reason with savages who wear funny clothes, talk a weird, regressive tongue, and live in troglodyte hovels. You need to control 'em, police 'em and put some stick about. Cromwell had set the precedent with a margin of success and the Williamite government had condoned or at least tried to ignore the excesses of the Massacre of Glencoe, which sent a chill message to the Tory Clans (not that any of them cared for McIan of Glencoe or showed any willingness to assist after the fires had burnt out).

Wade was the man for the job and a remarkable job he did. The general was first sent to assess the risk and then propose a remedy. He estimated that if the fighting clans could muster 22,000 broadswords overall, then at least half of those were inclined to mischief. His suggested plans were comprehensive and indeed sound. Like Caesar, he would divide and conquer by raising those previously mentioned independent companies of highland militia who, having taken the king's shilling, might be useful in stepping on more fractious and recalcitrant neighbours. He advocated rebuilding the old fort at Inverness (not today's Fort George but, as noted, its predecessor, where that Victorian court complex now stands in the centre). He wanted an armed sloop on Loch Ness. A new outpost at Fort William was to be thrown up, soon joined by another at Fort Augustus with police-style barracks in Glenelg and at Ruthven (the latter still stand).

These strongpoints needed to be linked by good all-weather arterial roads and so Wade set to work – for once the government, sufficiently alarmed by the recent troubles, was less inclined to quibble over costs, which were considerable. Much of the actual work was undertaken by the general's equally competent successor, Major William Caulfeild. Four major routes were pioneered: Fort William to Inverness, Dunkeld to Inverness, Crieff to Dalnacardoch and Dalwhinnie to Kiliwhimen (the new Fort Augustus). Wade, like the consummate professional he was, favoured the Roman view of long straight, surgical lines, overcoming gradients with hairpin bends. The most spectacular heights he conquered linked Ruthven to Fort Augustus via the Corrieyairack Pass.

Wade would live long enough to see his great feat put to very good use by Bonnie Prince Charlie's fledgling army in late summer 1745, when he beat stumbling Johnny Cope to the summit and obliged the Hanoverian General to scamper back to Inverness and seek a sea passage to the Lothians, and a defeat that September at Prestonpans that would lead to his court-martial. It's highly unlikely the irony didn't strike the ageing marshal.

Walking Wade's Roads and standing on the dizzying track of the Pass. There is no sound other than the suitably mournful curlew's cry, a skittish wind plucks at the heather, no single element of modernity intrudes:

John: George Macdonald Fraser describes the Elizabethan borderers as being free to a degree we can't comprehend.

Graham: Depends how you define freedom, they knew hunger, fear, pestilence and saw more death in the average season than any of us might ever see in a lifetime.

John: No damn electronic noises, 'this vehicle is reversing' as if you'd not twigged by virtue of the fact it's travelling backwards, no trilling and burping mobile phones ...

Graham: Just the sound of the pipes perhaps, enough to frighten off old Johnnie Cope.

John: Ah, the pipes of course, well nowhere is perfect...

The Shetland Bus

Every time I look at the daily weather map, I'm taken by just how far north Shetland is, more familiar now in many households thanks to Anne Cleves and her detective series. Otherwise, you tend to think mainly of big hairy Norse types beating up most places from their sleek dragon ships. Being remote doesn't mean you escape the war zone. From 1939–1945, the islands were heavily guarded, the fear being they might form a bridgehead for Hitler. He needn't have worried; the Shetlanders took the fight to him.

In March 1943, Leif Larsen, master of the *Bergholm*, a trawler from Lerwick was undertaking a most dangerous mission against Hitler's *Festung Europa.* His vessel was bound for Traena in the Nordland, delivering arms and equipment to resistance fighters based there. In all, Larsen and his crew were to make over fifty trips, missions fraught with risk from the fury of the elements and the vigilance of the Luftwaffe. This time, luck ran out. They made their delivery of vital supplies safely but, motoring back toward Shetland, two cruising ME110s dived and strafed the boat mercilessly. Of the eight-man crew, six were hit and one, a young lad named Nils Vika, died of his wounds. With the ship sinking, survivors, mostly injured, took to the lifeboat, and for four days rowed across the deadly Atlantic waters. Mercifully, they were rescued off Alesund by a motor torpedo boat (MTB) from Shetland.

Larsen was already something of a legend. He had escaped from occupied Norway in February 1941 in the fishing smack *Motig 1* and, after initial training, became a skipper on what was to become known as the 'Shetland Bus'. This was largely unsung but highly dangerous work, scratching at the flanks of Fortress Europe. Such brave men in small boats undertook a constant series of daring missions, sea passages contesting the worst of the elements and fury of the Nazis. The German invasion and subsequent occupation of Norway in 1940 is chiefly remembered for the fiasco of attempted allied intervention. Thousands of Norwegians, finding the yoke of Nazi tyranny intolerable, escaped in fishing craft and other small boats, many landing in Shetland.

Those who remained, in many instances, formed the core of a nascent resistance movement. By summer the following year, a base had been established at Lunna on the north-east flank of Shetland mainland, and civilian craft were pressed into service as transports for supplies and agents being infiltrated into Norway. Thus, the Shetlanders were in the forefront of a small but steady assault on Hitler's Europe. Many of those who fled this grim oppression did so in conditions of extreme peril. Kare Iversen, a fisherman who had joined the Norwegian resistance in 1940, pressed his father's craft

Villa 1, a 13-metre-long fishing vessel into service, when he and several companions, hunted by the Germans, sought refuge in Shetland. Their passage was eventful:

> The first two days passed in perfect weather conditions. But at 3.10 on our third day trouble arrived. I was coming out of the engine room when I saw through the wheelhouse door that a German flying boat was coming straight for us. At the time two of us were down in the engine room, two in the forward cabin. After the first burst of gunfire, I went out on deck and released all the halyards to let the sails drop to the deck, then we went back down to the engine room. The Germans continued shooting at us for twenty minutes. Their gunfire riddled the wheelhouse and holed the boat just above the water line with their shelling. Whenever we rolled, a big rush of water now came into the hold. The flying boat tried to land but the sea was too rough, so they gave up the attempt.[20]

After a fraught passage in high seas, his vessel damaged by Axis gunfire, Kare reached Shetland, anchoring at Sandwick. After initial de-briefing, he trained with fellow Norwegian exiles and then returned to Lunna where he sailed with Leif Larsen on *Arthur*, *Feie*, *Siglaos* and *Heland*. In December 1943, after many adventures, he became part of the crew of MTB *Hessa* and served until the end of the war, though he found time to court and marry local girl Cissie Slater.

Lunna was a reasonable beginning, but the place lacked adequate facilities so, by 1942, efforts were switched to Scalloway, where a local concern, William Moore and Sons, acted as engineers and repairers for the growing fleet of small boats. This was the very stuff of derring-do, a channel for the frustrated exiles, with local assistance, to strike back at the occupier. It was not undertaken without cost; *Sjo*, *Aksel*, *Feioy*, *Bergholm* and *Brattholm* were all lost with the lives of over thirty men. As unit strength was never more than double that number, such losses were dismal indeed.

When it was suggested that this was unacceptable and that the sailors could be more effectively employed in a conventional naval role, they simply replied, 'Give us better boats and we will continue the fight.' This proved a turning point. The Shetland Bus service was re-equipped with three US fast motor launches, *Hitra*, *Hessa* and *Vigra*. These sleek craft were built entirely for war, 100 feet in length and powered by twin 1,200 hp General Motors engines. Cruising at 17 knots with a further 5 in reserve at full throttle and bristling with machine guns, they could give a good account of themselves against marauding Messerschmitts. From November 1943 these replaced the ad hoc conversions to carry out many further operations. None was lost, nor did the crews sustain any fatal casualties.[21]

Shetland had been in the front line since November 1939 when half a dozen Heinkels came in low over Lerwick, aiming to attack vessels in the harbour. Eight bombs were dropped to no effect, but strafing attacks wrecked a flying boat, though all the crew were rescued by a brace of haddock boats. Happily, three trawlers each heavily loaded with marine munitions, in this case depth charge – *Northern Foam*, *Northern Isles* and *Northern Princess* – were spared. Once Norway fell, there was that looming possibility that Shetland could form the target beachhead for an invasion of Britain.

As a result, a large garrison, at its biggest some 20,000 strong, was drafted in and hut camps sprang up by Scalloway, Sumburgh and Sullom Voe. Sandbagged anti-aircraft batteries were positioned around Lerwick and new airstrips were levelled at Sumburgh and Scatsta. Sullom Voe also played host to a squadron of flying boats. Shetland became a militarised frontier with access strictly controlled; locals acting as Home Guard. Scotland, during the war years, became an embattled landscape: over 28,000 pillboxes, 414 roadblocks and 296 coastal battery positions were constructed, the most heavily defended part of Britain.[22]

Scalloway is located on Shetland's main island, once the capital with its fine natural harbour and commanding 16th-century tower house. In 1985, the first museum was opened by the local history group and kept going in its original format for two decades. The Shetland Bus was as it still remains, a prime exhibit, a wartime manifestation of ancient seafaring tradition, one the Vikings would surely have grasped. At the start of the millennium a Shetland Bus Friendship Society was founded, its objective to create a permanent memorial to the 44 men who died. A new monument was unveiled in 2003. Eleven years later and a whole new museum in larger premises was opened, the ceremony being performed by the Prime Minister of Norway. It tells this remarkable story simply but well.

West Highland Museum and Spean Bridge Commando Memorial, Fort William

My wife's godmother married Jack Whitmore Rasmussen (1922–1989), a Danish soldier and patriot who first enlisted with the Buffs then transferred to No. 4 Commando. He served both at Dieppe and on D-Day. I knew him. In 2004, I had the honour to attend the annual Armistice parade of surviving commando veterans at Spean Bridge, 9½ miles (13 kilometres) northeast of Fort William, with a march from the village uphill to the totemic memorial. There were lots of senior figures as this was to be the last such full parade, including HRH the Duke of Edinburgh. At this point most of those who had served were into their eighties, but when you saw them swing up the hill, you had to wonder just what they must have looked like in their twenties.

In 1949, sculptor Scott Sutherland won a competition to design the memorial, an Homeric bronze trio of lantern-jawed commandos, their thousand-yard stares directed over the verdant valley of the Rive Spean and across to the heights of Ben Nevis. Completed in 1951 and standing 5.2 metres tall, the monument has become a major visitor destination.[23]

Jack and his comrades were led by Simon Fraser 15th Lord Lovat, who, in real life, even managed to upstage his later screen persona as played by Peter Lawford,[24] piped ashore on Sword Beach on 6 June 1944 in fine highland style by Bill Millin. A consciously staged piece of legend-building, marvellous theatre but the enemy bullets were still real. The heroic piper was later and fittingly lauded in verse. 'Shimi' Lovat came from a long line of highland scrappers and looked the part, tall, Hollywood-handsome, who had moved from one conventional elite, the Scots Guards, to the rather less orthodox commandos. He had plenty of experience of raids in Norway and a distinguished role in the Dieppe disaster when his men, with superb sang froid, took their difficult objectives.

Now commanding 1st Special Service Brigade, he would land on Sword Beach. They'd had a typically rough crossing, oily sardines, and no rum for breakfast for those who had the stomach for it, then the waves of Allied aircraft screaming in overhead, a chorus to the symphony of naval shells engulfing the beach ahead. They were still eight miles out. As the flotilla closed, the big guns gave way to the lesser battering of destroyers, picking up enemy muzzle flashes. It took 45 minutes to close on the beach, Lovat had divided his landing craft into waves, he didn't want his entire brigade crowding the sands and giving the enemy an array of packed targets.

His commandos were going in behind 8th Infantry Brigade, their job to take assigned strongpoints – heavy infantry work rather than dash and grab raiding. One troop went ashore singing *Jerusalem*. Shimi himself was going ashore in a plywood-hulled craft; the first echelon had the use of sturdier steel-hulled boats. They felt very exposed. There was a great deal of smoke but not enough to completely shroud their landmark. A damaged LCT with wounded and dead aboard limped back in front of them. Now they were among the obstacles. Skeletal ribs of hedgehogs laced with mines, spiced with incoming fire. The enemy were firing armour-piercing rather than HE rounds, a bizarre blessing as these just punched through the flimsy wooden hulls and kept going, HE would have shredded them.

Despite damage, they made it in, five out of twenty-two launches were hit but most still made it ashore. Down went the bows on Lovat's boat and, as CO, like the clan chiefs of old, he led the charge, Bill Millin offered *Blue Bonnets* as a musical accompaniment. As befitted his raffish style, Lovat went ashore wearing a monogrammed white jumper beneath his battledress and carrying (allegedly) a strictly non-issue .45-70 lever action Winchester. John Wayne couldn't have done it better.[25]

They pelted up the beach, and men inevitably went down.

> Mullen – the artist – like a broken doll, lay with both legs shattered, at the end of a bloody trail. He was beyond speech but out of pain; a glance showed a helpless case and death was busy with him... David Wellesley Colley, a Downside boy with a cheerful smile, lay against the pack of a sergeant who had pulled him into shelter. He was shot through the heart. Tears were running down the NCO's face. 'Mr Colley's dead sir. He's dead. Don't you understand?'[26]

'Little' Ginger Cunningham was shot through the legs as he ran. 'Big' Murdoch McDougall yanked him up and carried the smaller man clear – Ginger loudly lamenting how the Germans managed to miss such a big bloke yet hit a far smaller one![27] Despite his wounds, Ginger, the unit medic, stayed in the game patching up wounded and hobbling gamely with the aid of Lovat's stick. A brave *resistant*, M. Lefevre, had sabotaged cables powering flame throwers on the beach. Lovat rather snappily felt he'd done better than 8th Brigade. Shimi was none too impressed with 3rd Division, 'muscle-bound mentally and physically, after four years training in the UK'.[28]

W. H. Jeffries served in No. 6 Commando and landed from an LCI:

> After sailing, below deck we made a very special study of our maps, checked our arms and ammunition, and had plenty of hot soup provided

> by one of the crew. June the 6th, soon after dawn, we were crouching low on the deck and to our left a battleship was firing, and above a few Spitfires to cover us in. At this point the enemy gunners were trying to get our range and shells were bursting all around us.
>
> Soon we were heading for our part of the Normandy coast, and at once all hell seemed to break out. As the enemy machine gunners opened up, very calmly the LCI crew dropped the landing ramps down, and with good luck from the crew we started on our way through the sea. Part of our task was to reach the airborne forces who in the night had taken and were holding the bridge, now named Pegasus Bridge. After leaving the beach we made our way through open grassland, and all around the Germans had placed notice boards warning of mines. But by a careful study of the ground, we found the way across a part where cattle had been grazing some days before. We moved so fast that we were on to one group of Germans drinking coffee in the edge of a field. Our instructions had to be carried out. Push on to the bridge, never mind the odds.[29]

One of Lovat's units, 1st Battalion, Royal Marines, 4th Franco-British Commando, was led by *Capitaine de Corvette* Philippe Kieffer. Their job, coming ashore at Rive-Bella, was to take the former casino, no longer a gambling den but a redoubt. Born in Haiti but of Alsatian stock, Kieffer had volunteered for the French Navy in 1939. He'd escaped the fall of France in June the following year and signed up for the Free French Navy in London. A fluent English speaker, he had been impressed by the fledgling commandos and despite already being forty he set about raising a Gallic contingent, the *Fusiliers-Marins Commandos*. Blooded at Dieppe, by 6 June 1944 they were veterans of numerous raids.

Keiffer had 177 men, 21 of whom were killed and 83 wounded. His commandos, many of them Bretons, were the only all-French unit to take part in the invasion. No. 4 Commando came ashore on Queen Beach Red. Lieutenant-Colonel Dawson was wounded, so Kieffer led the survivors towards Ouistreham. Luckily, a local gendarme was able to give some pretty precise information on the mini fortress, once the casino (rebuilt for defence in 1942). Kieffer himself was hit twice by shrapnel as he directed the attack, which became very up close and personal, a hard, bloody fight eventually won by the bayonet.

Despite his wounds, the Frenchman refused evacuation for two days and was back in action barely a week later, taking part in the breakout battles. He was amongst the first Free French to enter Paris, a bitter victory as his 18-year-old son, serving with the *Maquis*, was killed by the Germans almost at the same time.

Lovat himself, storming up the beach, passed through 8th Brigade busily digging in. Commandos don't really do digging in, they go and find more enemy to biff. With No. 6 Commando as vanguard, they charged. Bombing their way through the defences, stripped down to essential fighting kit, they chucked grenades into pillboxes, scorched the defenders with flame throwers and hosed everything in sight from Brens. This was what they had volunteered for. Snipers lurking in bombed-out buildings were systematically stalked and dealt with. Marksmen picked off any German daft enough to raise his head.[30]

A steady trickle of captured enemy was shepherded dazed down onto the beach and into captivity. 'We were almost through the Atlantic Wall.'[31] Even Shimi was impressed by the network his commandos had punched through. The Atlantic wall wasn't a line of static defences, 'grouse butts' as he calls them, but meticulously constructed sets of interlocking positions offering serious in-depth and all-round defence. The underwater obstacles fed onto minefields, back towards individual strongpoints studded with ordnance and MGs, behind them infantry in trenches and reserves of armour further back. 'Each pillbox was a citadel of reinforced concrete, sunk hull-down and half-buried in the ridges of the dunes. Walls two feet thick stood six feet above ground level, their height made up by a very solid roof, giving further feet of concrete head cover.' Lovat saw these German bunkers made the British equivalent look puny. They were sited in depth, say 100 or 150 yards apart, girded with strong wire entanglements, interlaced with more mines. They didn't face front but gave sweeping enfilade fire. There might be half a dozen defenders with a 75mm gun and automatics, pillboxes interspersed with MG cupolas.[32] Nevertheless, No 6 Commando 'moved like a knife through enemy butter'.[33]

Traditionally, the West Highland Museum was to a significant degree focussed on Jacobinism. It's a wonderful place, on the south flank of Fort William's modest town square. It's the classic 19th-century local museum that is now an endangered species. In 2014, the curator acquired an extensive collection of Commando kit; clearly, it's just down the road from Achnacarry of Special Forces legend. (British

Commandos, US Army Rangers and commandos from France, the Netherlands, Norway, Czechoslovakia, Poland and Belgium trained there). It's not a big display, the museum is short of space but hugely evocative and in exactly the right place. I talked to Vanessa Martin, the most helpful curator, who did intimate they're acquiring another significant collection and planning a special exhibition. That's one I don't intend to miss. A poll of visitors held in 2019 still put the Jacobites in first place as a visitor draw but with commandos coming a very close second.

At a coffee shop in Fort William:

Graham: Bloody marvellous, love that place. The commando stuff is spot on.

John: Like most places in the Western Highlands, I've been coming here for a distressingly long time and, being grey and cantankerous, love places that haven't really changed. The bulk of the museum, the Jacobite exhibits have been there for a very long while, but adding the commando element creates a whole new dimension, grand collection of Fairbairn-Sykes knives… 'Get tough, get down in the gutter, win at all costs… I teach what is called "Gutter Fighting". There's no fair play, no rules except one: kill or be killed.'

A digression on William Ewart Fairbairn. Graham and I are collectors of Fairbairn-Sykes (F-S) knives and Fairbairn is a hero of mine. Born in the south, he spent much of his youth in the tiny village of Lesbury in Northumberland. Lean and wiry, he was never the Arnold Schwarzenegger type, just a lot more dangerous and very much the real thing. He lied about his age to join the Royal Marines at 16 and spent four years, 1903–1907, in the Far East, based in Korea, then occupied by the Japanese (our allies at that point). He became gloriously obsessed by martial arts with an added twist of bayonet fighting. He'd spend his spare time sparring with Japanese opponents in 'friendly' bouts and, to be fair, the Japanese weren't frightened of using bayonets and were enthusiastic exponents.

After his stint on the Peninsula, he moved into policing work in Shanghai. As a regular beat, this sprawling metropolis made Al Capone's Chicago look like Emmerdale; organised crime wasn't just rife, it was an institution. Fairbairn honed his knowledge of martial arts (the first European black belt in Judo), developing his own bespoke hybrid version. He liked blades, just as well; he is said to have survived some 600 fights, building up a mass of scar tissue from defensive wounds but leaving many deceased opponents.

With his colleague Eric Sykes (not the comedian), he both trained his fellow officers and developed the famous F-S fighting knife. A modern stiletto that every renaissance assassin would have coveted, it has a 6.5-inch tapering blade leading to a needle-sharp point. It's not a utility tool, it is designed for killing, for thrusting into exposed areas, sliding between unsuspecting ribs and generally disposing of unwanted sentries and other inconveniences. The tapered handle balances the heft and facilitates a number of grips.

By 1940, Fairbairn was 55, but Hitler had plans and so did the Secret Service, recruiting these ex-police to train emerging Special Forces including British Commandos and US Rangers. Fairbairn never sugar-coated the pill. His creed was to do unto others as they'd like to do unto you, just do it first and better. Today's SAS still carry F-S knives.

Royal Scots Museum, Edinburgh Castle

I've mentioned the Royal Regiment of Scotland, founded on 28 March 2006. It initially comprised six regular and two Army Reserve Battalions, a grand amalgamation of Scotland's magnificent regiments. That wasn't the end, that August two of the six were further merged and a mere six years later one of the regular units was whittled down to just company size. At the end of 2021, another regular formation was removed from the order of battle to form 1st Battalion the Rangers.

This new, somewhat truncated regiment (the senior line infantry unit in the British Army) also has its museum in Edinburgh Castle. The Castle has the Scottish Cavalry Museum and indeed the Scottish branch of IWM (rather better than its London head office). Where else is more fitting for a Royal Regiment of Scotland Museum than in the ancient fortress on its rock.

In the summer of 2005, two young people purchased a house in Mount Vernon, a residential suburb of Glasgow. The property had previously been owned by an elderly couple who left the attic uncleared. The buyers had no expectation of undiscovered treasure, but behind a vintage Dansette turntable, they discovered a cardboard box which turned out upon investigation to contain the war diaries and memorabilia of Arthur William David Roberts, who had died over twenty years previously. At the core of Arthur Roberts' account of his life is his war experience, gained in the cauldron of Flanders during the Third battle of Ypres in 1917:

> For so short an army career, I think I may safely say, my life during that period was as varied, and eventful, as most private soldiers of a similar length of service. A soldier during war time if capable is pushed into many breaches whether for the front line or base. I have been fit for both; consequently, I have filled many breaches. The last sentence will perhaps lead the reader to think I am possessed of great capabilities, and this belief may be strengthened when I say that I have been company-runner, batman, guide, dining-hall attendant, bugler, dispatch clerk, aircraft-gunner, hut-builder, stretcher-bearer, and one or two other things. Now it has been unintentional, if I have seemingly blown my own horn about my military accomplishments, but I think this book, written as frankly as I could write it will exonerate me from any imputation of self-aggrandizement.[34]

Arthur Roberts was born in Bristol in 1897 of mixed-race parents. David Roberts (Jenkins), his father, a ship steward, had a Caribbean background, while his mother Laura Roberts (Jenkins), nee Dann, was from the West Country. By the time of Arthur's birth, the family had dropped the surname Jenkins. At some point in the early twentieth century, it appears Arthur and his father moved to Glasgow where the young boy was educated at Finniston Public School and would continue his studies at Kent Road Higher Grade School. He remained in full-time education till he was 18 and it is clear from the quality of his prose he was a highly intelligent and articulate young man.

For most of his adult working life, Arthur was employed by Harland and Wolff in Anderston and Duncan Stewart, which later became Davy United, in Bridgeton, Glasgow, as a marine engineer. He volunteered for military service in February 1917, firstly with the King's Own Scottish Borderers, then on 3 June with the 2nd Battalion Royal Scottish Fusiliers.

Relatively few Black soldiers served in a front-line role with Scottish regiments in WWI. Arthur Roberts does not focus on his race or cultural identity; he is very much an individual, both participant and wry observer, slightly detached whilst clearly not, in any way, excluded. Arthur Roberts is not just a military figure; his life is a window into the twentieth century. He was of mixed race and origins, yet he remained in education till aged 18, an achievement in itself. His grasp of language and grammar are exemplary, and he was as skilled with the brush as the pen. His war memoir is fluent and accomplished, delivered with humour and panache.

Accounts by Black soldiers in the First World War are very rare. His is very much a Scottish experience. It is extremely well written and beautifully illustrated; the writer was an accomplished artist. It is more than just a war memoir, it is the unique history of a man's life, a fully rounded account which has been ably researched and illuminated by Morag Thomson Miller and Roy Laycock. They know Glasgow intimately as natives and a great part of Arthur's story concerns his life as an adopted Glaswegian.

> I first actually entered the trenches on the dawn of 9 June 1917. I can tell you, after our gruelling march, I was a physical wreck. That night as I plumped down in a dugout, I was so tired that without taking off my equipment, I almost immediately fell into a trance. All the Kaiser's horses and all the Kaiser's men could not have put the wind up me that night. No, I was too far beyond the stage of self-preservation. Sleep and welcome oblivion were wanted. I believe I should not have cared if I had been told I should never wake up again.[35]

Arthur suffered recurring foot problems and did not see further active service after 1917, being de-mobbed on 5 December 1919. He returned to Glasgow, now his native city, to continue his apprenticeship. He resumed sketching, was confirmed into the Anglican faith, and became interested in music, particularly the banjo. He met his future wife in the 1930s, though they did not marry until many years later.

He continued to live and work in Glasgow during the Second World War. His post-war photos of beach holidays show him as still very much the dandified 'man about town'. His wife Jessie died at the age of 63 and he lived with Jessie's cousin Jean McDonald until she, too, died in 1977. Arthur's health declined and his later years were spent in care, where he died at the age

The house in Mount Vernon had been owned by an elderly widower and the attic appeared to have lain undisturbed for decades. The new buyers knew nothing of Arthur William David Roberts till they came across that large cardboard box.

The box contained His Great War diary and collected memoirs, his satchel, official documents placed separately, a number of photos, postcards and drawings with three albums and various Christmas cards. The young woman who with her partner had bought the house used some of the Great War material in her academic work and it was her course tutor who advised her to consult a military expert. The full significance of the diary was not appreciated until Morag and Roy, who had agreed to take on the research task, had consulted with Ian Martin at the KOSB Museum in Berwick. Thanks to Ian's experience and diligence, a clearer record of Arthur's Great War Service came to light. Morag and Roy were now fully possessed by that fever which all historians and researchers understand, the need to know more.

An overriding question was how the box had ended up in this particular attic, as Arthur Roberts appeared to have no connection with the address. A series of clues emerged from the cache of personal records. Tellingly, his death certificate (dated 15 January 1982), listed his executor as a Mr. J. Walker – it was he who owned the property and the link was now established.

Arthur's long anticipated baptism of fire came in late June 1917 in Flanders:

> Gradually as the day drew to a close, the German artillery fire increased, and by dark the stutter of machine guns had become more voluble. Up to now I had seen nothing but the Verey lights to excite my curiosity, and the significance of this exchanging of ammunition, held no warning for me. Eagerly I waited for the start of the expedition into no-man's land, so judge of my disappointment, and my comrades' relief, when the officer came with the news that the raid was off. That night we were manning the parapet with gas helmets on, and everything ready to give Jerry a warm reception. While the shells were whizzing and bursting overhead, I was much amused with the other chaps bobbing up and down, and jumping into corners at every near burst. I must say at that moment I felt much superior as I stood on the fire-step without a movement.
>
> This is no boast dear reader, for you must understand, that in my ignorance, guns and shells were only but sounds to me and I had never seen the effects of shellfire. Jerry eased off about dawn, and things were fairly peaceful until about six o'clock, then he showed us the results of

> his gun practice, and I can tell you when Jerry lets himself go in those days, he was generosity personified in handing out his shells.[36]

Arthur did his bit.

Stobs Camp, Hawick

Before we arrive at the camp, mention must be made of James Paris Lee (1831–1904), a native of Hawick whose achievements are commemorated in Wilton Lodge Park, the excellent museum there. His family emigrated when he was very young to Canada, where he quickly demonstrated a scientific flair and, like Lord Armstrong (see below), this was directed towards firearms technology. He constructed his first gun at the age of 12 and in 1861 designed a breech-loading cartridge conversion for the muzzle-loading Springfield percussion rifle.

His major impact, however, came when he devised a spring-loaded, column-feed magazine system for centre-fire rifle cartridges. Most of us who grew up in the 1960s learned to shoot with bolt-action rifles, usually a venerable Mark IV Lee-Enfield, chambered for .22. One of the prime innovations of Lee's system was using two columns or clips of five rounds, as opposed to only one, thus doubling the magazine capacity. More rapid loading was refined by using a five-round strip or charger, the famous 'mad minute'. The Lee Bolt, unlike the straight heavy bolt on a German Mauser, enabled far quicker shooting. The Short magazine Lee-Enfield (SMLE) proved a better weapon for trench warfare than the Mauser. The Germans found this out to their considerable cost at Mons in August 1914.

Andy Jepson is one of life's great enthusiasts, a man driven by passion (a much overworked expression) for what he does. In the early summer of 2019, on one of those glittering days that only the borders can produce, a team of us drove north from Northumberland to Stobs, just south of Hawick.[37]

We went north from Belsay where I live, up over the dour Ottercops, past Catcleugh Reservoir, up to the watershed of Carter Bar, the Reidswire, which opens up a panoramic vista and a broad sweep of the border hills, hunched and timeless, several crowned with Iron Age hillforts. Restless wind brushes the green and ochre grasses as though by a giant's hand. In my six decades and more, it has never failed to uplift. Then we branched left and westwards past Souden Kirk, via Chesters, towards Hawick.

Stobs is huge. It sits above the narrow valley below on the bare rise of the hill, and more hills around, swelling like breakers. It was magical in the warm sunlight, the school group we were taking as entranced as school groups are likely to get. This was an exceptional day of beauty and tranquillity, not, it has to be said, what you'd normally expect. The camp was a mini city, originally laid out as a Victorian training camp used by local regiments and Territorials for training. It's totally unique combining a historic facility used for training, for internment, and as a POW camp. Much of the structure remains intact or intact enough, and the visual impact is breathtaking.

During the war of manoeuvre in late summer 1914, the bulk of the 'enemy aliens' interned were civilian internees, 10,500 out of 13,600 in all. Edward Lowry from the US Embassy visited the camp in February 1916 and reported that the place was home to 4,616 inmates, of whom 1,829 were army, 504 Navy with 2,283 civilians. Most of these (2,098) were German; the rest were primarily Austro-Hungarian but with three Turks and one Bulgarian.[38] Only as the war progressed and the number of Central Powers' POWs swelled did the military/civilian balance begin to shift.

It is hard to give a real sense of the place, it kind of creeps up on you. When you begin the walk uphill, you pass a few buildings, admin and barracks piles, looking a bit post-apocalyptic where sheep have claimed back their birthright. It's unspoilt, not cluttered by tidy gravel paths, interpretation panels and the endless plethora of prohibitions, excessive signage and health and safety warnings we now tend to associate with heritage sites. Nor does it fit our view of the border landscape, no tall tower house looking as though it had sprung from bedrock, this is a manmade environment imposed on the landscape, a temporary expedient, even if it did last for three score years of occupation.

> In 1917 the POW camp consisted of a large compound measuring internally about 750 metres x 280 metres, surrounded by a heavy triple barbed fence which had electric lights at intervals. There was a sentry post at each of the corners and other posts at 70–100 metre intervals around the perimeter within the larger compound, there were four subsidiary compounds known [imaginatively] as 'A' Camp, 'B' Camp, 'C' Camp and 'D' Camp. These were separated from each other by fences. Each of the four camp compounds contained four sets of five

> huts measuring 120 feet x 20 feet... To the north of the built-up area each camp had a roughly rectangular open exercise yard.[39]

Stobs was pretty secure not least because of its remoteness, only the Waverly line linking Edinburgh to Carlisle passed close by, and the rest was hills. Some did make the break however, no Steve McQueen jumps on motorbikes, but in November 1916 three matelots, captured when the *Gneisenau* went down after the Falklands fight in the South Atlantic got out. Two were caught at Rutherglen and the third, Julius Schmidt, made it as far as Spittal near Berwick. He might have been in with a chance as he'd plied the route before the war as a sailor on commercial vessels trading timber. He had six shillings on him and plenty of rations in a 'strong wooden box'.[40]

The Devil's Porridge, Eastriggs

I always used to enjoy the road from Longtown to Gretna; it has an aura of its own. Running straight through woods, there were military looking buildings, blank but unmistakably MoD and wartime, standing in small and seemingly unconnected clearings. This was 1960s and if you were into Bond books and movies there was a distinct resonance of secret bases and forbidden places. I suspect most are now gone. At that time, I had not heard of the Devil's Porridge or how what happened in Gretna affected the outcome of the Great War.

Gretna is associated with romance and elopement of course. It hit the headlines for all the wrong reasons in May 1915, when the worst multi-train crash in UK history claimed 227 lives, including those of 215 servicemen from Leith Battalion Royal Scots *en route* to Gallipoli – somehow dying in a railway accident on your way to war seems more pointless than actually being killed in the fight itself.[41]

Those attempted break in battles of 1915 ended in bloody stalemate. Sir John French blamed a lack of shells (that excuse kept him in a job until Loos in September). In part though, he was right: Britain was not geared up for total war and the press, as ever, went crazy. The shell scandal, as it was called, led to Lloyd George being appointed as Minister of Munitions in Asquith's coalition government and things began to change. Certainly at Gretna, which was earmarked as the preferred site for a vast industrial complex dedicated entirely to industrial warfare.

From Dornock by the Solway shore to Longtown, a huge, nine-mile-long series of factories would be thrown up. Over ten thousand construction workers would undertake the work and they did it in nine months, a tremendous and largely unheralded feat. It's said it was Sir Arthur Conan Doyle who coined the unforgettable phrase 'devil's porridge', referring to the tricky mix of gun cotton and nitro-glycerine needed to manufacture cordite to propel the shells.[42] Artillery was fast becoming the dominant force on the battlefield and Gretna would be the Wayland's Smithy of the Great War.

It was women, up to 30,000 of them, who'd do this dangerous work. The cordite was mixed by hand in large vats, and the mix, inevitably, was very unstable. A single spark meant instant oblivion. The women wore a uniform of trousers, crossover tunic and belt, hair 'bobbed' or carefully covered in mob-style caps. This was, of itself, a game-changer. Young women had not had such opportunities before, teaching for the talented and educated, governesses for the eligible and domestic service for most. Whilst the graft was long, hard, exhausting, and dangerous (it also caused industrial diseases), it did mean a kind of liberation; it was vital war work and definitely better paid than domestic service.

By 1917, the complex was producing eight hundred tons of cordite a week[43] and earned a royal visit from George V and Queen Mary. Living accommodation was spartan, more *Great Escape* than *Homes and Gardens*. A timber town of huts was enclosed behind barbed wire, though with its own hospital, cinema, schools, and shops. These young women with cash in their purses were hard to corral and local hostelries thrived. This was shocking to some, tales of drunkenness and debauchery abounded, the press enjoyed yet another field day. This resulted in a rather unique and very British experiment in state control when the government effectively nationalized five breweries and some four hundred public houses. As you'd expect from the state, the beer was noticeably weaker.[44]

Now the factories are gone, most went at the end of the war but some of the ancillary buildings, better quality housing and the (former) St Ninian's RC Church survive. The charming, rather idiosyncratic small museum has now been replaced by a much larger purpose-built structure dedicated to telling this remarkable story. The women of Gretna made a significant contribution to winning the war and their efforts also influenced views on women's suffrage. The Devil's Porridge is an important chapter in the region's history, though it concerned a very different kind of war from those of the border reivers!

Monica Ramirez Lopez is the education officer at Devil's Porridge. Monica sees this remarkable collection as the treasury of local memory. For years, the history of the complex was largely forgotten until rescued by a group of enthusiasts at the start of this millennium. Now, the museum, a bespoke centre, tells the whole story. Monica writes:

> At one of our school visits, a little Ukrainian girl, after listening to our war stories and reflections, ended up not depressed (as her country is at war and she is a refugee), but actually hopeful. She said that seeing how well Scotland and particularly this area looks after the war, and after being bombarded, makes her feel hope that Ukraine one day will also recover from the war. Seeing how good Scotland is doing gives her hope. That's what I mean when I say that the museum reflects the resilience and strength of the Scottish people: A museum that shows the horrors of the world wars also gives hope, inspiration, and the best learning about actions and reactions of human beings, so we don't repeat the past.

I'd say that says it all.

The Johnny Armstrong Museum of Border Arms and Armour

> I curse them walking and riding; I curse them standing and sitting; I curse them eating and I curse them drinking; I curse them waking and sleeping, rising and lying; I curse them at home; I curse them away from home; I curse them inside their houses and outside their houses; I curse their wives, their children, and their servants who help with their evil deeds. I curse their crops, their cattle, their wool, their sheep, their horses, their swine, their geese, their hens and all their livestock. I curse their halls, their rooms, their kitchens, their stalls, their barns, their cowsheds, their barnyards, their cabbage patches, their plows and their harrows. I curse all the goods and every building necessary for their sustenance and well-being.
>
> Archbishop of Glasgow: *Monition of Cursing* 1565

In the early 1990s when I was working on a book about Scottish battles, I went to the remarkable Johnnie Armstrong Museum at Teviothead, west of Hawick, run by the dynamic Moffat family whose main business was and is goldsmithing and jewellery. This was open from 1986–1998 and housed a stunning collection of primarily 16th-century arms and armour, such as might have been used by the justly infamous Anglo-Scottish border reivers, a lifetime obsession of mine.[45]

The 'steel bonnet' of the Border rider was a close-fitting, peaked helmet of the type described as a burgonet, which covered the cheeks but left the face exposed, thus protecting the rider against a blow delivered to the rear or sides. The foot soldiers of this period frequently wore the morion, which, in profile, offered a semi-oval outline with raised comb and flatter rim. Body armour comprised, for those of rank, a simple breast and back; lesser mortals had to rely on a leather or canvas jacket padded for comfort and reinforced with iron plates sewn on. A more sophisticated version was the brigandine – defensive plates and chains were sometimes stitched onto doublets or breeches to provide rudimentary protection. Leg harness was rare, most riders preferring thigh-high leather boots.[46]

These Borderers were essentially business-like in appearance – after all, they rode for profit and not for glory. Ostentation was shunned and in war the combatants were distinguished by a device sewn on, or worn as an armband. This lack of display had a disadvantage in battle; many a Border laird was killed alongside his tenantry, as at Pinkie where the dead of the Scottish armies proved difficult to distinguish by rank.

A horseman's chosen weapon was the lance; mounted on an ash shaft some 13 feet (4 metres) in length, used couched for thrusting or hurled over arm like a javelin. Swords, axes and daggers were in regular use, as was a handy form of crossbow known as the latch. A particularly fearsome weapon was the Jedburgh axe, or Jedhart staff, which receives its first mention early in the 16th century. Craftsmen working in the ancient burgh produced these formidable arms. The heavy elliptical blades averaged 4 feet (1.3 metres) in length, set upon an oak handle, the head socketed and secured by long flanges or languets extending down the shaft. These served to prevent the head from being lopped off in close combat and the fighter's hands were protected with iron stops or vamplates, similar to those featured on lances for the tourney.[47]

The cumbersome matchlock received little favour as it was essentially unsuited to the cavalry arm. Wheel-locks, when introduced, proved more practical though expensive, and a gentleman might carry a brace of pistols secured in holsters worn each side of the saddle, in addition to sporting a carbine or caliver which, as the 16th century wore on, replaced the latch as a missile weapon. In military terms the Borderers operated as light cavalry, incomparable scouts in their own marches, though far less reliable in any other role.

Throughout the Border wars the marchers evinced a distinct reluctance to see blood shed, particularly their own, in the cause of national honour. It limits the incentives when there is a good chance a victorious king will deploy his power against his own subjects; and the complexities of clan loyalties, marriage alliances and sheer pragmatism combined to limit or defeat any show of patriotism. Lord Home's celebrated reluctance to commit his own, largely unscathed, division to the final carnage at Flodden was just hard-headed calculation. Defeat by the English would see the Scottish border ablaze, and a prudent man husbanded his resources to defend his own.

To a Borderer, his mount was everything and the shaggy little garrons or 'hobblers' were prized for their surefooted stamina. The raiding season got under way in the autumn as the nights drew in, when the mosses remained dry from the summer sun. The reivers' paths across the trackless wastes were numerous and devious. A raid could cover formidable distances, far in excess of those to which conventional cavalry might aspire. As good mounts were so valued, horse breeding became a local and national pastime. The Stewart kings imported stock from as far afield as Hungary, Poland and Spain. Racing was popular amongst the Borderers, and race meets – like truce days – were an escape from the relentless grind.[48]

One truce degenerated into the fracas that became known as the Raid of the Reidswire, during the summer of 1575. The Reidswire lies on the line of the Border, some three miles from Carter Bar, with the whole magnificent sweep of the Border hills as a backdrop, the 'swire' being a narrow neck of land linking two hunched and rolling crests. The English were represented by the middle march warden, Sir John Forster. In his mid-seventies and destined to hold sway until his tenth decade, Sir John was a man who thoroughly understood the wickedness of Border ways, having practised most of them enthusiastically since his youth. Sir John Carmichael, Keeper of Liddesdale, appeared for the Scots.

For three hours the meeting continued amicably, though there was little love lost between the two or between Liddesdale Crosers and the Tynedale men. An argument over a particular bill or some other matter degenerated into a slanging match between the wardens. Their respective compatriots, fuelled by drink, followed suit and insults were quickly followed by blows. A chance arrow killed Fenwick of Wallington and first blood seems to have persuaded Forster and Carmichael to calm themselves and their followers. A precarious peace was almost restored when the Scots suddenly fell upon the English warden and his party, capturing the outraged Forster and leaving his deputy, Sir George Heron, dead on the turf.

Now the Tynedale men surged forward, their longbows creating havoc amongst the Scots, who wavered but, at the last, rallied and charged again, bolstered by a timely contingent arriving from Jedburgh. It was the turn of the English to flee, hotly pursued by the victors. The Scots were left to celebrate their win, but the matter was most embarrassing to the government seeking to foster good relations with England, and Forster, old rogue that he was, soon gained his freedom and regained his wardenship.[49]

Blood feud. No more bitter rivalry existed between any of the riding families than the enmity between the Maxwells and Johnstons. In 1528 Dacre was able to report that the state of open warfare which existed between the two kindreds had resulted in the devastation of the Debateable Lands and much of the west march. The prize most sought was the march wardenship itself, and the office passed like a bauble between the factions with almost bewildering speed. In the 1580s each kindred possessed an active and aggressive chieftain, Johnnie Johnstone clashing frequently with John, eighth Lord Maxwell. Maxwell had two spells in office around 1580, being dismissed once and latterly imprisoned, an experience he came to understand well. When Johnstone was outlawed in 1581, Maxwell entered upon a third term of office. The following year he was once again removed, and Johnstone reinstated. Maxwell was not the man to submit lightly, however, and he rebelled against his detested rival.

The Chancellor, Arran, was kin to Johnstone and purported to punish Maxwell for his impertinence by stripping him of the office of Provost of Dumfries, a lucrative sinecure. Enraged, Maxwell raised the stakes by importing English Grahams to harry and burn. By the spring of 1585, administration of the march had degenerated into anarchy; 80 Johnstone towers had been sacked or razed, but far from being intimidated Johnstone allowed the tide of violence to swell by laying waste the Maxwell village of Duncow, and they naturally responded in kind. The body count continued to rise steadily.

By the summer, Maxwell's star was firmly in the ascendant, and he had taken the march by storm; by August Johnstone was captive. At this point King James, bowing to what appeared inevitable, acquiesced to Maxwell's reinstatement. Unable to keep clear of mischief, Maxwell's political adventures, linked to his unfashionably ardent Catholicism, landed him back in gaol the following year. Johnstone was swift to make capital from his enemy's discomfort and he laid waste Maxwell territories. For this he was in turn incarcerated.

Though he was subsequently freed, the years of strife had taken a heavy toll and Johnstone died in 1587, a broken man.

Maxwell, as ever, proved more resilient. In exile now, he busied himself in Spain abetting plots against England and in the year of the Armada was stirring up resentment in the Catholic cause. Not unsurprisingly, he was once again imprisoned. Somehow, he managed to talk his way out again and by 1592 he was reappointed as warden, this time for life. Fully in control, he arranged for the young Johnstone chieftain to marry a Maxwell girl and the feud seemed, if certainly not forgotten, at least under control. Matters might have rested thus, at least for a while, but a petty feud led to a violent clash between the Johnstons and the Crichtons, which left 15 of the latter dead. Maxwell, for once, appears to have been reluctant to become involved but his promises of succour to the offended Crichtons and their Douglas allies came to the ears of the Johnstons, who yet again drew swords.

With the Johnstons once more antagonistic, Maxwell decided upon a draconian solution to the problem of his quarrelsome neighbours. Formally summoning Johnstone to surrender, the warden moved towards Lochwood with a force of 2,000 men. Desperate for allies to bolster his thinner ranks, Johnstone was joined by Elliotts, Scotts, Irvines and the English Grahams, who were always available for mayhem. Despite these reinforcements he could barely muster 400 riders.

On 6 December 1593, as Maxwell's force lumbered past Lockerbie, Johnstone appeared at the head of forty-odd riders. Maxwell countered by unleashing twice as many of his 'skurrouris' or scouts, who gave chase to the now retiring Johnstons. This apparently precipitate retreat was no more than a ruse, for Johnstone had concealed his whole force, 'Lyland Darnit', in a wood and the warden's vanguard were thrown into wild disorder as the main body swept from the fringe of trees and fell upon them with a vengeance. It was the Maxwells' turn to flee and the ensuing chase threw the centre into confusion.

A hacking, sprawling mêlée spread over the plain, known as Dryfe Sands, and knots of fighters spilled into the streets of Lockerbie. So many fearsome wounds were dealt by the axes and staffs wielded by the Johnstons, fighting with the fury of utter desperation, that such injuries were thereafter referred to as 'Lockerbie licks'. The Maxwells were cut to pieces and the warden himself was among the fallen, who numbered over 1,000. Dryfe Sands was the bloodiest 'clan' fight in Scotland's history.

Johnstone was outlawed for his victory and though the wardenship continued to change hands he gained a pardon and the appointment, thus becoming the last warden of the west march. In 1608 a formal truce was arranged, and Johnstone met with the new Lord Maxwell, son of the formidable father killed at Lockerbie. Though the meeting was intended finally to heal the rift, Maxwell proceeded to empty a brace of pistols into the warden's back, a crime for which, on 21 March 1613, he went to the block, the last of four 'heidmen' to die in the vendetta.[50]

After the Union of the Crowns in 1603, James VI (James I of England), clamped down on his unruly marchers with a level of severity that would find him accused of crimes against humanity now. Standards then were more flexible, and harsh as it might have been, the treatment seemed to work. The reivers were consigned to the margins for a couple of centuries until Sir Walter Scott resurrected and re-branded them as Hollywood Robin Hoods. It was really only when the late, great George Macdonald Fraser published his seminal *Steel Bonnets* in 1971 that the revised image was revised again, nearer to reality. The BBC had a go with a TV series in the 1960s with Michael Gambon and Iain Cuthbertson, fun but pretty dire overall.[51]

At the Borderers Fair, March in Hawick 2022:[52]

Graham: I don't think this place has changed since I was at school, no recent makeovers.

John: The wool trade is pretty much dead, plenty of retail outlets where the mills used to operate, probably all made in Taiwan now. Heritage is becoming the new industry. Much as I love the borders it's not marketed like the Lake District, despite having just as much to offer.

Graham: You can thank Scott for bump-starting the revival, Macdonald Fraser did the remake and you've done a fair bit yourself. I assume the reivers weren't really your Robin Hood types?

John: God no, just the opposite, they were uplands mafia, took no notice of the official Anglo-Scottish divide. Why would they? It made no odds to them unless it suited and both governments heartily despised them when they weren't being useful. The Liddesdale names made common cause with the Tyndale banditti to pillage all and sundry, never a hint of racial discrimination, they'd rob and murder anyone.

Graham: Sounds like our local council. Mind you, if you ever played rugby against Hawick you wouldn't need any lessons on border warfare....

3

North Wales, Northwest and West of England

'War is a womb, big with many miseries.'

Civil Wars pamphleteer

'Tried and Valiant' – Cumbria's Museum of Military Life

Dusk on 13 April 1596 was thick and murky, god-fearing people weren't going out, but the rain-soaked night cloaked an elite group of eighty riders, moving like ghosts through the near pitch black and the well-named Debatable Land, the threap that separated the two ancient enemies of England and Scotland. Sir Walter Scott, known as 'Bold Buccleuch', was coming to Carlisle to spring his infamous client Kinmont Will Armstrong from the clutches of English justice. Will deserved to hang but he'd been on a Day of Truce, which was bad form and counter to border law and custom. The English West March Warden Lord Scrope simply refused to let the old villain go. Well, if diplomacy didn't work, Buccleuch must fall back on force, a remedy he was accustomed to adopting. In fact, the whole brilliantly executed operation, the commando raid par excellence, was an inside job. Scrope was betrayed by his shifty subordinates the Carleton brothers, Kinmont was duly sprung and Scrope had egg all over his outraged face.[1]

Carlisle Castle would never be a pushover. The Scots had tried hard on numerous occasions to take the place and failed at every turn. With a fine sense of historical irony, it wouldn't be until 1745 that Prince Charlie's Jacobites browbeat a few ancient defenders into submission. The daft Italian insisted on leaving a garrison there after his retreat from England as a balcony for imaginary further incursions. This doomed party was mainly made up of men from the Manchester Regiment, some of those few luckless souls who'd been stirred to volunteer for the white cockade in England. Against Cumberland and the big guns he brought up they had no chance whatsoever, no hope for clemency either, a number suffered the full horror of a traitor's death.

Now, a main ring road cuts across the face of the red sandstone walls and squat keep that rises up like a cliff. George Macdonald Fraser memorably described wild Hermitage Castle in Liddesdale as representing 'Sod off in stone'. He's not wrong and Carlisle gives the same impression, as though its bulk was deposited there by an immense and angry act of nature.

You can see straightaway why it was always such a tough nut. Through past the impressive 14th-century gatehouse and the parade ground interior is divided between largely Victorian barrack piles and the inner ward with keep. The pleasingly austere buildings proudly bear the names of past battles, and the military museum is housed in the imposing Alma Barracks. I remember its previous location within the inner ward, wonderfully eclectic. It has had a substantial makeover and Cumbria's Museum of Military Life (MOML) is really first class, a well laid out chronological tour through the regiment's long and varied history. It truly commemorates and celebrates the county's proud history of long service.

Jules Wooding leads a very dedicated and enthusiastic team and 2022 was the museum's 90th anniversary. The collection was fortunate to take delivery recently of a significant artefact, George Macdonald Fraser's bush hat, one he wore while serving with Slim's 'Forgotten' 14th Army in the Burma Campaign, 1944–1945. We'll come back to that.

For me there are two very different exhibits that really resonate. The first is a life-size mannikin of a Yeomanry trooper[2] and horse, just on the left as you come into the ground floor, detailed down to the last strand of forage, it's a brilliant recreation. Rather less gung-ho and very sobering is a child's bonnet recovered from the horror of Cawnpore during the Indian Mutiny[3], it's one of those everyday objects that tells a terrible story more eloquently than any narrative.

It was in 1881 during the era of Childers' Reforms that the Border Regiment was formed, an amalgamation of the 34th Cumberland Regiment of Foot with the 55th Westmorland Regiment. The Cumbrians saw extensive service in South Africa and through both world wars and in 1959 were merged with The King's Own Royal Regiment (Lancaster); other amalgamations followed, and today's successor unit is the Duke of Lancaster's Regiment.

The Redan[4] – the name is synonymous with expressions like suicide attack. The Crimean War 1854–1856, has itself become synonymous with futility, horror, suffering and incompetence. Homeric failures such as the Charge of the Light Brigade have been transformed into epics, but most of the work was done, as ever, by the poor bloody infantry and much of that involved digging, holding, and advancing trenches, a precursor to the more abundant horrors of the Great War. The war was a punitive stroke against the expansionist aims of Czarist Russia, keen to gobble up tempting tracts of Ottoman territory. The campaign was something of an afterthought, blitzing the great Muscovite naval base at Sebastopol.

Early opportunities to more or less walk into the place in the wake of the Allied success at the Alma (20 September 1854), were allowed to slip by, and the great Russian military engineer Todleben threw a ring of fortifications around the city. Despite subsequent battles, Balaklava (25 October) and Inkerman (5 November), autumn mist and mud saw a deepening conflict of attrition as the Allies (Britain, France, Turkey with Piedmont-Sardinia), dug in for a protracted siege. That winter was awful, chronic mismanagement and dire logistics saw the threadbare redcoats freezing and starving on barren high ground, while supplies rotted in the harbour at Balaklava because nobody knew how to move them up.

Inkerman is often called 'The Soldier's Battle', a savage slogging match where a depleted and exhausted corps of British defenders saw off a series of massive thrusts launched by successive Russian columns deployed in vastly superior numbers. The redoubtable Minie Rifle, despite soaked powder from dank overnight, did good service, as did the 55th Foot. Ground and key features changed hands with bewildering rapidity, casualties were very high, but the battered defenders held or recovered their ground. It was here that Private Beech won his VC, the newly instated award for extreme gallantry.

He dashed out to rescue Colonel Carpenter leading the 41st who'd been cut off in No-Man's-Land, assailed by enthusiastic Russians with very sharp bayonets. Beech saw off the attackers and rescued the colonel. Overall, the 34th were more fortunate, avoiding the worst rigours of that awful winter, only arriving in the spring. They were very soon in the thick of it when the Russians on 22 March set to with an all-out blow against the sodden trenches, a confused close quarter mêlée in the darkness of night, which was seen off but only with heavy loss. This wasn't good, but much worse was to follow.

Cracking the ring, as the siege deepened that spring, occupied all Allied commanders. The Russians had stolen a march on the French (who held the right of the siege lines), by capturing a strongpoint to their front known as the Mamelon. This defended the much mightier Malakoff Redoubt, the key to the whole defensive line; if it fell, Sebastopol was lost. Todleben was not unaware of this. In front of the British lay the infamous Redan, fronted by an outwork Tommies had dubbed the Quarries. On 6 June, the allied guns opened up a furious bombardment, overpowering the defenders' batteries, though gaps torn in the lines were soon repaired at night. Next day, after a further preliminary bombardment the French assaulted the Mamelon, which, after a vicious to and fro, finally fell. With this objective in allied hands, British storming parties now leapt forward to seize the Quarries. They succeeded and saw off a whole series of frenzied counter attacks that night.[5]

Now there loomed the Redan itself, and nobody could doubt this would be a very tough nut indeed. But it didn't actually need cracking. If French forces could successfully assault the Malakoff, the Redan didn't matter. Pride didn't see it that way. For the French to win all the glory by a winning stroke against the Malakoff just wouldn't do. Britain expected her boys to do their bit and that bit was the Redan. Built on what had once been a vineyard, the structure formed an arrowhead, each face some 70 yards (64 metres) long, flanked by outlying defences, fronted by a deep ditch. This meant the outward face rose for 30 feet (9 metres). It lay around 450 yards (412 metres) ahead of British parallels pushed forward after the fall of the Quarries.[6]

Both the British and the French believed overwhelming firepower was the key to success against these twin strongpoints. The plan was for drenching fire that would pulverise the works so thoroughly they'd be ripe for an all-out assault right along the whole line. Raglan assigned 8,000 soldiers to the job, a third of the size of the French contingent, who would assault in three columns, one against each of the faces and one at the hinge. Columns would be accompanied by skirmishers drawn from the Rifles, sappers and scaling parties would follow on, the bulk of the attackers ready behind, with reserves in waiting. By way of a diversion, 3rd Division would advance to their left.

On that bright dawn of late spring, almost Mediterranean, at precisely 04.00 800 allied guns thundered, a massive cannonade, unprecedented. Russian counter-battery fire seemed to be easily suppressed, but in fact their gunners were merely conserving valuable ammunition. A massive blunder followed. The French decided to bring forward their assault by two hours. Nobody told Raglan. When the British commander did become aware, he agreed. Most of his contemporaries agreed this was a dire mistake. He was too much of the gentleman to disingenuously hold back his own men until he could see if the French blitz on the Malakoff had succeeded. If they failed, there was nothing to be said for throwing men's lives away by having a solo go at the Redan. In the event, this shifting timetable just added to prevailing uncertainty in the jumping-off trenches. Giving orders from HQ is one thing, but translating these to such cramped masses of men on the ground was another thing altogether.

It's often said that no plan ever survives contact with an enemy. 18 June 1855 was no exception. A nervous French officer mistook a mortar round for a signal flare and sent his men off half an hour too soon, many to their deaths, the whole scheme unravelled, and Russian guns did the rest. The defenders had learnt some very costly lessons and used their hard-won knowledge to great effect. By dawn, the French had retreated, piles of dead and writhing wounded attested to the bravery of their attempt. What should have happened was that Raglan would postpone his attack, allow the bombardment to resume and his allies time to get their breath and regroup. Raglan ordered the British attack to go ahead. Meanwhile the Russians, having guessed that was about to happen, shifted their substantial reserves to meet this new threat.[7]

Advancing over that distance in the teeth of accurate and sustained enemy fire was a nightmare. The left-hand column was quickly in difficulties, scrambling up out of their trenches, cohesion lost before they even began to move. They were shot to pieces and driven back. There was no banquette or firing step to help the men get out of their trenches and the 34th, making up the bulk of this attacking column, became disordered even before they had taken a step toward the enemy and his bristling cliff of booming defences, roundshot screaming. Morale was badly dented already, and onlookers reported the advance as 'hesitant'. Very soon the regiment had taken nearly 200 casualties and their attack stalled. Colonel Lacy Yeo of the Royal Fusiliers, commanding a brigade of the Light Division, sent in his supports, leading himself. He got as far as the abattis (an improvised field fortification of sharpened outward pointing timber stakes), before being shot through the head. Despite this mortal wound, he crawled forward a few yards, clutching his sword before dying.

Despite such a magnificent example, the attack sagged like a burst bladder, men going to ground as best they could, all forward impetus lost. Neither officers nor NCOs could get men back onto their feet, most who attempted rallies were immediately shot down. One of the officers from the 7th Fusiliers recorded:

> We had some hundred yards to advance across an open plain with guns loaded with grape and canister shot blazing away into us. As I advanced, I thought every second would be my last. I could hardly see for the dust that the grape shot made in ploughing up the ground all around us – before – behind and on each side – shells bursting over my head and fellows rolling over right and left. I seemed to bear a charmed life for nothing would hit me! My haversack was covered with blood from men who were shot near me and so was my sword. When we got to the abattis, which is at least 50 yards from the redan, the fire was so heavy that no mortals could stand it and there was nothing for it but to retreat as rapidly as possible… In fact, we were regularly beaten back, and I saw the rascally Russians taking off their caps and jeering at us.[8]

It was carnage that could only end in defeat, but that was not the fault of those men of the 34th or any other regiment. Lord Cardigan told his diminished brigade after the earlier debacle, 'It was a damn fool plan but no fault of mine.' As his lines recoiled, Raglan ordered his gunners to reopen their fire. This caught masses of Russians lining the parapet, inflicting severe loss. It was still a defeat though, and Sebastopol didn't fall until September.

Harry Flashman, the notorious bully and bounder and villain of Thomas Hughes' *Tom Brown's Schooldays* was resurrected as a proper Victorian anti-hero by George Macdonald Fraser (1925–2005).

Born in Carlisle of Scottish stock, he was expected to follow his father into medicine but rather frittered away opportunities through lack of interest and enlisted with the Border Regiment in 1943. He might have stayed a soldier, having gained a commission, and transferring into the Gordon Highlanders. By 1947, he'd decided against a military life in favour of journalism. The first *Flashman* novel appeared in 1966 followed by a regular list of sequels. One of his seminal works is *the Steel Bonnets*, a history of the Anglo-Scottish border reivers, Flashman would have fitted in quite nicely! His wartime memoir *Quartered Safe Out Here* (1992) is rightly regarded as one of the best, and this is where the bush hat comes in.

Macdonald Fraser's family gifted this totemic headgear, featured on the cover of the 2000 paperback edition, to the museum, a very personal and tangible link to the great man's service in Burma. Few have come close to his evocation of the stifling reality of the jungle:

> I peered across the bunker's top, wishing there was a moon; the sky hadn't begun to lighten, and ten yards away the landscape was just a blur; a Jap fighting patrol could get within a stone's throw undetected, if they were quiet enough ... was there something out there beyond the shadows, or was it just my imagination?[9]

At the Cumbria museum:

Graham: This is what I call a museum, bloody brilliant, stuffed with kit and no apologies.

John: For a long time, military history was the poor relation, academics scoffed at soldiers. In school we only studied cause and effect, the actual conflicts were brushed aside in a few sentences, nobody seemed interested in the detail. I got going with Liddel Hart, Fuller and Colonel A. H. Burne – all soldiers and all disregarded by the more rarefied circles of academe.

Graham: Higher education's long been a bastion of the left, according too much weight to the glories of Empire and the colours wouldn't suit the prevailing dogma.

John: That general contempt has now morphed into either apologising or re-writing. I kicked off a few years back in the new museum by Liverpool Docks, which is damn good otherwise, but the military section which was small stayed half hidden away behind a staircase and consisted of a weak regret over 'The Indian First War of

Independence' which, after a while I realised was what we used to call The Indian Mutiny (1857–1858), re-branded as a heroic struggle for independence from the Raj.

Graham: And here's me thinking it was an internal industrial dispute that got a bit out of hand?

John: It was. Independence as such wasn't really on the agenda and only one of the three presidencies was affected. You can see a whole weave of strands when you look closely but just re-packing the whole thing for the sake of PC is bollocks, never mind the dreadful syntax.

Graham: The presence here of a child's lace cap taken from the well at Cawnpore probably wouldn't be allowed on show in several other places, too emotive and the wrong brand of emotion.

Blue Streak; Spadeadam RAF Range

Night has fallen/And the barbarians haven't come
And some of our men just in from the border say
There are no barbarians any longer...

C. P. Cavafy: *Waiting for the Barbarians*

They called it Blue Streak, though it was more 'Beast from the East' when I last led a group there in 2018, so no better day for a band of intrepid Explore Students to visit RAF Spadeadam.[10] North of the River Irthing, across the strategic gap that marks the line between Northumberland and Cumbria. Once past Gilsland, the land rises beyond the twisting stream to the bleak moorland of Spadeadam Waste.[11] It's serious country, bleak, raw and untamed where ancient history's so close to the surface you can just about breathe it. Phil Abramson, MoD archaeologist, was our guide and his enthusiasm for this wild place with adds immeasurably to the experience, as does his vast store of knowledge.

Spadeadam, there's a name, you wouldn't get somewhere like this is Surrey. It's northern and it's a frontier; the russet-tinted fell stretches north towards Bewcastle and the Debatable Land. There's a lot of history here, much of it violent; the Three Hundred Years War followed, far more recently and perhaps unexpectedly, by the Cold War. And it was cold, an arctic, knife-edged wind drove snow flurries into whipped furies and, if you weren't careful your ears would drop off.

In 1955, the Waste was chosen as the site for the UK's Intermediate Range Ballistic Missile Test Centre, our UK contribution to NATO and the Cold War (the original that is). Post-1945, we relied on the formidable 'V' Bombers[12] to deliver nuclear bombs, should we need to have them delivered. Politically though, we wanted our very own deterrent, an intercontinental ballistic missile (ICBM). We couldn't just buy these in wholesale from the US owing to the restrictions imposed by the Atomic Energy Act of 1946, and besides, as a major power still, we felt entitled to our own Union Jack branding.

During spring 1954, our American allies proposed a joint venture. They'd build an ICBM, and we would construct a medium range weapon of, say, 2,000 miles (3,700 kilometres) capability. That August, both powers agreed terms (the Wilson-Sandys Agreement) and the programme got going. De Havilland Propellers won the contract to produce the kit, which would use liquid oxygen and kerosene. Now you could load up with kerosene, cheap as chips, pretty easily but the sixty tonnes of liquid oxygen were far trickier, had to be bunged in just before launching or it might, probably would, freeze up. This wasn't so good – the process took 4½ minutes, far, far too long. The missile was extremely vulnerable to a pre-emptive strike, launched with no warning, typical sneaky Russian trick, so we'd have been toasted before we'd even filled the tank.

Rockets never did fly from Spadeadam; an actual test site was set up at Woomera in South Australia. The initial budget in 1955 was £50m. Within four years this had increased six-fold. Both the US and USSR were galloping way ahead of us and it was going to cost us £1.3bn to finish the job. Blue Streak got the chop in April 1960, and we bought in the US Skybolt system instead.[13]

If anyone's looking for a post-apocalyptic landscape or a new version of *Dr. Strangelove*[14] – Spadeadam's your ideal location. The Waste is bleak, the MoD road winds through commercial woodland to the RAF guardhouse and, once on the range itself, there's an anorak's heaven of old Soviet gear, multiple, tracked rocket launchers, anti-aircraft quads and trucks. The snouts of old 'Bloodhound'[15] missiles point at the sky. The engine testing area with water reservoirs and squat, grey bunkers is an uncompromisingly brutalist complex, no pretty ponds or alpine flowers, the modern equivalent of peels or bastles, built for purpose and nowt else.

At Greymare Hills stand the remains of two rocket-firing stands; massive, infinitely ornate, imposingly ugly structures that fit like the devil's mark into the fold of hills. They are redolent with menace; age has not diminished them. They appear as timeless as the rocks and heather. Age, as we say, shall not wither them. Nothing short of a fifty-megaton blast could hope to and, certainly as far as I'm aware, there's no such comparable site in the whole of NATO's demesne.

On the range:

Graham: I don't know of anywhere quite like this, bloody unique and I spent years on every kind of range you can imagine, still have nightmares about the Brecon Beacons.

John: The amount of ex-Soviet vehicles scattered around as handy targets does give another weird twist. And it's in the middle of the wastes of the old English West March, nearest outpost is Bewcastle and it's amazing how many times the garrison there was looking the wrong way at critical moments. God knows what the reivers would have made of this.

Graham: Giving our riding ancestors' temperaments, access to missiles wouldn't have been ideal. Remember that series of books the *Sterkarm* trilogy, time-travelling reivers, pre-dates *Outlander*, I think. At least once they get hold of AK47s and other modern kit, it doesn't end well. [Susan Price, *The Sterkarm Handshake* (1998); *The Sterkarm Kiss* (2003) and *The Sterkarm Tryst* (2017)] In the first book, as I recall, the hero is transported forward to today's Carlisle General to be stitched up after a scrap and sees modern warfare as depicted on the news and decides 20th-century folk, 'elves' as they call them, are far more vicious and destructive than the borderers.

John: You couldn't really say he was wrong either, could you?

Lancashire Fusiliers Museum, Bury

Death is not an adventure for those who stand face to face with it.

Erich Maria Remarque

The Lancashire Fusiliers served their country for 280 years. But they weren't raised in Lancashire. In November 1688, Colonel Richard Peyton raised a regiment that, being his, was named after him as colonel, as most formations were. From 1740–1746 and through the Jacobite wars they were designated as Bligh's, then after 1751 till 1782 as 20th Foot, the 20th East Devon between 1781 and 1881, and as the Lancashire Fusiliers after 1881 until Haldane's Reforms of 1908, when the regimental structure was reformed and the Territorial Army came into being – they were still Lancashire Fusiliers.

On 25 April 1915, the 1st Battalion took part in the initial stages of the Gallipoli Landings at Cape Helles, carnage and glory followed like blood brothers, their 'six VCs before breakfast' represent an enormous sacrifice. 'W' Beach became known as Lancashire Landing. Bury is now the home of the Regimental Museum, made up of two main galleries: the Lancashire Fusiliers (LF) and the Royal Regiment of Fusiliers (RRF). The LF gallery tells the story of the 20th Regiment of Foot and the Lancashire Fusiliers from 1688 to 1968, when they then became part of The Royal Regiment of Fusiliers.[16]

Forcing the Dardanelles wasn't really such a bad idea, though perhaps not so good an idea as it appeared in 1915. On the Western Front, the trench lines had spread like impenetrable fissures across Northern France and Belgium, a total of 25,000 miles (42,250 kilometres) across a front of over 400 miles (650 kilometres). Both sides attacked that spring; in general terms, nobody could advance more than a few blood-soaked miles. The lines had solidified and the war that was supposed to have been over by Christmas wasn't going to be over anytime soon, in fact it had barely started. Russia in the East, part of the Triple Entente with Britain and France, had taken a fearful hammering and desperately needed re-supply. Meanwhile the Ottoman Empire had, rather disappointingly, sided with the Central Powers. On 2 January 1915, the Czar's C in C, Grand Duke Nicholas, appealed to his allies to intervene by supporting Russia's efforts in the Caucasus. The obvious remedy, which had been looked at before hostilities began, was to force the famous 38 miles (61 kilometres) passage of the Dardanelles, beloved of antiquity, and seize Constantinople. This was a pretty tall order, the narrows were studded with forts and layered with minefields. Still, an ally in need had to be supported and Churchill, as 1st Lord of the Admiralty, was an enthusiast. This was just his kind of thing, all too often derring-do masquerading as strategy.

On 28 January, the Dardanelles Committee approved a seaborne attack, put in primarily by the Royal Navy, deploying out-of-date battlecruisers rather than new, monstrously expensive dreadnoughts. In mid-February, the Committee had another think and decided the Navy would need soldiers in support to occupy key strongpoints once they'd been subdued. General Sir Ian Hamilton was sent to command troops, supported by a French contingent mustering in Egypt. The big naval guns first boomed on 19 February, followed by a hiatus, first of several, blamed on bad weather. When the ships went in, losses were alarming; three ageing battlecruisers went down, and another trio were damaged.

Was an amphibious assault a good idea? Not really, the nature of the narrow ground vastly favoured the defenders and most of Britain's best formations were serving elsewhere. Only 29th Division, a first-class unit, was on home service and available for Gallipoli. Four Fusilier battalions making up 86 Brigade would land on 'V', 'W' and 'X' Beaches, the Lancashire Fusiliers drew the short straw of 'W' at Tekke Burnu, to be immortalised as Lancashire Landing. They'd pay a very high price indeed. On the day, they fielded 27 officers with 1007 ORs. By the end of it they would only be able to muster 16 officers and 304 soldiers.

Cape Helles lies at the southern end of the peninsula, a rugged finger of steep, riven rock. Brigadier S. W. Hare whipped his men on with a rousing appeal to regimental honours and tradition: 'Behave in such a way that the men of Albuera and Minden, of Delhi and Lucknow may hail us as their equals in military achievement.'[17] Hare was soon badly hit and his successor was subsequently killed. The troops were conveyed aboard HMS *Implacable* and *Euryalus* before being loaded into launches, towed in by steam-pinnaces. Sitting ducks. For 2nd Royal Fusiliers coming ashore on 'X' Beach, opposition proved mercifully light. This was not the case at 'W' Beach. The Turks were well dug in and had plenty of machine-guns. They returned to their trenches as soon as the covering fire from ships subsided. Those men of 1st Lancashire Fusiliers would, on that day, both match and surpass any of their forebears famed successes, but the cost would be terrible.

In the sparkling light of a golden Mediterranean spring morning, men crowded into the boats and set off. At 0.500 HMS *Albion* fired her first crashing broadside. The hope was that this drenching fire would kill or demoralise many of the Turkish infantry and blow apart the wire. Both hopes proved pious. Defenders held their fire until their massed targets were fully in range, then let rip. Rifle and machine gun fire swept the boats, a hurricane of lead, thick as hail. Men were massacred as they crouched, soldiers and sailors drifting lifeless or writhing in rudderless craft. One airman described a sea of blood below. Still, they struggled ashore, the storm never abating, and struggled up the beach, men dropping as though 'mown down by a scythe' as General Hamilton lamented. He described the attack as displaying levels of courage and endurance 'unsurpassed' in British military annals. Support was provided by 4th Worcester's from 88 Brigade, but this was Lancashire's Landing and very many families from across their home county would come, in their sorrow, to know that name.

Soldiers had boarded their boats when some 2,000 yards out from the beach. No enemy could be seen but the barbed wire and parapets of his trenches were clearly visible. A and B Companies were to storm the right-hand side, taking trenches and seizing Hill 138. C Company was to make for the centre and Hill 144. Captain Richard Willis commanding C Company, who would win a Victoria Cross that morning, noted:

> It might have been a deserted land we were nearing in our little boats. Then, crack! The stroke oar of my boat fell forward to the angry astonishment of his mates, and pandemonium broke out as soldiers and sailors struggled to get out of the sudden hail of bullets that was sweeping the beach and the cutters from end to end. Men leapt out of the boats into deep water, encumbered by their rifles and 70 pounds of kit, and some of them died right there, while others reached the land only to be cut down on the barbed wire.[18]

Willis's company would win four of the six VCs awarded for 25 April. He showed inspirational leadership, keeping his men moving despite the tornado of fire hurled against them.[19] Just to add to the mix, Captain Shaw leading B Company found his damned wire cutters weren't up to the job. Lance-Sergeant William Stephen Keneally, aged 28, who'd been down the mines since he was 13, volunteered for the same task with the same frustration, but he kept on trying despite Johnny Turk's best efforts. The wire was thick and untouched by the opening bombardment. Men were rammed against it like lemmings and enemy machine gunfire never slackened. Keneally, too, won a VC but died two months later on 28 June fighting in Gully Ravine. (The Battle of Gully Ravine was fought between 28 June–05 July 1915 and resulted in a limited tactical win for the Allies but at a cost of over 3,000 casualties.)

Sergeant Alfred Joseph Richards got to shore but was hit three times in his right leg, which was almost severed, just as he reached the beach, but somehow he kept going, pushing his men on through the infernal wire, leading, driving, and encouraging until he passed out. Evacuated to Egypt he lost the leg but gained a VC.[20]

Rifles jammed with sand; bolts had to be stamped on to get them to function. Men fought on despite serious wounds. The narrow beach became a slaughterhouse. Dead men jostled for space like limp marionettes, hung up on the cursed wire. Sergeant Frank Edward Stubbs spurred his platoon on towards Hill 144 where they were to RV with 'D' company, the battalion reserve. He got them there but never made it himself, shot clean through the head on the slope. Of the Lancashires VCs, he was the only one who died in the beach battle.

Despite the storm of steel and lead hurled at them, the Lancashires battled on, clawing their way, yard by bloodied yard, up the beach; by 01.15 the landing ground at fearful cost, had been secured. Lieutenant (later Major) Cuthbert Bromley, acting as adjutant was wounded in the back shortly after wading ashore. He ignored his injury (didn't in fact report it for three days). He again showed magnificent leadership in Gully Ravine where he was hit a further three times in the foot, but again refused medical aid until his objectives were secured. His wounds were severe enough for him to be evacuated back to Egypt for surgery and recuperation. He wanted to get back into the fight and embarked for the return journey on *Royal Edward*. The ship was torpedoed and he perished.

Lance-Corporal John Elisha Grimshaw was 22 and a signaller, part of the beach landing operation liaising with *Euryalus* offshore. He played a full and distinguished role in the assault, inspiring and motivating the men around. 'In boats we got within 200 or 300 yards from the shore when the Turks opened a terrible fire. Sailors were shot dead at their oars. With rifles held over our heads we struggled through the barbed wire in the water to the beach and fought a way to the foot of the cliffs leaving the biggest part of our men dead and wounded.'[21] Miraculously, he came through unscathed, his webbing, pack and water bottle peppered with shrapnel. Evacuated in 1916 with frostbite, he returned to active service the following year. Not only did he survive the war but enjoyed a long army career, retiring as a Lieutenant-Colonel. He was the sixth VC winner and lived to 87, dying in 1980.

Six VCs in a single regimental action over so short a time says everything.[22]

Royal Welch Fusiliers, Caernarfon

Ich Dien[23]

'Good-morning, good-morning!' the General said
When we met him last week on our way to the line.
Now the soldiers he smiled at are most of 'em dead,
And we're cursing his staff for incompetent swine.
'He's a cheery old card,' grunted Harry to Jack
As they slogged up to Arras with rifle and pack.
But he did for them both by his plan of attack.'

Siegfried Sassoon, 'The General'

In the early summer of 2016, I took my son-in-law's support company from 1st Battalion Royal Gurkha Rifles on a Great War Battlefield tour to where their great-grandfathers had fought within the Ypres Salient. Sam was able to go to the location on Pilckem Ridge where his own great-grandfather had won a Military Medal on 31 July 1917 during the opening moves of the Third Battle of Ypres, often called Passchendaele. This terrible battle degenerated into a sea of sucking, heaving mud, which symbolises many people's impression of the War. He had served with the Royal Welch Fusiliers.

The regimental museum is in the mighty Caernarfon Castle. It seems very fitting that so renowned a regiment should be remembered in so famous a fortress. Famous indeed, though not necessarily welcomed by the medieval inhabitants. Edward I decided to build on what was the site of an earlier castle and a Roman fort. This was imperium writ in stone and writ large. His vast new citadel would be surrounded by a civilian settlement and would form a bustling administrative centre for North Wales. The cost was prodigious, so much so it was never in fact finished, though less prodigious than the resentment this authoritarian imposition represented. It was built mainly between 1283–1285 and the first Prince of Wales, the future Edward II, was born there in April 1284. The locals struck back: Madog ap Llewelyn trashed the place ten years later, and later still, Glyndwr undertook a lengthy siege.

It was some time afterwards, in 1689, that the Welch Regiment of Fuzileers was raised in the wake of the previous year's Glorious Revolution. In 1713, they were able to add 'Royal' and stayed as such until the Childers reforms of 1881. Then they became officially ', but the old name stuck. Since 2006 and amalgamation, they have formed

1st Battalion Royal Welsh. With a long and distinguished history of service, their regular battalions served continuously on the Western Front 1914–1918. The Great War produced a significant body of literary output, poetry, narrative, and memoir, the Royal Welch more so than most; aside from Robert Graves and Siegfried Sassoon there were others. David Thomas, for example[24] and John Bernard Pye Adams, who is barely remembered but after being wounded in 1916 wrote a memoir, *Nothing of Importance*, whilst convalescing, the only such record actually published during the war. He didn't live to see it go to press, killed in February 1917.

One of the first Great War books I ever read was *Goodbye to All That* and I have re-read it numerous times since; for me, perhaps the best of all the memoirs. Born in Wimbledon in 1895, Robert Graves was the son of a celebrated Irish poet, A. P. Graves, an inheritance he maintained. He is perhaps best remembered for his versions of the Greek myths and novels such as *I Claudius*, *Claudius the God* and *Count Belisarius*. For me, Graves was the perfect example of the officer class of 1914, middle class background, public school educated (perhaps more of a 'temporary gentleman' as he won his place through a scholarship) and with a distinct romantic bent. The war would see to that. By the end:

> Very thin, very nervous and with about four years' loss of sleep to make up, I was waiting until I got well enough to go to Oxford on the Government educational grant. I knew that it would be years before I could face anything but a quiet country life. My disabilities were many: I could not use a telephone, I felt sick every time I travelled by train, and to see more than two new people in a single day prevented me from sleeping. I felt ashamed of myself as a drag on Nancy but had sworn on the very day of my demobilization never to be under anyone's orders for the rest of my life. Somehow, I must live by writing.[25]

This paladin's mother Amalie von Ranke was in fact German, married to Graves' father when being Teutonic was still fashionable. Robert nearly died age seven when he contracted double pneumonia following a bout of measles. He carried the scarring, and it nearly did for him again after he was wounded on the Somme. The terrible influenza epidemic of 1918 came close to finishing the job Kaiser Bill had started, but he lived to see his 90th year. In 1909, he had won a scholarship to Charterhouse, a prestigious public school founded in 1611. He wasn't gentry so suffered some bullying on that account, worse because his mother was a Hun and the Boche were, even then, not well liked. Nonetheless, he was both intellectually distinguished and good at sports, a school welterweight and middleweight champion. Legendary climber George Mallory was one of his most influential mentors. From Charterhouse, he won a place at St John's College Oxford, but the outbreak of war intervened.

That August, he volunteered and was commissioned as a 2nd Lieutenant in 2nd Battalion Royal Welch. His abilities were recognised and by October 1915 he was commanding a company. Equally rapidly, he began to gain a reputation as a poet, forming a firm friendship with Sassoon. Severely wounded during the Somme Offensive, he spent much of the final years of the war in England. At one point Sassoon (who'd turned against the whole business of the war), and Wilfrid Owen were patients at Craiglockart under the brilliant, pioneering psychiatrist W. H. R. Rivers, beautifully portrayed by Pat Barker in her masterful *Regeneration* trilogy. Graves was in fact instrumental in getting his friend admitted, which sidestepped the possibility of a court martial.

He first wrote *Goodbye to all That* in 1929, though he revised the narrative extensively for the 1957 edition. His marriage to Nancy Nicholson eventually failed and he went into voluntary exile in Spain. He did take up his place at Oxford post-war where one of his closest companions was T. E. Lawrence. Clearly, he, like so many others who came back apparently in one piece, was suffering from post-traumatic stress disorder (PTSD).

At the Chantilly Conference in December 1915, Generals Haig and Joffre had discussed ideas for joint offensives in 1916. Haig, an ardent Francophile, was by no means confident of his ally's capacity to do more and was increasingly (and correctly) of the view that Britain must shoulder the greater burden.[26] A week before Verdun furiously erupted, the two commanders met again. The concept of a major joint offensive astride the River Somme was, at least in general terms, agreed with a proposed start date of 1 July. Haig was not a fan of the Somme area. Strategically it had little to offer. As ever, he preferred to look toward Flanders.

Verdun changed all this. Without Haig's full support it was feared the French would crumble. When on 26 May Haig informed his ally that his New Army divisions would not be ready until perhaps August, Joffre was appalled. The French, he expostulated, might very well not survive that long. Haig rather drily observed that a fine vintage brandy calmed Joffre's Gallic nerves most admirably. That the British would attack in the Somme *Department* was now a given. Haig's misgivings were secondary to his need to prop up Britain's ally before the weight of German blows proved too much.

General Henry Rawlinson, leading British 4th Army, now held the right of the British Line and the brunt of the coming battle would be borne by divisions under his command. The fight would be spread over a front of some 20 miles (32 kilometres), across the deceptively mild chalk downs of Picardy, from Foncquevillers in the north to Mericourt in the south, where the British and French sectors would intersect on the line of the River Somme. Much of the British sector lay on the line of the Ancre and some felt the battle should have been named accordingly, but by then it was too late. The Somme featured too prominently in the public consciousness. It still does.[27]

Those who have heard anything about the Great War will probably have heard of the Somme. As with Verdun for the French, it symbolises the loss of a generation, a frightful gobbling up of blood and manhood for seemingly trifling gains. Some writers, even those as rightfully revered as the late Sir John Keegan, have asserted that after the Somme, Britain declined irreversibly and was ruined as surely, if less evidently, than France.

Along the otherwise insignificant ridge running from Thiepval to Ginchy the Germans had turned sleepy hamlets into bristling fortresses: Serre, Thiepval, La Boisselle and Fricourt. Their front line was studded with strong redoubts, with successive belts riding the crests of ridges behind, like spume on breakers, each a major tactical problem in itself. Chalk, firm and clear after the mud and slop of Flanders, was ideal for carving deep dugouts, the enemy line was immensely strong and utterly formidable. Rawlinson himself was an advocate of 'bite and hold'; a series of limited assaults, each quickly consolidated and then built up in readiness for the next. Haig needed rather more. He had to have a breakthrough, a smashing of the enemy line and a clear run for his cavalry to Bapaume, up the dead straight Roman Road from Albert. This would signal victory and his political masters demanded no less.

Rawlinson diplomatically, if with reservations, bowed to his superior's wishes but insisted on massive, drenching bombardment as a necessary prelude. A continuous deluge of fire so utterly overwhelming it would crush the enemy front-line defences to dust and decimate the troops holding it. Kitchener's New Army men would be the main attacking force and it was feared their training was insufficient to instil complex 'fire and movement' tactics. Instead, they would move forward from their start lines in four orderly waves, rifles sloped and walk towards the enemy front line, which would simply be theirs for the taking.

It didn't work out like that. Despite the weight of a million and more shells, much of the German wire remained uncut. The desperate 'race to the parapet', generally won by the enemy, who for the most part was safe if traumatised in his deep bunkers, resulted in unspeakable carnage. The attacking battalions, many being blooded for the first time, suffered a staggering total of 57,470 casualties, of whom, very nearly a third (19,240) were fatalities.

Graves was sent to the Front in early July, every trained officer was needed to make good the dreadful losses. Despite the grimness, he found opportunity to fulfil 'a childhood dream' when he helped stoke the French locomotive lugging its leisurely way forward (the rolling stock tended to be very leisurely in its progress, but why hurry to meet death?). He'd been hoping for 1st Battalion but found himself in trenches at Givenchy, serving with the 2nd. They were across the canal from the Cuinchy Brick Stacks, great surreal towers.[28] He got there on 5 July just as the haul from a successful trench raid was being herded down the communication network: 'Saxons, just back from a divisional rest and a week's leave to Germany, with new uniforms and their packs full of good lootable stuff. One prisoner got a stern talking to from 'C' Company sergeant-major, a Birmingham man, shocked at a packet of indecent photographs found in the man's haversack.'[29]

This was retaliation, as only a few days before the enemy had exploded a mine beneath 'B' Company; The 'Bs' were proverbially unlucky.[30] Casualties had been high, and the inevitable German attack only just seen off. The raid was meticulously planned. Fire and smoke would drench the enemy flanks while a creeping barrage pounded both the front line and supports. The intention was to force the enemy to continually disappear into his deep dugouts, then reappear swiftly and stand to. They'd get slower, giving the raiders, shielded by smoke, the chance to dash across No-Man's-Land and catch them off balance. It had worked.

In his report, Graves noted with dry irony that a regiment which had abandoned the puissant pike in the late 17th century now, in the age of industrial warfare, attacked with knives taped onto broomsticks. He and his comrades were amused at a journalist's rather Hollywood account, which alleged the men went over the top yelling 'Remember Kitchener!' and 'Avenge the *Lusitania*!' 'What a damn silly thing to shout. Old Kitchener served his purpose as a figurehead, but nobody wants him back at the War Office… As for the *Lusitania*, the Germans gave the Yanks full warning; and if her sinking brings them into the war, that's all to the good.'[31]

Graves had expected a reasonably warm reception, even though he'd originally been commissioned into the 3rd Battalion, but old jealousies, which even the horrors of the Somme battlefield could not overcome, re-emerged, and rumours spread around his 'tainted' parentage. Four days after the raid, the battalion began moving up towards the Somme, through the skeleton of ravaged Bethune, entraining for Amiens, reaching the front line late on 14 July, the spot where so many former comrades had died. By now, the line had shifted forward through 'Happy' Valley, where even Graves, already hardened, was shocked to see so many dead or maimed horses and mules. By the time they reached the vicinity of Mametz Wood, they were under fire.

Dregs of gas remained like tendrils of mist (tear gas rather than chlorine), which left a foul taste. They passed a captured German gun, chalked up to their 1st Battalion, where Graves was reunited with his old pal Sassoon. Their next and difficult objective was the staunchly defended 'Quadrangle' fronting Mametz Wood. Siegfried Sassoon single-handedly stormed an enemy front-line position, one which had earlier resisted a full battalion attack. With just a couple of riflemen giving cover, he'd bombed the occupants out of their bunkers, then calmly sat down to read amongst the debris. His CO wasn't impressed, as this maverick brilliance, however compelling, had delayed a planned assault. He was called 'Mad Jack' by now.[32] Nights were cold despite the season and Graves foraged through the emaciated and defoliated woods for German overcoats. The place was a graveyard, dead Prussians, Royal Welch, and South Wales Borderers were heaped in contorted piles: 'Going and coming by the only possible route, I passed by the bloated and stinking corpse of a German with his back propped against a tree. He had a green face, spectacles, close shaven hair; black blood was dripping from the nose and beard.'[33] He also saw a dead South Wales Borderer with his bayonet buried in the corpse of his killer from the Lehr Regiment. On 18 July, the battalion moved up to new ground just north of Bazentin-le-Petit to relieve the Tyneside Irish. As they marched through what was left of the village, blackened gables and a confusion of charred rubble, they were again hit with gas shells: 'Orders were to ignore these, don't halt to fumble with respirators, as they're only lachrymose.' These weren't, 'so we lost half a dozen men.'[34]

On 20 July, Graves' battalion were in reserve during an attack on Bazentin Churchyard, hunkered down with French 75s blazing away overhead from a nearby gun line. It was at this point, as was often the case, that orders came up from division. These were normally banal and irrelevant to the game of life and death being played out at battalion level. In this instance, a private was to be placed under guard and escorted back to Albert, together with an NCO who was summoned as a witness at his court-martial.

This was serious, the soldier was charged with the wilful murder of a French civilian in a café at Bethune a few months before. Both parties appear to have been the worse for drink, but the Frenchman had suspicions about the squaddie and his own spouse. Tempers flared, cognac was consumed, and the outraged husband's Gallic temperament got the better of him. Not only did he insult the Englishman and all others of his ilk, but he also cheered on the Germans, hoping they'd win. This was too much for a loyal Englishman who forthwith drew his bayonet and permanently cut his tormentor off in mid-flow. He got off with the enthusiastic support of the French civil representative at the trial, who commended his actions in combating defeatism.

It turned out the accused was lucky in more ways than one. When the Royal Welch were finally committed to a desperate fight, they lost a third of their number to German shellfire before they advanced a foot. Graves and his company moved back 50 yards at the double toward a less exposed position. He didn't make it: 'I heard the explosion and felt as though I had been punched rather hard between the shoulder blades but without any pain.'[35] At first, he didn't realise he'd been injured, but blood was running into his eyes, and he felt he was losing consciousness. Bizarrely, he had sustained two minor cuts to his left hand in the same positions as two he'd had to his right during the opening stages of the battle of Loos, September 1915. This time it was rather worse: 'One piece of shell went through my left thigh, high up, near the groin; I must have been at the full stretch of my stride to avoid emasculation. The wound over the eye was made by a little chip of marble, possibly from one of the Bazentin cemetery headstones.' [36]. More seriously, 'A piece of shell had also gone in two inches below the point of my right shoulder-blade and came out through my chest two inches above the right nipple.'[37]

For Graves the war, for quite some time, was over.

It's hard to know if you're alive or dead
When steel and fire go roaring through your head.

One moment you'll be crouching at your gun
Traversing, mowing heaps down half in fun:
The next, you choke and clutch at your right breast–
No time to think–leave all–and off you go...
To Treasure Island where the Spice winds blow,
To lovely groves of mango, quince and lime–
Breathe no goodbye, but ho, for the Red West!
It's a queer time.

Robert Graves, 'It's a Queer Time'[38]

The Royal Welch had several literary luminaries serving in their ranks besides Graves and Sassoon, titans among the pantheon of Great War poets. One of the more remarkable was the celebrated Welsh bard Ellis Humphrey Evans (1887–1917), better known by his assumed name of Hedd Wyn. Not necessarily warrior material, Wyn worked on the land as a shepherd and his romantic verse, much lauded even during his short lifetime, was rooted in pastoral or faith-based themes. He did write some war poems after 1914, was conscripted into the regiment and died on 31 July 1917 at Pilckem Ridge. His supreme accolade, the Bard's Chair at the National Eisteddfod, was awarded three weeks after his death in battle.

Colonel Nick Lock OBE is a very effective and inspirational advocate for the regimental museum. In many ways it's a daft question to ask a trustee to identify his or her favourite objects. As I know from experience, we all have many! Nick feels very strongly about the medal room, which, he observes, is the heart of any regiment and the human stories the awards unlock are infinitely varied and often deeply poignant. The Royal Welch are in the midst of a project to refurbish their display and offer more such biographical detail.

His selection of items from the collection is eclectic. The museum holds Sassoon's privately purchase Browning pistol which he carried as a back-up gun as well as his .455 Webley. Contrary to myth, the great poet, when he repudiated the war, didn't ditch his MC into the river, only the ribbon, the medal itself resides in the collection. There's also a de Walden trench knife. These are perhaps the antithesis of our understanding of an industrial war. Howard de Walden 8th Baron is reputed to have modelled his design on a form of short sword carried by Welsh archers as far back as Crecy (1346), issued to machine-gunners and bombers for very close quarters. It's a very elegant but purposeful weapon with a broad, prominently ribbed, leaf-shaped blade, if it was in bronze, you'd expect to see Hector or Achilles wielding one.

There is a silver cigarette case, bearing the regimental crest, the property of Major Geoffrey Lee Compton-Smith 10th Battalion (1889–1921) who was kidnapped by the IRA in 1921 and subsequently murdered. A Londoner by birth, he preferred army life to the law, his father's profession, and joined the Royal Welch. Serving on the Western Front, he was wounded at Arras and earned a DSO. After the war he was posted as a local area commander to Cork in southwest Ireland where a vicious 'hedgerow war' was being waged against the IRA, led by Michael Collins. By the spring of 1921, this wasn't going too well for the rebels, quite a number of whom had been recently captured and four condemned to death.

Compton-Smith was active as an intelligence officer and as such was inevitably on the IRA's watch list. Collins wanted a viable hostage to, as he hoped, secure the release of his condemned men, or at least a commutation of sentence. The major unwittingly drew the short straw and was kidnapped on 16 April, moved from safe house to safe house over a period of days. During his captivity he showed no animus towards the IRA and wrote a series of letters, although these weren't found until after his death. His fate was sealed by 28 April when the four volunteers were executed. Two days later, by way of reprisal, it was his turn.

> Compton-Smith then wrote his final letter to his wife, Gladys. He tells her that he will die with her name on his lips and her face before his eyes and that he will 'die like an Englishman and a soldier'. He is not thinking about himself but of his wife and daughter, his father, and his regiment. He is even thinking about the man who has the job of

> executing him, Jackie O'Leary, to whom he gives his watch—an act of forgiveness, perhaps? The second letter was to his regiment, and in it he says: 'I should like my death to lessen rather than increase the bitterness which exists between England and Ireland... My cigarette case I leave to the mess. I shall die with it in my pocket.'[39]

And he did, after the final customary cigarette.

Bosworth Battlefield Centre/Richard III Museum, Leicester

> It has been a hard day's night,
> And I have been working like a dog,
> It has been a hard day's night
> And I should be sleeping like a log,
> But when I get home to you, I find the things that you do
> Make me feel all ... right.
>
> Peter Sellers (pace Laurence Olivier and William Shakespeare)[40]

It used to be impossible to get through schooldays without seeing Laurence Olivier's 1955 film version of Shakespeare's *Richard III*, the classic scheming villain, plotting everywhere, murdering everybody in his way. His portrayal, royally over the top and in glorious technicolor, influenced a whole generation of actors since. Of course, there have been very many, such as Ian McKellen and Anthony Sher, who have re-defined the role and added their own stamp. Why not, and there's space for plenty more. Richard remains an enigma, history's eternal maverick. To his near idolatrous fans, he can do no wrong. His statue outside Leicester Cathedral looks more Michelangelo than Machiavelli and the debate over whether he killed his nephews will run and run. (I'm sure he did, but very many disagree.) A decade after the movie, smack in the heady days of Beatlemania, Peter Sellers did a brilliant spoof, blending Olivier – the hump, the staccato delivery – the Bard and John Lennon into a sixties mix.[41]

The real thing was rather more serious. That first bloody scrimmage in the streets of St Albans in the spring of 1455 unleashed a generation of intermittent warfare, tinged with the bitterness of blood feud. Losers, especially those about to be attainted, received no mercy and their loss inflamed successors. Richard was born in 1452, so he grew up in an age of extreme violence. His father, brother, uncle, cousin, and a slew of others were all butchered, heads grinning sightless above Micklegate Bar. He survived Warwick and his older brother Clarence's treachery and emerged as King Edward's ultra-loyal servant. He avoided the fractious politics of a Woodville-dominated court, ruled the tumultuous north wisely and well, wrested Berwick back from the Scots and established an English hegemony on the marches. So far so good.

The key question was and is, did Richard move to take the throne in 1483 as a survival tactic, a knee-jerk response, or was this a long-prepared plan? Nobody expected Edward to die at such a young age, but Richard would have been aware his brother was failing. He was equally aware that the Woodvilles would move fast and hard to consolidate their grip and they wouldn't be taking prisoners. Even though he wasn't at court when the king succumbed, Edward's will did name him as guardian of the young prince, then aged only 12. His riposte was a masterpiece of disinformation and timing that would have set Machiavelli beaming. He intercepted and effectively kidnapped his nephew, rounding up Earl Rivers, the Queen's brother, and others, hoodwinked them and sent them north to Pontefract. Pretty soon they were dead, as very soon was Lord Hastings who, despite having supported Richard, might have begun to sense his darker intent.

We know the rest. Richard, aided by the Duke of Buckingham, executed a carefully stage-managed coup, the young Duke of York was almost literally dragged from his mother's arms and sent with his brother to the Tower, neither ever emerged and sparked one of the longest whodunnits in English history. Richard must be prime suspect, but no compelling evidence has ever come to light, nor will the Crown allow DNA testing on the supposed remains.

Richard's short reign was an uneasy one. Buckingham had to be dealt with and the spectre of rebellion hung like a malevolent cloud. Henry Tudor, the last surviving quasi-Lancastrian candidate, had been written off as a nonentity but became the emerging focus of Lancastrian/Yorkist/Woodville resentment. The Dowager Queen, despite the decimation of her family and the almost certain murders of her sons, remained a significant player. Richard's line faltered with the death first of his son, Edward of Middleham, and then of his wife Queen Anne. Whether or not Richard had in fact done away with his nephews, most contemporaries believed he had, and the taint of their taking off followed him like a contagion. Medieval opinion was pretty robust when it came to disposing of political opponents, but the deliberate killing of innocents, and by the very man who was their uncle and protector, provoked widespread revulsion.

> The cat, the rat and Lovell our dog, Rule all England, under an hog.[42]

High and uneasy summer 1485 and Henry Tudor was no longer a rank outsider. In fact, on 7 August his very modest invasion fleet dropped anchor at Milford Haven. The king, who'd been at Nottingham, ordered a general muster to assemble at Leicester. Richard nominally possessed a pair of very slippery aces in the brothers Stanley. Thomas Lord Stanley and his sibling Sir William were traditional Yorkists but had reason to dislike Richard as he had failed to back them in a land dispute. Lord Stanley himself was Henry Tudor's stepfather – his son Lord Strange was with the king, as much a hostage as supporter. Both the campaign and battle of Bosworth are unique in that there were three armies involved, the powerful Stanley contingent forming a third, and neither Richard nor Henry quite knew which way they'd jump, possibly at the outset neither did they.

King Richard III, last of the Plantagenet monarchs, marched his impressive army out of Leicester on 21 August. By dusk all three armies were only a few miles apart, next day would decide the matter. King Richard's short and troubled tenure would be vindicated in trial by battle, or Henry Tudor's gamble would pay off. Nobody at this stage was sure of the Stanleys, they'd given assurances to both sides and reassured neither. Richard had probably already decided to move towards the spur of Ambion Hill to gain the advantage of height, around 400 feet (122 metres), above the plain below (provided we accept that's where the ensuing battle was fought).

Moving up in line of columns, Norfolk led the vanguard of the Royal Army, Richard the centre and Henry Percy IV Earl of Northumberland the rear. On paper, the king had greatly superior numbers, the advantage of ground and he was a far more experienced and highly successful commander. A tantalizing question emerges, the chroniclers hint at a pre-battle address – Henry V style – to his troops, where he asks them to judge him not by what he's done but by all that he might do if God grants him victory. Meanwhile, the shadow army of the Stanleys deployed north and south in two contingents, Sir William took the former and Lord Stanley the latter. Nobody had a clue what they were up to, but the rival commanders would be casting continual nervous glances, their intervention, either way, could prove crucial. And so it did!

Before the armies could engage, John de Vere, Earl of Oxford, leading the pretender's vanguard, had to skirt an inconvenient water feature at the base of Ambion Hill. Guns banged in the fine summer morning, roundshot flew from both sides but there's no evidence any did much damage. Assuming Dr Glen Foard has the right location (see below), then the wealth of cannon balls found clearly suggests an intense opening artillery duel. Oxford launched his men up the slope against Norfolk into a fierce, swirling mêlée with both generals feeding in reinforcements. The rebels began to gain the upper hand, despite the gradient being against them. At some point Norfolk went down and his son Thomas Howard (the victor of Flodden 28 years later), was wounded. Percy, rather tellingly, stayed supine, totally out of character for so famously aggressive a bloodline.

Legend asserts Tudor left the line to try and speak personally to Sir William Stanley, whose men still stood idle. It does seem most unlikely that Henry would canter off at such a critical moment, given Oxford was doing rather well and the Welshman needed to display strong leadership. Precisely what prompted King Richard to launch a downhill mounted charge at the head of his household men, we don't know, though obviously he intended to end the battle with a single bold stroke. We might surmise he feared what the Stanleys might do and perhaps what Northumberland might not do. It seemed at first he might just pull it off, smashing through Tudor's own company and coming so very, very close to dealing with the pretender himself.

Too close for Sir Wiliam Stanley, who could see how short his own future might be. His bows and bills finally surged forward, turning the tide in an instant, overwhelming Richard's knights. Whatever is said of Richard, his courage and fortitude were never in doubt, and he fought bravely to the last, but with his death, the Battle of Bosworth was over, and the earth had shifted. My old school history book on the Tudors, referred to the day at the last of the Middle Ages and the dawn of a new era. It was in many ways, but I very much doubt if any of the bloody, bruised, and dazed survivors thought so. Still, drawing one breath after another was quite enough.

Another legend claims it was Thomas Lord Stanley who discovered the discarded crown hanging on a hawthorn bush and placed it upon Henry's head. Nice story, and indeed Thomas did very well out of it, gaining an earldom, though William later lost his head. One doubts there were too many mourners. Thomas Howard survived and clawed his way back, but only his momentous win at Flodden in 1513 assured the return of his father's dukedom. Percy didn't do too badly either, but he was killed during a tax riot four years later. As for the late usurper's stripped and mutilated remains, well that turns out to be quite a long story...

Was King Richard III hero or villain? The answer must lie somewhere in between. He was a ruthless man in a ruthless age. Did he murder his nephews? Almost certainly yes in my view. He comes over as rather dour and humourless, tending to both deviousness and self-righteousness but, had he been a nice upstanding sort of fellow, he would not have made it past puberty. The Woodvilles, when Edward died, would have done for him for sure, and for his young son. Richard made choices we can easily condemn, but that doesn't mean he did so out of malice, merely self-preservation and expediency. Was the coup of 1483 planned in advance? To a degree, yes it must have been, but only a fool would not have prepared a contingency plan, seeing the king's health decline and the Woodville vultures circling.

Even if Richard had just asserted himself as protector and not taken the throne, he was only buying time. Once the young monarch came of age, under his mother's baleful influence, he'd have no need of his uncle. Then what would happen and what of the fate of Richard's own son? A bit like Macbeth, 'in blood steep'd so far' and all that. I doubt either he or the Bard could have quite foreseen how the telling of the tale would unfold.

I rate the Bosworth Battlefield Heritage Centre as one of the best in England. The café is situated in a reconstructed 15th-century tithe barn and I really like the museum, which has a whole raft of interactive stuff, excellent models and a gallery of facsimile arms and armour from the era. The site is well laid out, with Richard's white boar banner floating above Ambion Hill (that's assuming you agree this is the correct site). It's the hitherto agreed historical location for sure, the remaining features correspond and, for me anyway, it has the right 'feel'. Others disagree.

As far back as 1985, an obvious anniversary year, archaeologists began to express doubts, as recorded by the UK Battlefield Trust:

> Until 2009 Bosworth was the most contentious of English battles, because at least three alternative sites had been proposed for the

> battlefield. We now know that the battle was fought not far from Stoke Golding, some two miles or so south-west of Ambion Hill, the 'traditional' location of the action. In addition to locating the battlefield, the archaeological survey has unearthed the largest group of cannonballs ever found on a medieval battlefield – no fewer than 34 lead shot of a variety of calibres – ranging from bullets fired from handguns to roundshot from substantial artillery pieces.[43]

This ground-breaking work was undertaken by Dr Glen Foard of Huddersfield University who was appointed by Leicestershire County Council to lead the investigations in 2005. This discovery or re-discovery of the true site did unearth a trove of artefacts, one of the most telling and poignant being a silver gilt livery badge in the distinctive shape of a boar. The late Professor Richard Holmes described this find as the most important Bosworth discovery of his lifetime. It's hard to disagree.[44]

This could have been embarrassing, given English Heritage's significant investment in the traditional location, but the museum incorporates an excellent exhibition of finds from the newly identified site that neatly ties the whole controversy together. Aside from regularly re-enacting the battle and taking tour groups around, I have spent time there with the Richard III Foundation Inc. and its patron, the late, great Robert Hardy. A lot of my obsession, as I've said previously, is his fault. It was watching him as Henry V in the BBC's *Age of Kings*, back in the sixties, that got me addicted to history. In 1996 at Durham Johnston School, which spreads over the field of Neville's Cross (1346), I watched him deliver a stunning lecture on the English warbow at the end of a long and intensive day, when people's attention spans were shortening. Yet Hardy mesmerised and energised, one of the most brilliant lectures I've ever heard. This taught me an awful lot about the art of teaching as performance.

On 22 August 1485,[45] Richard's mangled corpse was taken to Greyfriars, Leicester, where he was crudely interred in a makeshift pit. Losers don't get state funerals. There was a contemporary account which described his remains as being dumped in the River Soar, but this is erroneous. Following the dissolution in 1538, the site of the actual grave was lost. For some time during this century the Richard III Society steadfastly pursued their 'Looking for Richard' project, one that involved detailed scholarship with just a hint of mysticism. Most people didn't give them much credence. But in August 2012, a dig led by University of Leicester Archaeological Services, working under the auspices of the City Council, got started.

By way of seriously confounding their many sceptics, the very first day's work uncovered the skeleton of a male, identified as being in his thirties (Richard was 33). This man's remains displayed clear evidence of a condition known as scoliosis, a spinal deformity. His end had not been peaceful. Immediate cause of death was most probably a massive slashing injury to the back of the head. This had likely been inflicted by some form of polearm, a bill or halberd, slicing through the skull like an egg, exposing his brain. If that hadn't done the job, a thrust though the head certainly would have. His battered bones revealed details of many additional wounds probably inflicted postmortem. From Richard's perspective, you'd hope so.

But was this Richard III? Circumstantially, it appeared a strong possibility and analysis of mitochondrial DNA extracted from two lineal descendants of the last Plantagenet king's sister, Anne of York, supported the archaeological evidence. On 4 February 2013, the University of Leicester announced that the remains, beyond reasonable doubt, were those of the king. Richard, despite his 500 years of skulking beneath a municipal car park, was able to cause yet more conflict. He never struck me as having had much sense of humour, but he would surely have smiled if he'd seen the storm that blew up over his funeral games.

It had been agreed between the diggers and the city council that, should the bones prove their provenance, they'd be regally re-interred in Leicester Cathedral, only a stone's throw from the original grave. This seems reasonable enough, but the new tomb had the potential to generate significant tourist income, so the City of York weighed in with a claim. Litigation, the benefactor of lawyers' pensions, ensued, but Leicester won, and Richard was entombed in a splendid new monument inside the Cathedral, the Archbishop of Canterbury and a slew of church dignitaries officiating with the world's media looking on. Eat your heart out Henry Tudor.

Over the position of his long-lost grave, 'Crookback Dick' now has his own attraction – the King Richard III Visitor Centre. This showcases the original pit with a ground-floor exhibition offering a journey through the king's life which is nicely done and comes with a curator who is also a re-enactor, keen at the drop of a groat to show off his impressive kit. He did so for several tour groups I've taken there, and it really added to the visit. Upstairs is the story of the archaeology, with a nod to the Bard. Ian McKellen's 1930's *Schutzstaffel* type uniform, an arresting sight, is on display there.

In the agreeable converted tithe barn café/bistro:

Graham: Did he then, murder his nephews?

John: This would get me lynched in certain circles, but I'd say almost certainly yes. Means, motive and opportunity. His nephews, even if anyone bought the story about them being bastards, were a constant threat to Richard himself and his own proto-dynasty. Leaving them alive would inevitably expose his own son Edward of Middleham. Doesn't make him a psycho, just a ruthless pragmatist.

Graham: He did have form?

John: Ask Earl Rivers, Grey, Vaughan, his old ally Hastings and new one Buckingham. There's no real evidence he killed Edward of Lancaster at Tewkesbury or poor old Henry VI after, as Shakespeare would have us believe.

Graham: Olivier has a lot to answer for.

John: I know we've neither of us seen it but there's a new film, *The Lost King* about the finding of the body [a BBC film production, directed by Stephen Frears]. It's a comedy of sorts I understand, but the actor playing Richard seems taller, altogether more genial and is wearing interesting armour, meticulously researched by Dr Tobias Capwell [formerly curator at the Wallace Collection and a leading specialist in high medieval arms and armour]. I'm not sure you could accuse the real king of ever having demonstrated a sense of humour. Doesn't necessarily make him a bad person of course.

Graham: I read about that, there's been quite some controversy between Philippa Langley the Ricardian and Richard Taylor the archaeologist. He was clearly outraged at his portrayal: 'bullying, cynical, double-crossing, devious, manipulative and even worse, sexist'.

John: The real Richard III would have found him a job then, I'm sure.

As we leave Bosworth Field, a fond memory. At a Richard III conference there some years ago where Robert Hardy, who'd then be 90 or very close, was ready to take the podium, the sound engineer was fussing with tech stuff and exclaimed with horror he couldn't get the microphone to work so nobody would be able to hear the noble thespian when he spoke. 'No need to fret dear boy,' the great man calmly replied. 'If the folk in this room can't hear me as I am' – there'd be about 200 – 'then I've wasted the last sixty years.' Needless to say, they heard every word.

The Soldiers of Gloucestershire Museum

The Glorious Glosters, already a famous county regiment with more individual battle honours than any other, gained yet more fame for a magnificent stand against impossible odds during the Battle of the Imjin River (22–25 April 1951) as part of 29th Brigade (with 1st Battalion, Royal Northumberland Fusiliers, 1st Battalion Royal Ulster Rifles and a Belgian Battalion). My late uncle by marriage, who served with the Northumberland Fusiliers as a very young subaltern, suffered severe wounds during the closing stages of the fight and spent 18 months recovering in Japan. He always said the Northumbrians felt a bit cheated of their own share of the glory on account of the Glosters! This glory was very dearly bought, and nobody could deny, well-earned.

The regimental museum is located in Gloucester's refurbished docklands district. I've dragged my grandsons, not entirely unwillingly, around the galleries a few times, happily they're not shy of weaponry and some modest treats usually suffice as a half-decent bribe. They also like to remind me of the time I went 'arse over noddle' outside on the ice there in winter!

Raised in 1694 as Colonel Gibson's Regiment of Foot, the Glosters (1881–1994) were an amalgamation of 28th (North Gloucestershire) and 61st (South Gloucestershire) Regiments. They had the unique distinction of wearing their badge on both the front and back of their headdress. This dates to the Battle of Alexandria fought on 21 March 1801, when the British were out to frustrate Bonaparte's attempt to subsume Egypt. French cavalry succeeded in fracturing the line, wheeling around behind the Gloucestershire men, ready to charge. For infantry to be caught when deployed in line without time to form square was potentially calamitous. Due to the quick thinking of their officers and NCOs, the rear rank faced about, so the regiment fought with their line back-to-back and repelled the attack.

In April 1951, during the Korean War, Chinese forces launched a major offensive against UN positions along the lower Imjin River, aimed at crashing through the thin crust and moving through to recapture Seoul. The war had begun nearly a year before when the North Korean Army had invaded its southern neighbour, the two states arbitrarily divided along the 38th Parallel. This move was backed, after a fashion, by Joseph Stalin and Mao Zedong.

At first, Kim Ill-Sung's troops drove all before them, scrambling UN units piecemeal and herding what was left of the South Koreans and UN allies back into a last-ditch defence of the Pusan Perimeter. General Macarthur, US Supremo in the Far East, brilliantly outflanked the North Koreans with a bold amphibious landing, one which turned the tide, and the UN began to drive the North Koreans back towards the Yalu River. Now the Chinese waded in, and the pendulum swung back. UN forces, often with indecent haste, scattered back and by the closing days of 1950 the Chinese had penetrated the south. Seoul changed hands several times. The spring offensive along the Imjin was part of China's 5th Phase Offensive, Macarthur had been sacked and Mao saw final victory as potentially within his grasp. He was wrong.

The 'Battle that saved Seoul' primarily involved 29th Brigade at the tip of the UN spearhead, together with supporting guns and armour, facing massive odds in a desperate three-day battle. Chinese objectives were to destroy US 3rd Division and South Korean 1st Division as well as 29th Brigade. The four battalions had to cover a 12-mile (19-kilometre) frontage, essentially comprised of battalion 'boxes'; there simply weren't enough men to form a continuous line.

Brigadier Tom Brodie had to deploy focussing on defensible hill positions. On his left the Glosters were set to guard a ford over the Imjin a mile east of the South Koreans. Some two miles northeast of them, the Northumbrians were to hold the centre while the Belgians, positioned on Hill 194 on the right, were the only UN unit formation holding ground north of the Imjin. Their comms depended on retaining two rather fragile pontoon bridges about ½ mile apart. These crossings linked the entire brigade to their main arterial supply line, Route 11. Such dispositions were far from ideal, but as it seemed the ground was only to be held for a short period, nobody had really set to and dug in to any depth; no wire and no mines.

For the offensive, Marshal Peng Dehuai could deploy a total of over 300,000 soldiers. His order of battle would ensure that his 63rd Army group, comprised of three divisions totalling 27,000 men in 27 battalions, would be flung against 29th Brigade's front. These odds were not propitious.

During the hours of darkness on 22 April, Chinese patrols began infiltrating from the north, avoiding the Belgians posted on Hill 104, pushing towards those vital pontoons. Elements from Brigade reserve,

1st Royal Ulster Rifles, were sent forward to counter these probes, but as Chinese strength increased, they were forced back, leaving the Belgians dangerously exposed. Now it was the Fusiliers' turn. Pressure mounted against the Northumbrians' arc of company positions and ground had to be ceded, two vital hills, nos. 194 and 257, were lost and counterattacks failed. The brigade's position was fast becoming critical.

It would soon be the Glosters' turn. First to engage was a 17-man patrol from C Company guarding the riverbank. In increasing waves, massed forces from Chinese 559th Regiment began bumping up against this comparative handful. It was only shortage of ammunition that obliged the Glosters to fall back, having inflicted significant enemy casualties. Next, it was A and D Companies positions that were blitzed. Despite odds of six to one and climbing, the Gloucestershire men held on until finally driven off their perch on Castle Hill. Lieutenant Philip Curtis made a valiant attempt to retake the position, singlehandedly grenading and knocking out a Chinese MG position before being cut down by another. He was awarded a posthumous VC.

One officer who served in this fight was Captain (later General) Sir Anthony Heritage Farrar-Hockley (1924–2006). Originally from Warwickshire, he'd run away from school at the outbreak of WWII to enlist with the Glosters while still only 15. He was soon rumbled and had to wait another two years, but he was an NCO by 17 and commissioned a year later. Much of his wartime service was with 6th (Royal Welch) Parachute battalion, winning his MC during allied intervention during the fratricidal Greek Civil War in 1944. After the war, he received a permanent commission in the Glosters and fought at the Imjin, where his fearless leadership proved inspirational. Captured during the attempted breakout, he spent two terrible years as a prisoner of the Chinese. Despite this ordeal, he went on to command all NATO forces in Northern Europe. His account of the fighting and subsequent captivity, *The Edge of the Sword*, first published in 1954, remains a classic work of military autobiography.

Imagine, now, the scene on Castle Hill and that long hill on its right flank where the men of D Company stand to. Riflemen, machine gunners, mortar and gunner Observation Posts, signallers, medical orderlies, stretcher-bearers wait now, alert and ready; those under cover smoking; those outside scanning the moonlit slopes that lead to their positions, searching vainly the black pools of shadow that mark re-entrants, folds, depressions in the ground. Often those eyes are drawn towards the river and the crossing from whence the short, fierce engagements are apparent, marked by light and sound.[46]

Next morning, St George's Day, the Fusiliers supported by elements from US 3rd Division's reserves strove to retake ground lost during the night-time battles. The Americans attacked by Hill 257 to ease pressure on the isolated Belgians and facilitate their withdrawal. A tricky operation as the Chinese had now massed in force, but one successfully completed with only the battalion transport destroyed. Once they'd escaped the noose, the Belgians dug in again south of the beleaguered Glosters and Fusiliers.

> Now, almost hand to hand, the Chinese and British soldiers meet. Figures leap up from the attacking force, run forward to new cover and resume their fire upon the men of the defence who, coolly enough, return their fire, as targets come to view, as the attackers close with them. Occasionally an individual climax in an encounter between two men when, only a few feet apart, each waits to catch the other unawares, sees a target, fires and leaps across to follow his advantage.[47]

Relentlessly, regardless of casualties, the Chinese just kept on coming. By 20.30 hours A Company had been reduced by almost 50 per cent with all its officers knocked out, and survivors withdrew to Hill 235. D Company was now vulnerable and suffered as night drew on. With one platoon particularly badly hit, those who remained also fell back to the Hill. Though B Company hadn't suffered severe attacks, they were now badly isolated; D Company on their left had withdrawn, as had the Fusiliers on their right.

They, in turn now moved back to occupy Hill 316 which lay 800 yards (730 metres) east of C Company's new position. That night B Company, with odds of 18 to 1, felt the full weight of the enemy's might. They withstood six charges, piling dead around them, but the Chinese just kept coming, and only a stonk called down on their own positions broke up the final wave. At 08.10 next morning, grey light filtering over a scene of utter carnage, the Chinese came back for another go. Finally, the survivors had to break out and attempt to get to Hill 235. Only a score of men made it. What was left of the battalion now clung to the summit. 'Watching from the Command Post, I saw the men withdrawing, step by step, down the reverse hill slopes: D Company and A Company leaving the ground they had fought for so well, that had cost the enemy such a price.'[48]

It was no easier over on the right. A fresh communist division, the 188th, swarmed across the river to strike both Fusiliers and Ulstermen. They joined 187th hammering the right whilst 189th Division kept up their relentless, if very costly, pressure on what remained of the UN centre. This was bad, but what was worse was the Chinese spearhead, which had wedged between the Glosters' right and Fusiliers' left. The former were now at risk of complete encirclement. To retrieve the situation, UN armour was deployed to counterattack. An ad hoc group comprised of US Philippine 10th Battalion Combat team in agile M24 Chaffee tanks twinned with 8th Hussars in Centurions was pushed in to eliminate this rupture. The tanks got to within 2,000 yards of Hill 235 but were eventually forced back. Now the Glosters were on their own.

By dawn on 25 April, the survivors remained isolated on Hill 235 – soon to be known as 'Gloster Hill'. If the brigade must withdraw, and it's plain this had to be soon, before the huge vice finally closed, they needed to fall back down Route 11. The Belgians, dug in near Brodie's CP, covered the road whilst the Fusiliers, Ulstermen and Hussars held on further north. US elements were nearby to provide support, but the unending pressure from Chinese attacks frustrated an attempt by US 65th Infantry to punch through to Gloster Hill. Further attempts by UN armour also foundered. Time was running out.

Brodie allowed the Glosters' CO, Lieutenant-Colonel James Carne, to make his own decision as to whether his survivors should accept surrender or try to break out. By 08.00 the die was irrevocably cast as US 1st Corps decided on a general withdrawal to a new defensive line further south, Plan 'Golden A'. Prospects for the surviving Glosters were very far from bright: 'I remember watching the slow, wind-tossed descent of a helicopter that came down for some casualties to whom the winding, bumpy road back south would have meant certain death. I saw Bob and the padre standing back as the plane lifted, their hair blown wildly by the slipstream from the rotors as she lifted into the sky.'[49]

As the exhausted elements of 29th Brigade began to filter back down Highway 11, the Belgians gallantly maintained their blocks. It was a grim process, as Chinese forces continued their relentless pressure, worse as they now held several high vantages dominating the line of escape. It was, as one survivor stated, 'one bloody ambush'. Casualties among UN forces were high and included Lieutenant-Colonel Kingsley Foster, Fusiliers' CO, killed when his jeep was mortared. B Company from the Ulstermen brought up the final rearguard. Meanwhile, the

Glosters weren't going anywhere: 'But now it is dark. Already their stealthy infantry will have left the little holes in which they have kept hidden from the sun and our observers.'[50]

By now, B and C Companies had been amalgamated into one. Resupply by air failed, as did the break-in attempts. Despite unceasing attacks and with stocks of ammunition fast dwindling, the Glosters tenaciously clung to their improvised positions all through 24 April, during that night and into morning of the 25th. By then supporting artillery had also been withdrawn and Colonel Carne ordered survivors to attempt a break out. Most didn't make it, and 459, including Carne (who would later be awarded a VC), were captured, an unenviable fate. Only the remnant of D Company led by Major Mike Harvey made it back to UN lines.

Farrar-Hockley was one of the unlucky ones:

> All the way up the valley I had heard machine-gun fire sounding and resounding among the hills; but none had been directed at us. Only now, as we drew near to the saddle, as the walls of the valley seemed to close right in upon us in this dark and cheerless spot – now almost a ravine – did we feel the breath of the enemy's fire ... there can be no doubt, had they wished to, they could have mown us down like grass before a scythe... After all that we had done, after all the effort we had exerted in fulfilling our task, this was the end: surrender to the enemy![51]

At the end of the battle, the Chinese thrust had been severely blunted, their total losses probably amounted to the equivalent of a whole division. British casualties amounted to 141 killed with 1,169 wounded, the Belgian Battalion suffered a dozen fatalities. Seoul was, however, safe and no one in Britain at the time would forget the name Gloster Hill. It resonates still.

It was a great pleasure to chat about regimental history to Matthew Holden, the curator. This younger generation faces the problem that many military museums share, it's been a while now since most of the old cap badges disappeared through disbandment or a pyramid of amalgamations, so personal family contact fades and the custodians of heritage have to move forward. The Glosters are indeed looking to the future, planning a new 360-degree form of interpretation, looking at diversity and ethnicity, countering myths about the nature of empire and regimental families. This, thank God, isn't an apology, we see far too many of those, but a reasoned exploration, I hope my grandsons will love it.

'Glorious Glosters' was in fact a name devised by a US journalist with a typical yellow press flourish. Those who have served prefer their old nickname – 'The Slashers'! This rather dubious monicker was apparently a souvenir from Canada, where a highly critical local governor, who'd had many harsh words for the men of the Glosters currently stationed thereabouts, awoke in the wee small hours to find an intruder slicing off one of his ears. Whilst the assailant was never actually identified, he was seemingly wearing a redcoat and was suspected of being a Gloster. The name stuck.

It's invidious to ask a curator which is his favourite artefact, as a breed we tend to revere them all, but Matthew likes a watercolour executed by one of the Glosters in the 19th century of Bucephalus' Tomb. Alexander found time to ensure his famous steed was given a good send off, but the monument has since disappeared. The museum also holds Colonel Carne's VC, which is rightly treasured. But, for Matthew, he finds the back-to-back badge as the most telling reminder, a totally unique distinction. A very fine regiment.

Devonshire Regiment Museum, now the Keep Military Museum

> By all the glories of the day
> And the cool evening's benison
> By that last sunset touch that lay
> Upon the hills when day was done,
> By beauty lavishly outpoured
> And blessings carelessly received,
> By all the days that I have lived
> Make me a soldier, Lord.
>
> Noel Hodgson: 'Before Action'[52]

Formerly the Devonshire Regiment Museum, The Keep Military Museum incorporates the Dorset Regiment, the Devonshire and Dorset Regiment, The Dorset Yeomanry, Queen's Own Dorset Yeomanry, The Dorset Militia, The Royal Devon Yeomanry and 94 Field Regiment RA.

If you are touring the Somme battlefields, Major and Mrs Holt's definitive guide in hand, and travel to the southern sector by Mametz, then at the crossroads follow the C2 towards Carnoy you'll find yourself driving just behind the line of the old German forward trenches from July 1916. When you get to the crucifix, innocuous enough, it was here the enemy with their usual thoroughness, eye for ground and contempt for local sensibilities, dug out an MG pit in the base of a venerable roadside shrine. You're in the Vallee St. Martin and opposite you, on the far side in rising ground you'll see a line of trees and that's Mansell Copse. It was from there that the 9th Battalion of the Devonshire Regiment (part of 20 Brigade) attacked on 01 July.

One company officer, Captain D. L. Martin, had enjoyed some home leave prior to the 'Big Push'. An artist in civvy street, he took a large-scale map of 20 Brigade's operational area and at home crafted a plasticine terrain model of the sector. The lie of the ground convinced him that this MG post would create havoc among his battalion on the day of battle. Deeply perturbed, he took his model back into the line with him and showed it to brother officers, expressing his doubts. Superiors were impressed by the model as a useful tactical briefing tool whilst ignoring the point he was striving to make. Brigadiers weren't encouraged to have doubts, it smacked of faintheartedness and was likely to result in rapid reassignment, being *stellenbosched* as it was known, or *dégommé* (literally 'unstuck') as our French allies had it. Nobody wanted to come unstuck, so nobody expressed doubts.

Captain Martin was right. On the day, 9th Devons succeeded in gaining their objectives. Not many did. That fine summer morning they strode into battle 775 strong and sustained 463 casualties including virtually all officers, only one of whom escaped uninjured. Martin was among the dead and now he lies just behind Mansell Copse in Devonshire CWCG cemetery.

He keeps good company. With him are 162 others, including Lieutenant William Noel Hodgson (1893–1916). For me, Hodgson was a particularly fine war poet and deserves greater recognition. Born in southwest England, his father was Bishop of St Edmundsbury and Ipswich, yet much of his growing up was done in that great border *bastide*, Berwick upon Tweed. Educated at Durham School, he was both a distinguished scholar and fine sportsman, as handy at the oars as in a steeplechase. He won an exhibition to Christ Church where he was a contemporary of Rupert Brooke. In 1914, he immediately volunteered for service with 9th Devons, a Kitchener battalion.

'Smiler', as he was known from his affable nature, served in France 1914–1915, winning an MC at Loos. His unit recuperated in England before being sent out again to take part in the forthcoming Somme offensive. They would be attacking in the south, opposite Mametz, and by April the unit was dugin around Mansell Copse. Hodgson was battalion bombing officer and like Captain Martin died on the first day, shot through the neck as he tried to resupply comrades who'd secured one of the German forward trenches. His body was only discovered by his batman next day. He lies in Devonshire Trench (Grave 3, Row A).[53]

Of all the evocative cemeteries on the line of the Somme offensives, Devonshire Trench is, for me, the most resonant. At the time or shortly after a wooden placard proclaimed: 'The Devonshires held this trench, the Devonshires hold it still.' At the 70th anniversary in 1986, a new stone tablet bearing this inscription was unveiled by the Duke of Kent. I cannot pass this without feeling emotion and I cannot conceive of a finer or more poignant tribute.

After a few false starts, Henry Somerset Marquess of Worcester raised his regiment, The Duke of Beaufort's Regiment of Foot, or Beaufort's Musketeers, in 1685 – by which time he'd been elevated to his dukedom for the specific purpose of defending Bristol against the rebel Duke of Monmouth. At one point it did seem the Somerset men might be in a position to threaten Bristol, had they gained the city the whole dynamic would have shifted, and Monmouth's ill-conceived adventure might have begun to look a genuine threat. True to form, Monmouth bottled it at the decisive moment, which, having passed, never reappeared and Bristol as well as King James' throne was safe (in the case of the latter this was only a three-year reprieve, see below).

In 1690, the regiment saw action at the Battle of the Boyne and then at the Siege of Limerick. It fought throughout Marlborough's wars, including the less glamourous and unfortunate campaign in Spain. The Devons battled Jacobites in 1715 on the confused field of Sherrifmuir, then fought again on the continent at Dettingen (1743) and Fontenoy (1745), with extensive fighting through the Seven Years War (1756–1763). In 1751, they were numbered 11th Foot, then thirty-odd years later as the 11th (North Devonshire) Regiment. They battled in the fever-ridden West Indies, then through the savage grind of the Spanish Peninsula. Tough fighting at Salamanca earned them the name 'The Bloody 11th'.

Long years in far-flung imperial outposts followed the peace of 1815. Because of the Childers Reforms in 1881, they became the Devonshire Regiment, one that already in fact had two battalions. They went to France in 1914 and battled in October at La Bassée, storming Givenchy Ridge. Inevitably they were sucked into the attrition of 1st Ypres and during the following April in the terrible fighting for Hill 60, losing ⅔ of their officers and ⅓ of ORs. After such extensive service through both world wars, the Devons became amalgamated with the Dorsets in 1958 to form the Devonshire and Dorset Regiment. Finally, in 2007, the combined regiment was itself further amalgamated with the Royal Gloucestershire, Berkshire and Wiltshire, the Royal Green Jackets and Light Infantry to become the Rifles.

Calvary on the Somme would be followed in 1918 on 27 May by the epic last stand of the 2nd Devons at Bois des Buttes, the opening day of the 3rd Battle of the Aisne. The Devons were attached to 8th Division, which had already suffered badly in the earlier *Kaiserschlacht* battles. From 12 May they had been re-assigned to General Duchene's French 6th Army, occupying a hitherto quiet sector. A week or so later they were withdrawn completely from the line and sent back to Roucy, while numbers were filled from freshly arrived drafts. This was a short respite. By the 26th, rumours of another German blow were circulating. These were gleaned from POWs and deserters, though the picture was very far from complete. Bois des Buttes is a wooded sandstone knoll southwest of the village of La Ville-aux-Bois, which had changed hands sveral times. Besides the normal network of trenches and communication, the position benefited from a series of *creutes*, ancient quarries whose cavernous galleries made perfect dry and airy dugouts, complete with electric light.

Initially, the position lay some 1,200 yards (1,100 metres) behind the then front line. The Devons were intended as 23rd Infantry Brigade reserve and by midnight on the 26th had marched the 4½ miles (7.25 kilometres) up from Roucy, a soft, quiet night in warm spring air. So far so good. At 01.00 things changed as the German guns opened up, stabbing flames lighting up the night air. Sheltering in their deep caves, the Devons were relatively immune, though gas shells caused casualties. At dawn the first storm troopers appeared. Tactics on both sides had shifted since the early days of mass infantry attacks. These new elites preferred infiltration, picking up weak spots. They possessed new weapons, light portable artillery and the first sub-machine guns, which added enormously to section firepower. The Allied front line began to fracture, and very rapidly Bois des Buttes became a sole bastion north of the Aisne. This didn't augur well for the men of the 2nd Devons.

As the enemy guns fell silent and the front began to crumble, the battalion emerged from sanctuary underground, taking up position as best they could, clawing defensible spaces in their blasted trenches. As ever, field telephones were early casualties and comms were lost. What resulted, inevitably, was a soldier's battle. German infantry from the 50th Division pushed forward but were seen off, and the hilltop vantage allowed the Devons to bring fire down on the road from La Ville-aux-Bois to Pontavert, interdicting the proposed line of German advance. They'd become the cork in the bottle. Airstrikes, still in their relative infancy, followed. For anxious German officers, mindful of the need to exploit the gap they'd blasted in the Allied front, this was all very annoying. They made their annoyance felt.

From 07.00 the intensity of their rushes intensified. Within the next hour and a half, the battalion was progressively encircled, so Lieutenant-Colonel Anderson-Morshead pulled his HQ and such supports as he could muster onto the reverse slope in order that they could still keep firing at the road. By 09.30, barely 50 men of the battalion were still on their feet and unwounded. More Germans with artillery support were pushing up from Juvincourt towards Pontavert. The Colonel divided his meagre survivors into two assault teams to form a pincer movement to counter the enemy's fresh push. The odds were not good, but the Devons still succeeded in breaking up the attack, knocking out some guns.

This could only be a respite. Their Thermopylae was close to its inevitable denouement. Fresh German forces were thrown into the attack and more poured down over the hill the Devons had fought so hard to hold. Anderson-Morshead was killed, but the survivors fought on grimly. Ammunition ran perilously low as survivors staged a breakout towards the Aisne. Against all odds, a roughly company-size unit made it out. Some were captured as the Germans had already succeeded in gaining the south bank, but a handful did succeed in catching up with the British withdrawal. The cost of their Alamo had been very high: 29 officers and 552 ORs failed to return.[54]

Their sacrifice was by no means in vain. Time, the commodity the Germans did not possess, had been lost. Their magnificent stand had disrupted the attack to the degree it never recovered momentum and ground to a halt less than a week later. From here on in, it would be steadily backwards for the Kaiser's army.

Bois des Buttes was included in the first of ten lists of battle honours for the Great War, published in February 1924. The Devons were the first to be awarded a Croix de Guerre with palm, on 5 December 1918. Noel Hodgson could not, of course, have known this, but his 1916 poem would apply equally tellingly to his comrades at Bois des Buttes:

> I, that on my familiar hill
> Saw with uncomprehending eyes
> A hundred of thy sunsets spill
> Their fresh and sanguine sacrifice,
> Ere the sun swings his noonday sword
> Must say good-bye to all of this;
> By all delights that I shall miss,
> Help me to die, O Lord.[55]

The Regimental Museum is in the Keep, Dorchester, formerly barracks for the Dorsets. Built between 1877 and 1881 on the site of an old militia compound with the kind of epic Gothic revival façade so beloved of Hollywood directors, who imagine this is what a medieval castle looked like. It was the product of Cardwell's Reforms, which preceded Childers and marked the period when local connections were being promoted and encouraged.

Sedgemoor

> We need not now to feare the aspiring ambition either of a Cromwell or a Lambert. Nor need we now feare of being kept in bondage and slavery by an inconsiderable part of a Parliament, or a Phanatick Committee of Safety under the notion of a free state, or Commonwealth.
>
> The Reverend Ames Short: Restoration sermon

Sedgemoor is a disappointment. For such an important action with a marvellously dramatic plotline, there's very little there to quicken the blood. R. D. Blackmore in *Lorna Doone* gets the atmosphere just right, but the place is effectively forgotten. It deserves much more. The human tragedy which unfolded there commands our attention and led to a savage reprisal.

King Charles II died on 6 February 1685, his Catholic brother James, Duke of York succeeded as James II. Charles left no legitimate offspring but a large family of illegitimate sons and daughters from his stable of mistresses. The eldest of these was the Protestant James Scott, Duke of Monmouth, who had served his natural father ably enough as a soldier.

James II of England and VII of Scotland (1633–1701), the last Stuart king. James succeeded his elder brother Charles II, but this new king was very much out of step with the times as an ardent Roman Catholic and appeared to be trying to steer his joint realm back to papacy. This did not sit well and gave Protestant Monmouth his chance, though he made his attempt far too early, before James' unpopularity had evaporated. The king learned nothing from Monmouth's rebellion and the savage persecution of those who had taken part inevitably backfired. He was final ousted by his Protestant daughter Mary and her husband, the Stadtholder William of Orange, during the Glorious Revolution three years later. Marlborough, hitherto a staunch supporter and friend to the king defected and James had to flee for France. His exiled faction became known as the Jacobites.

Monmouth, handsome, charismatic, and shallow, claimed his father had actually married his mother, Lucy Walter, though there was no evidence to support this. At the time of Charles II's death, Monmouth was in self-imposed exile at The Hague in the Netherlands, following his alleged participation in the Rye House Plot, a conspiracy to murder both his father and the Duke of York. Monmouth's mother Lucy Walter was possibly Charles II's first mistress (though there is doubt) and she was only 18 when she gave birth to her son, apparently by Charles, in Rotterdam. As king, Charles recognised the boy James Scott as his illegitimate offspring and created the boy Duke of Monmouth at age 13. Monmouth served his father as a soldier during the Third Anglo-Dutch War commanding an Anglo-Dutch Brigade, earning plaudits for his conduct at the Battle of St. Denis in 1678.

The following year he commanded a royal army that defeated rebellious Scottish Covenanters at Bothwell Brig in June. Monmouth was inclined to be merciful, which earned the raging scorn of fiery Tam Dalyell of the Binns, the King's older, choleric C in C Scotland. During his ill-fated rebellion, however, he showed no spark of outstanding political or military talent; nor did he receive any clemency from his uncle after capture. James did grant him an interview, unheard of in circumstances where no clemency was likely, but was further disgusted by Monmouth's abject behaviour. The Duke met his appointment with the executioner on 15 July 1685. He implored the headsman Jack Ketch to do the job swiftly and with a single blow, but his end was messy and protracted.

Monmouth had toured the southwest of England in 1680, posturing ably as a popular figure and was roundly cheered. Taciturn James, wedded to the old religion, was far from popular. The Duke planned his ill-starred rising in Holland. He had coordinated his plans with a northern effort in Scotland commanded by the exiled Archibald Campbell, 9th Earl of Argyll, who had fallen foul of James when he was governing the northern kingdom on behalf of his brother.

The Duke of Monmouth set sail for England in May 1685 with a small force of adherents and landed at Lyme Regis in Dorset on 11 June. A lengthy manifesto drafted by Robert Ferguson, a fanatical Scottish Presbyterian Minister, denounced Catholic James II, going so far as to claim he had poisoned his brother. Monmouth, dressed for the part in royal purple and wearing a silver star, declared to the assembled crowd, 'For Religion and Liberty.' Meanwhile, Gregory Alford, Mayor of Lyme Regis, raised local militia for the Crown while exciseman Samuel Damsell rode swiftly to London to warn the authorities. Monmouth declared himself 'now head and captain-general of the protestant forces of this kingdom' with a 'legitimate and legal' right to the crown, which he promised he would exercise only with the agreement of a free parliament. He seemed a popular figure, referred to as 'King Monmouth' by many of the commoners; this outward show was impressive, even though it hid a vacuum.

John Churchill, later Duke of Marlborough, was appointed as commander of the regular Royal infantry, while overall command was eventually handed to Louis de Duras, Earl of Feversham. Over the next few days, Monmouth continued to recruit men, his Protestant appeal was strong among the rural population of Somerset and Devon and by 15 June he'd recruited around a thousand ill-trained, under-equipped but enthusiastic volunteers. Two of his officers who had joined him in Holland, Thomas Hayward Dare and Andrew Fletcher, however, fell into a disagreement about a horse, which led to Dare shooting Fletcher, Dare was promptly placed under arrest. This could be said to have rather set the tone for the campaign.

Monmouth, on receiving news of the approach of royalist reinforcements, left Lyme Regis, heading north into Somerset. Just a few days later, Royal Navy vessels reached Lyme Regis, effectively

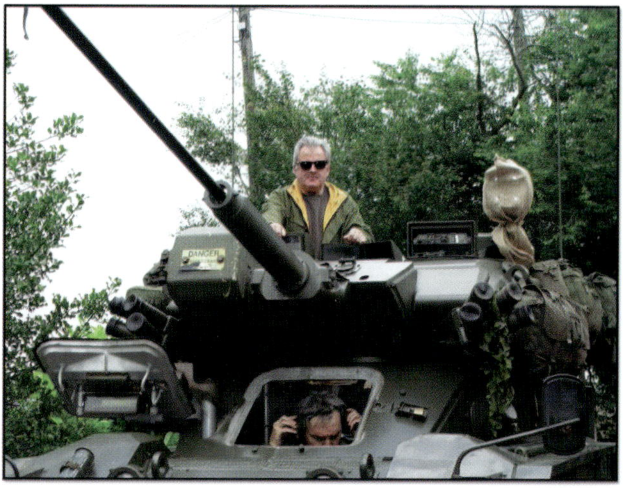

The authors in Graham's Fox armoured car, somewhere in Mid-Northumberland *c.* 2009. The FV721 Fox Combat Vehicle Reconnaissance in a 4x4 armoured car mounting 30 mm Rarden Cannon and built by ROF Leeds, in service from 1975 to 1994. (Author's collection)

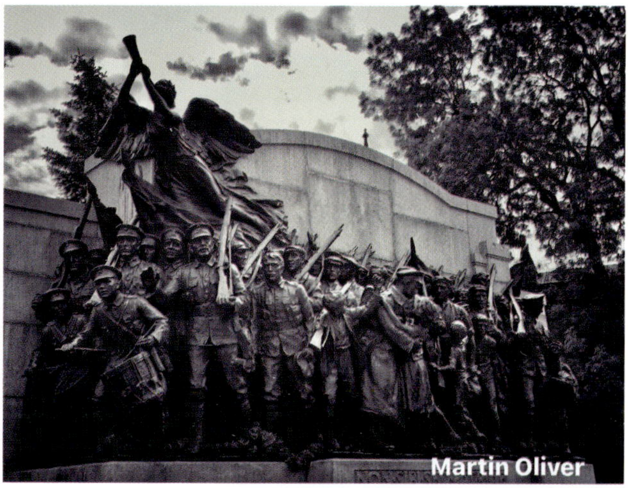

The Response, Haymarket/Barras Bridge, just north of the Church of St. Thomas the Martyr, Newcastle upon Tyne; commissioned by Sir George Renwick 1st Baronet and designed by Sir William Goscombe John, unveiled in 1923. (Martin Oliver)

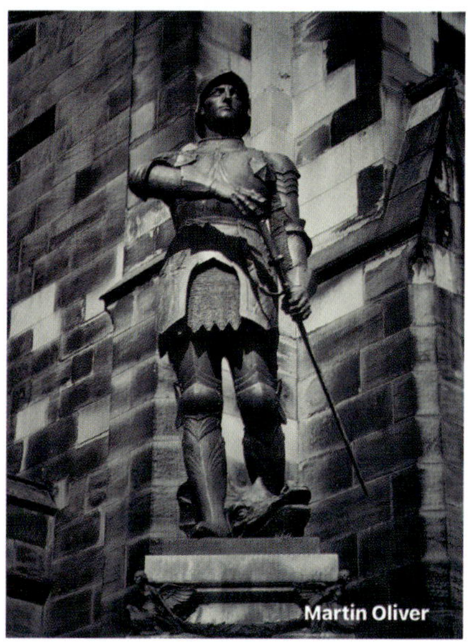

St George – a bronze memorial to the men of 6th Battalion Northumberland Fusiliers, 43rd & 49th Battalions of the Tank Regiment, designed by J. Reid and unveiled in 1924, standing by the Church of St. Thomas the Martyr, on the corner of Barras Bridge & St. Mary's Place, Newcastle upon Tyne; the Great War as chivalry. (Martin Oliver)

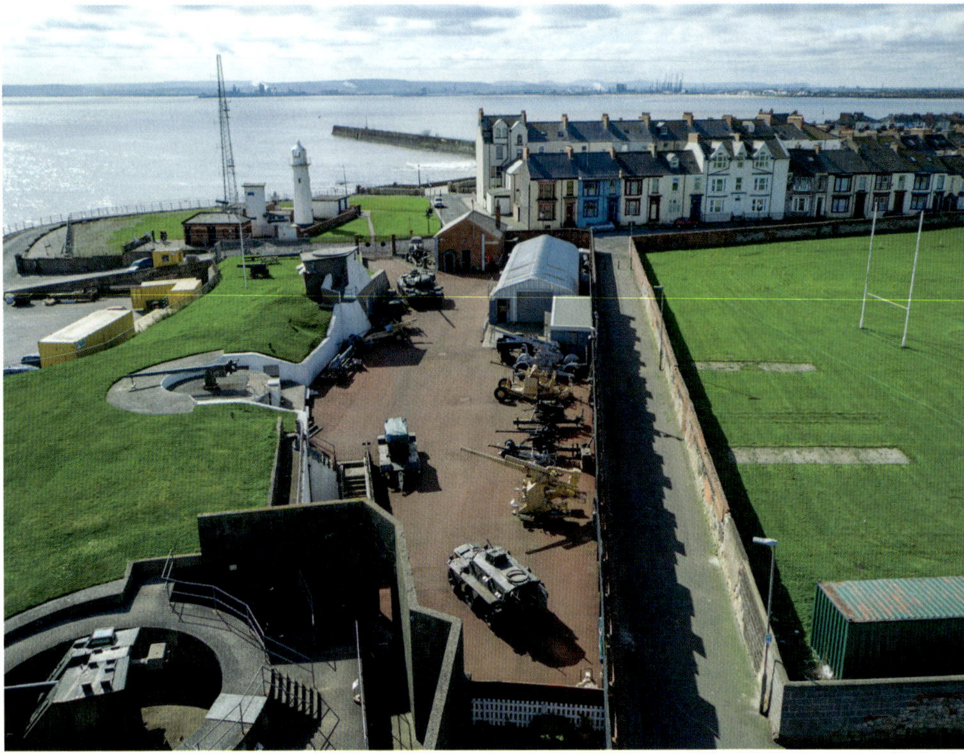

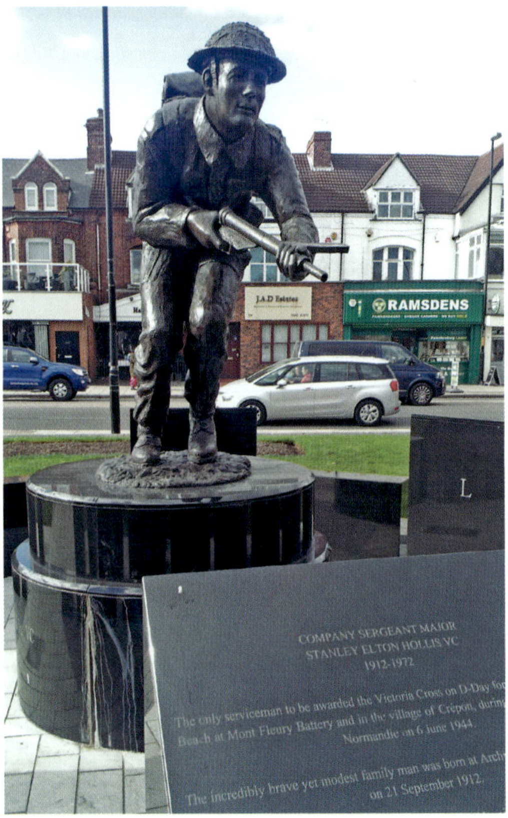

Above: Heugh Battery Museum, Hartlepool, the old Victorian fort that withstood the fury of German naval bombardment in December 1914. (Heugh Battery Trust)

Left: The Green Howards Museum, Richmond – bronze statue of Stanley Hollis VC, sculpted by Brian Alabaster ARBS and unveiled in November 2015. It stands close to the Middlesbrough Cenotaph and in front of the Dorman Museum. (Wikicommons)

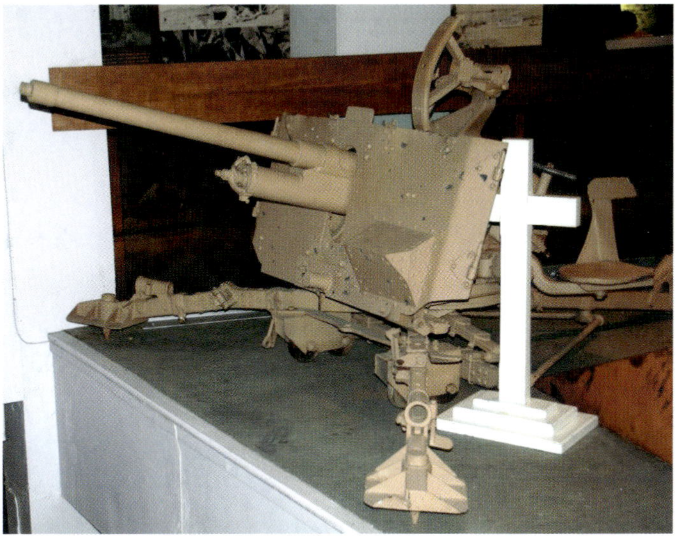

The Durham Light Infantry Museum, Durham – the Wakenshaw Gun. This used to be in the former DLI Museum premises and there are strong hopes this may soon reopen. (Durham County Council)

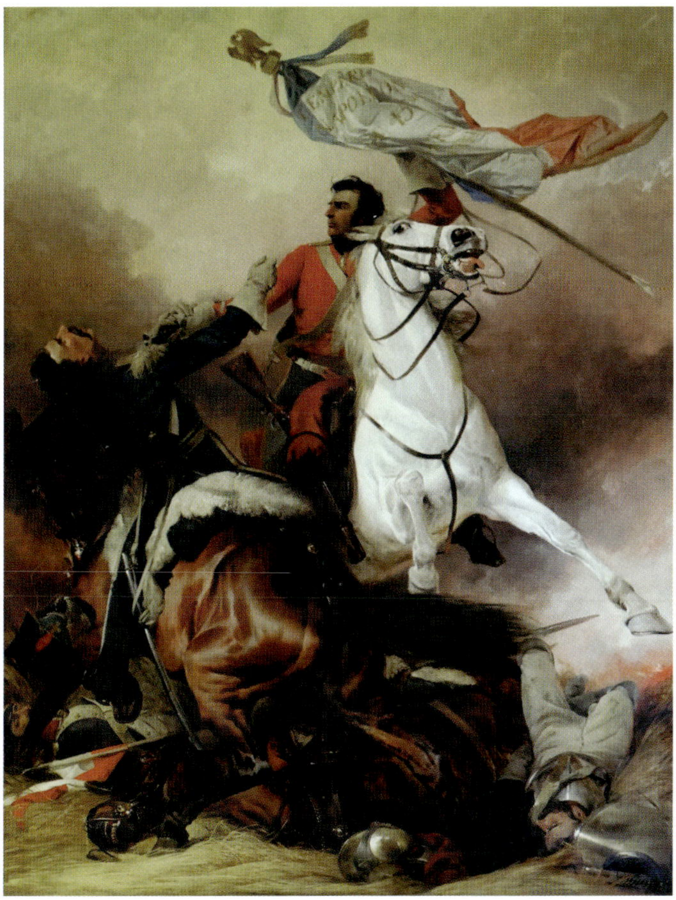

The Royal Scots Dragoons Museum, Edinburgh – Ensign Ewart wins the French colours, an epic, if perhaps rather romanticized image. (Eric Gaba Photographie)

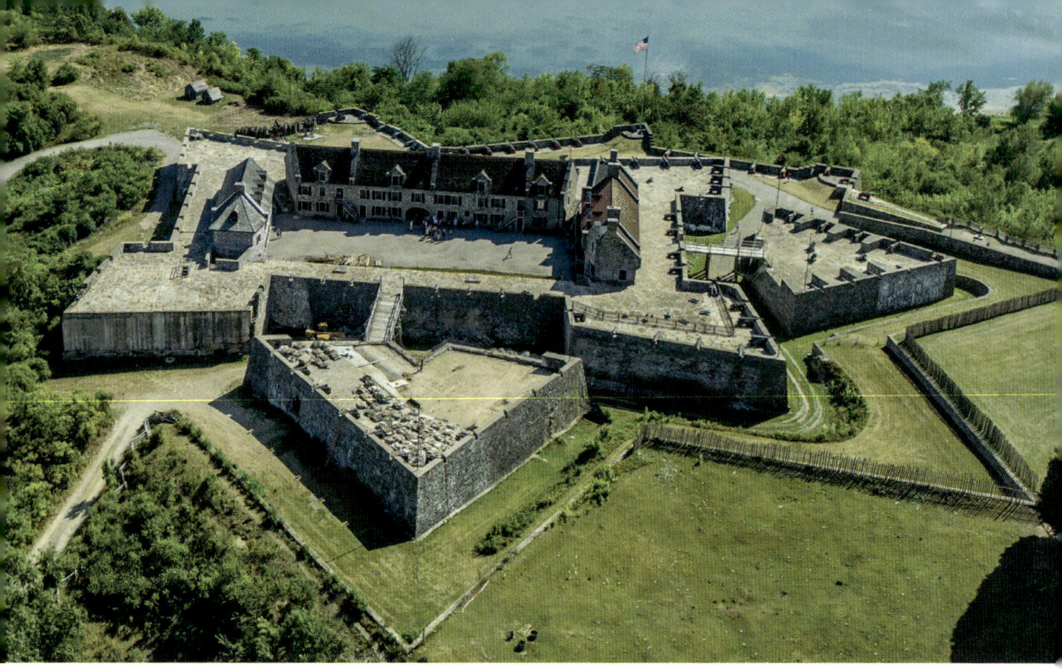

Above: The Black Watch Museum, Perth – aerial view of the rebuilt Fort Carillon, scene of the heroic failure. (Adobe Stock/Tamas)

Below: The Prince Olav slipway used by the Shetland Bus, those fishermen who took on Hitler and survived. (Hanc1)

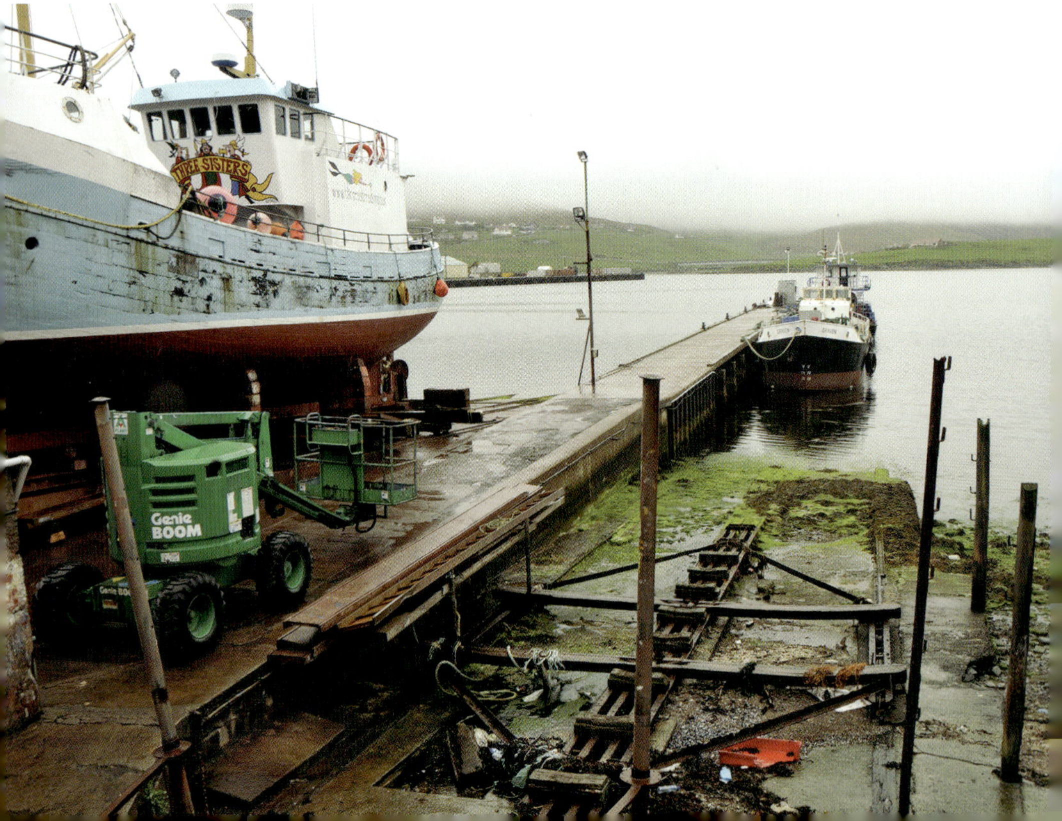

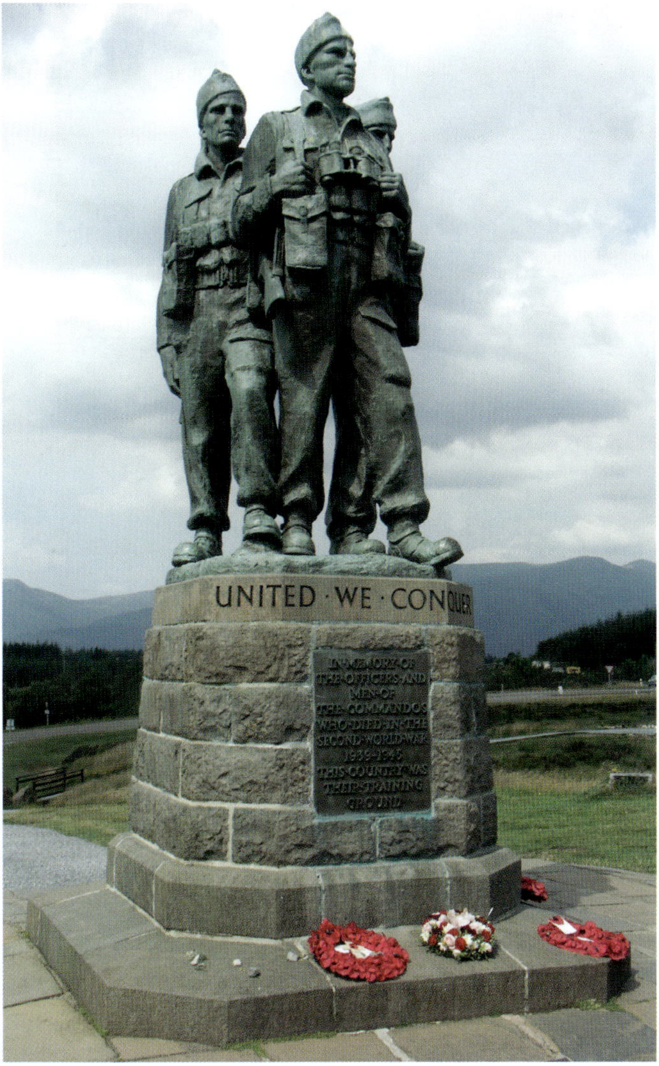

West Highland Museum, Fort William & Spean Bridge Commando Memorial, completed in 1951. One sculpture (designed by Scott Sutherland), which is truly iconic, placed in a dominating position above the town with sweeping views across Lochaber, 17 feet (5.2 metres) tall. (Wikicommons/JMB)

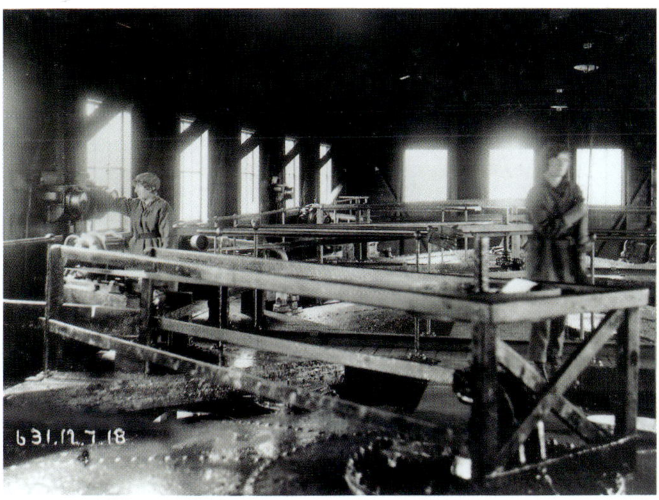

The Devil's Porridge Museum, Eastriggs; acid-mixing station – blending tanks. (Devil's Porridge Museum)

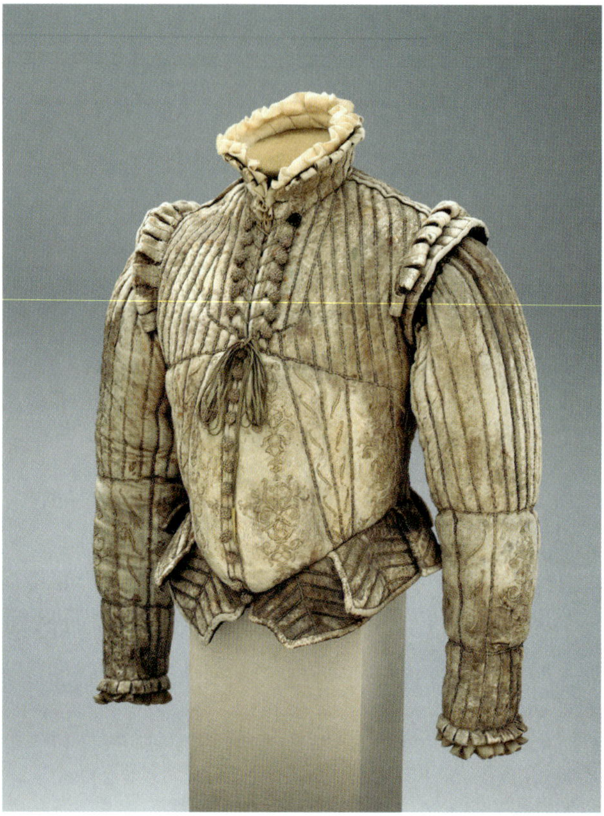

The Johnnie Armstrong Museum of Border Arms & Armour, Teviothead – a rather splendid example of a protective 'jack', this one most surely the property of a gentleman reiver. (Metropolitan Museum of Art)

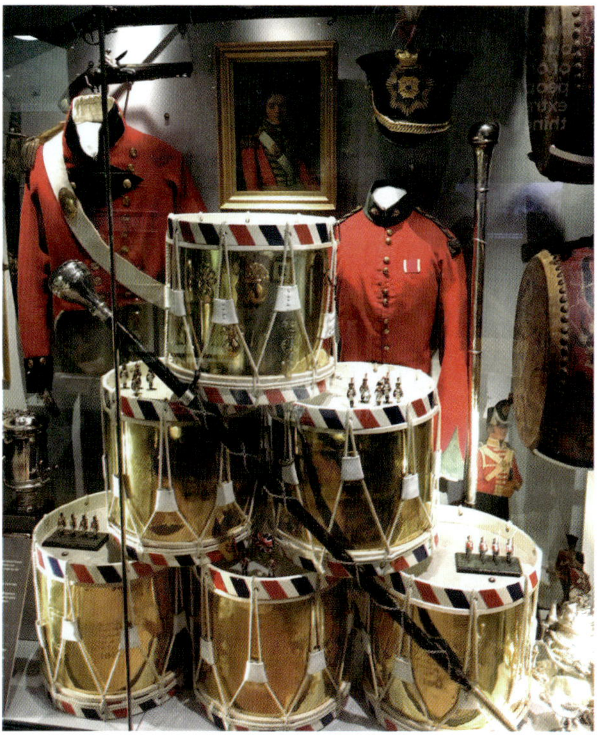

Cumbria's Museum of Military Life, Carlisle – the Arroyo Drums. These were captured at the Peninsular War Battle of Arroyo do Molinos, fought on 28 October 1811, when General Rowland Hill decisively beat a French force under General Jean-Baptiste Girard. (CMoML)

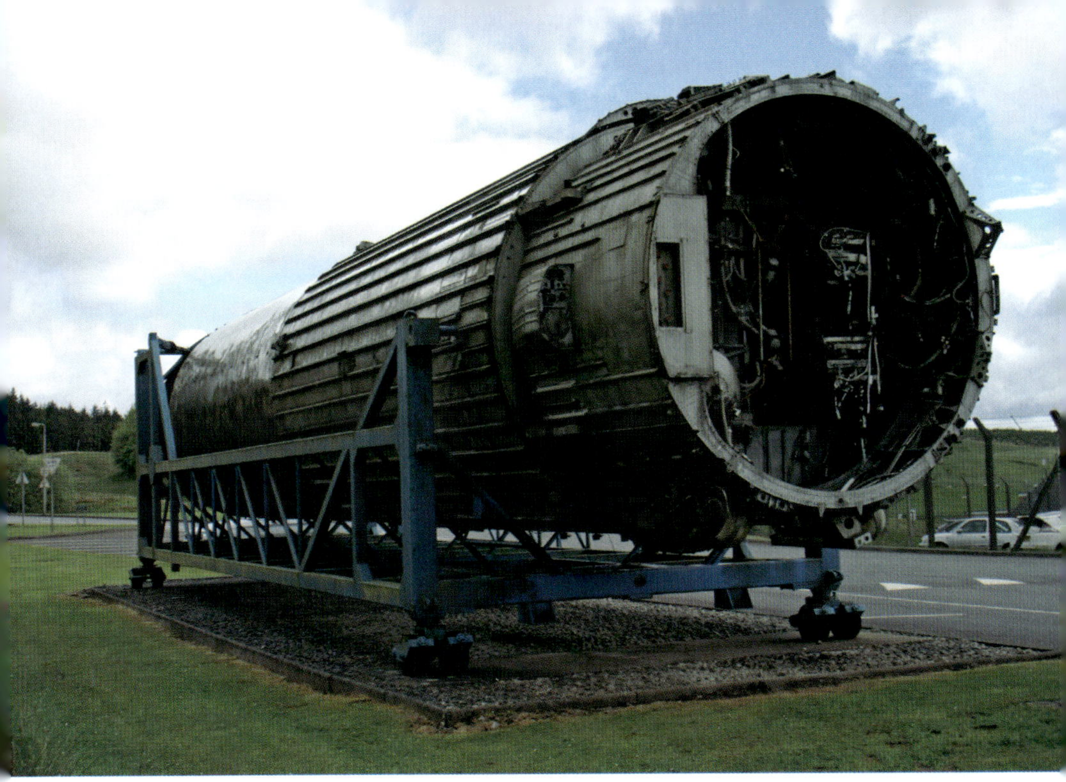

Above: Blue Streak, Spadeadam RAF Ranges. (Phil Thirkell)

Right: George Arthur Roberts.

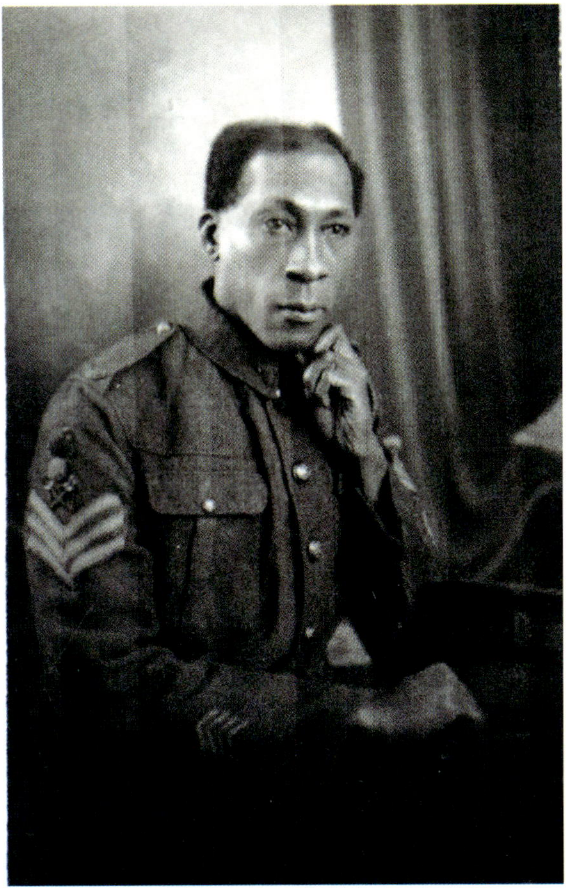

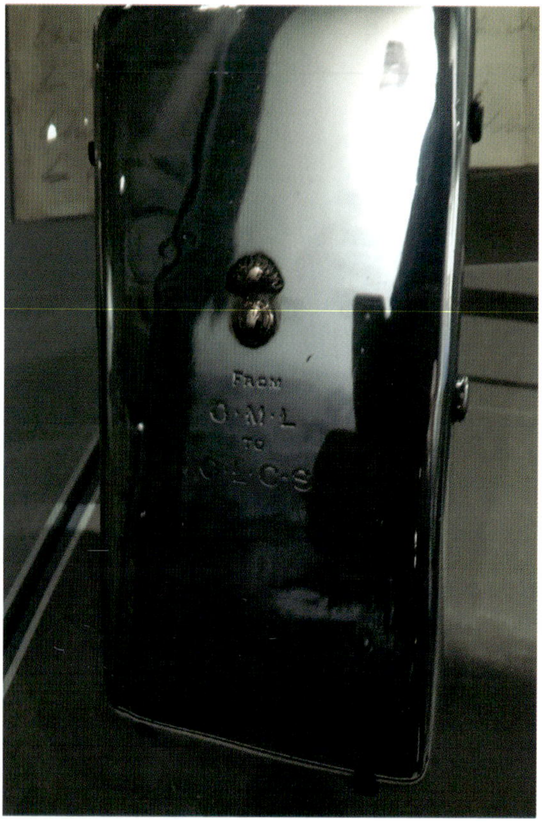

Royal Welch Fusiliers Museum Caernarfon - a silver cigarette case, bearing the regimental crest, the property of Major Geoffrey Lee Compton-Smith 10th Battalion (1889 – 1921) who was kidnapped by the IRA in 1921 and subsequently murdered. (Royal Welch Fusiliers Museum)

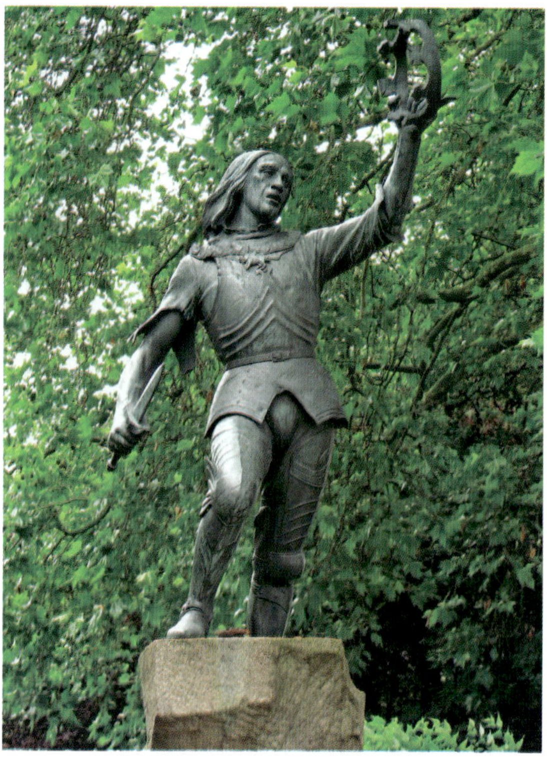

Bosworth Battlefield Centre, Market Bosworth, and Visitor Centre, Leicester – a bronze life-size statue of Richard III, commissioned by the Richard III Society in 1980, it originally stood in Castle Gardens but was moved to its present location, Cathedral Gardens, just outside the visitor centre in 2014. (Wikicommons)

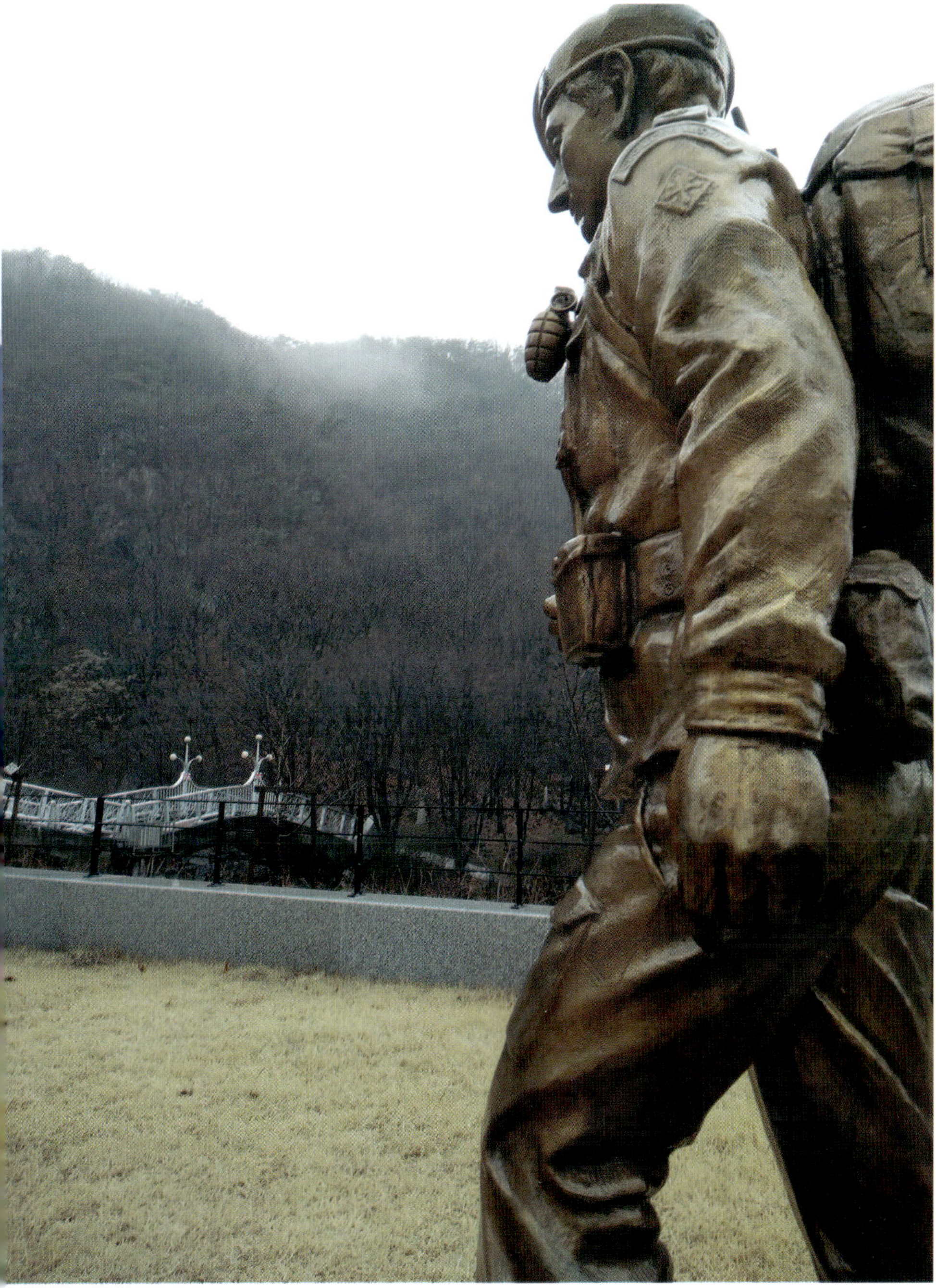

Soldiers of Gloucestershire Museum, Gloucester; the Gloster Hill Battle Monument. (Wikicommons)

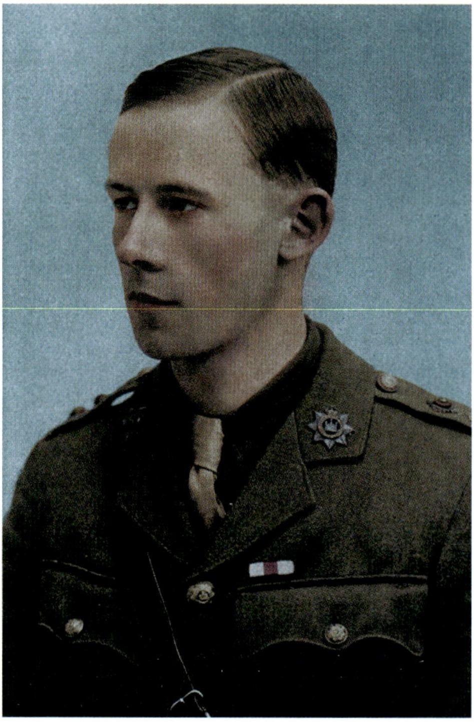

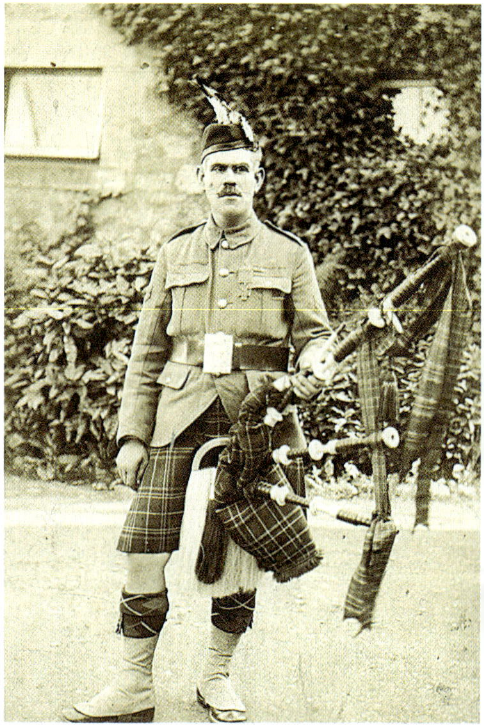

Above left: The Devonshire Regimental Museum, Dorchester – the poet Noel Hodgson, photographed in 1915. (Wikicommons).

Above right: Daniel Laidlaw VC, the Piper of Loos.

Below: Gurkha Museum, Winchester – Gurkhas in action. (Sam Meadows)

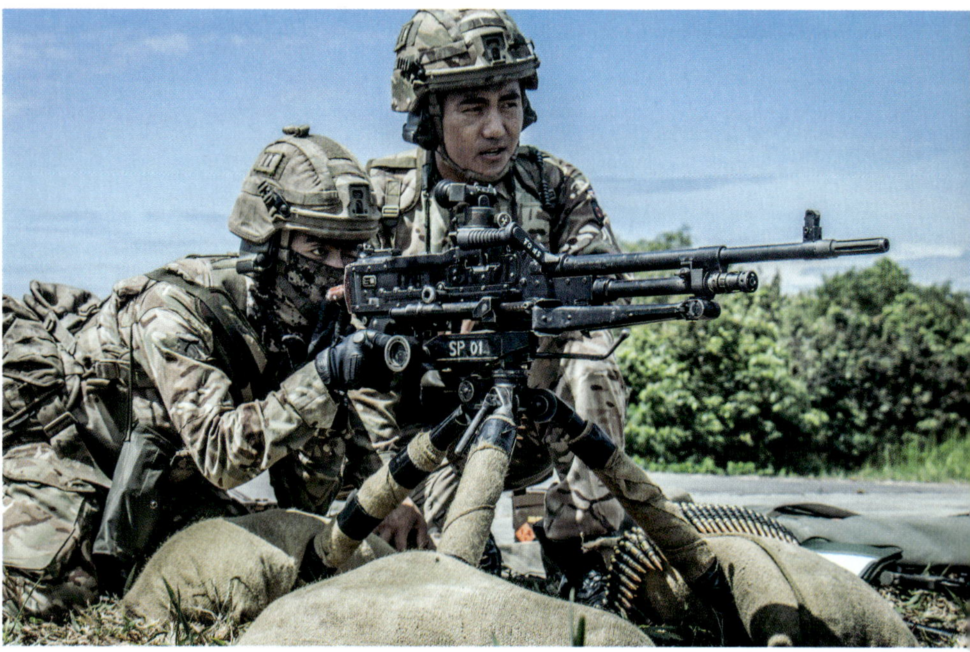

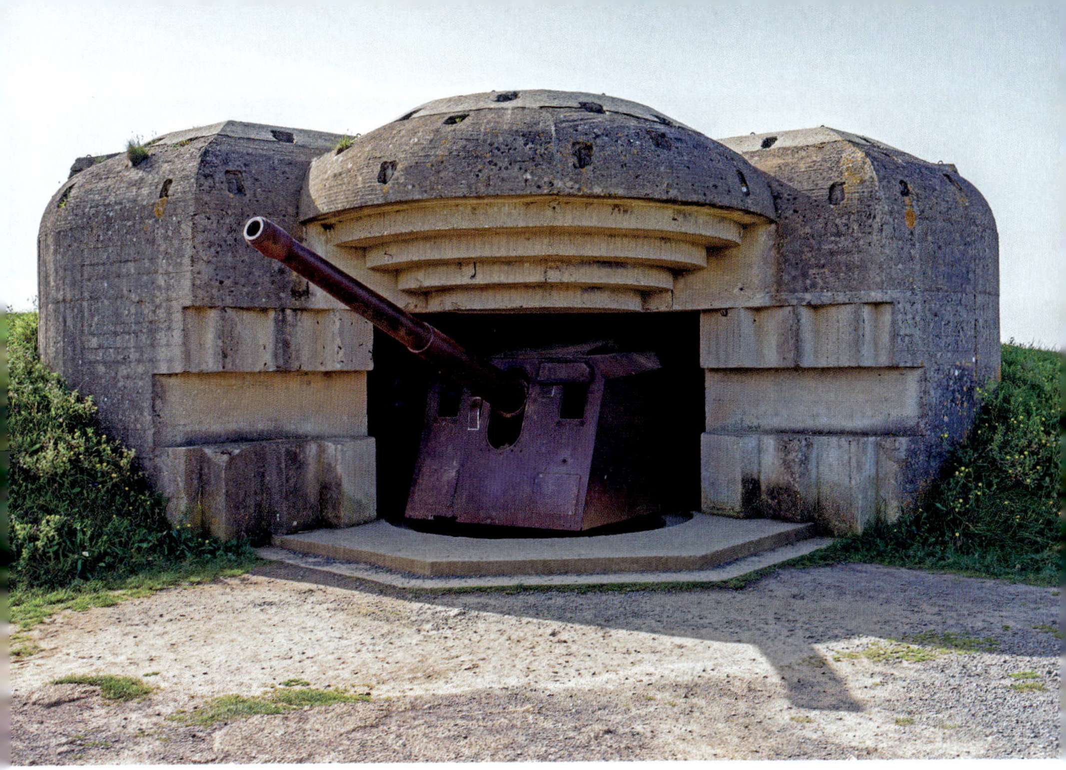

D-Day Museum, Portsmouth – preserved Atlantic Wall artillery position. Though not complete, Axis defences remained formidable. (Adobe Stock/olrat).

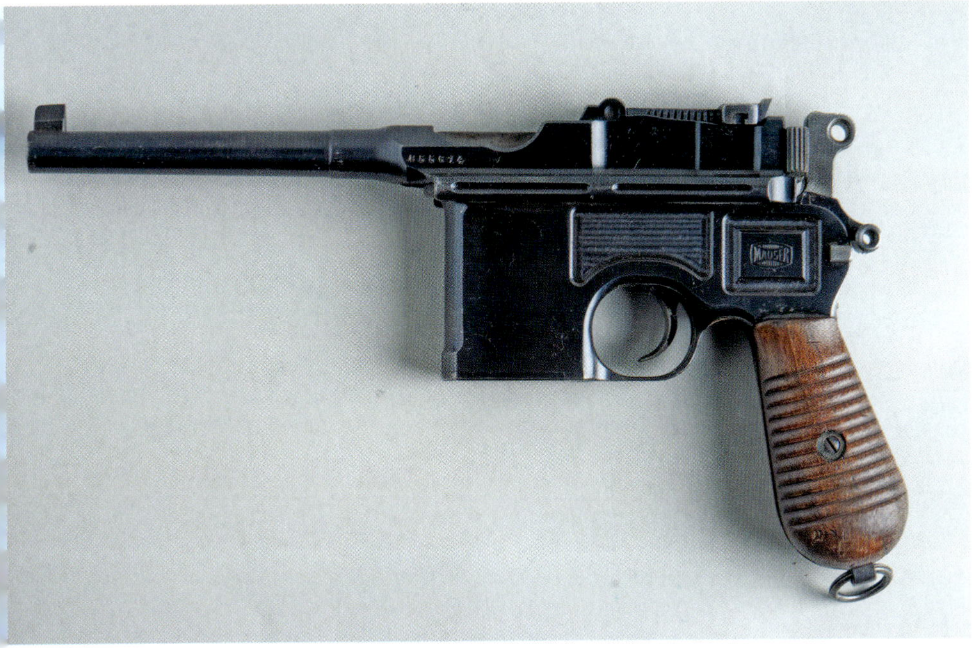

Churchill's War Rooms – Mauser C96 semi-automatic pistol, like that carried by Winston Churchill at Omdurman in 1898. (Wikicommons)

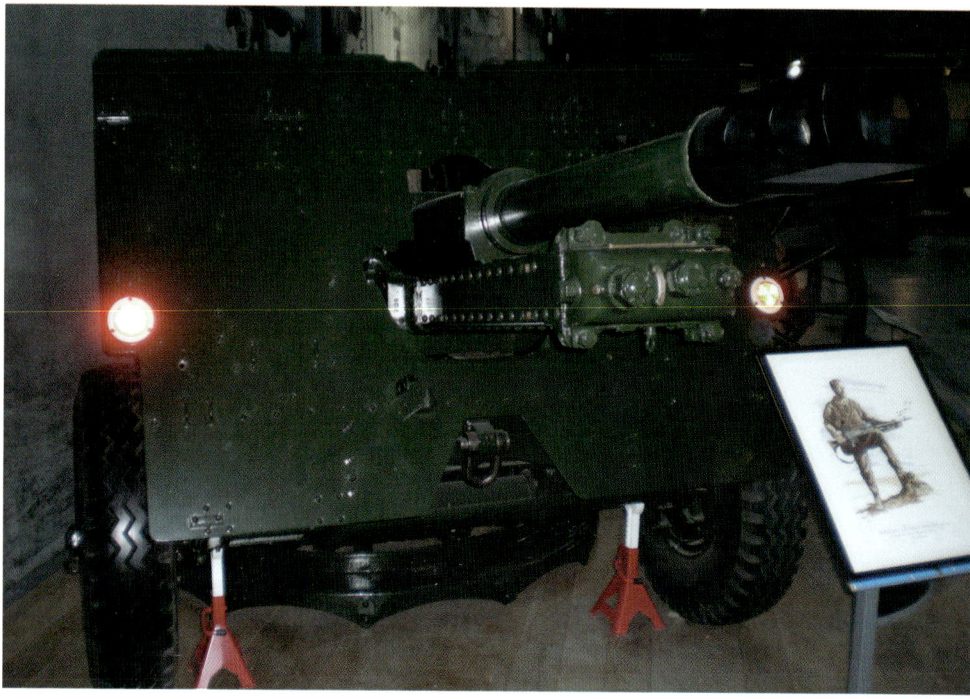

Above: Firepower, Royal Artillery Museum – the Mirbat 25-pounder gun which played so important a role in the defence. (Firepower, Museum of the Royal Artillery)

Below: Bofors gun at Dover Castle, guarding the tunnels begun during the Napoleonic Wars and used by Admiral Ramsay and his Dover Command staff to plan and execute the 'Miracle' of Dunkirk. (Adobe Stock/lukasseirl).

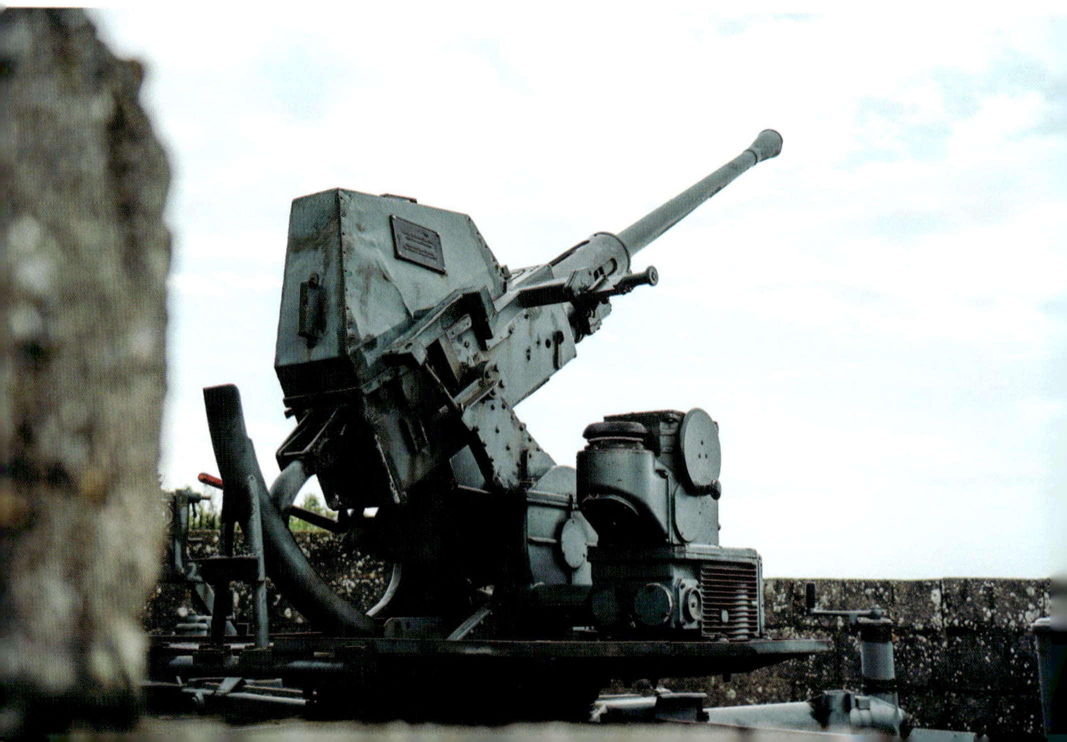

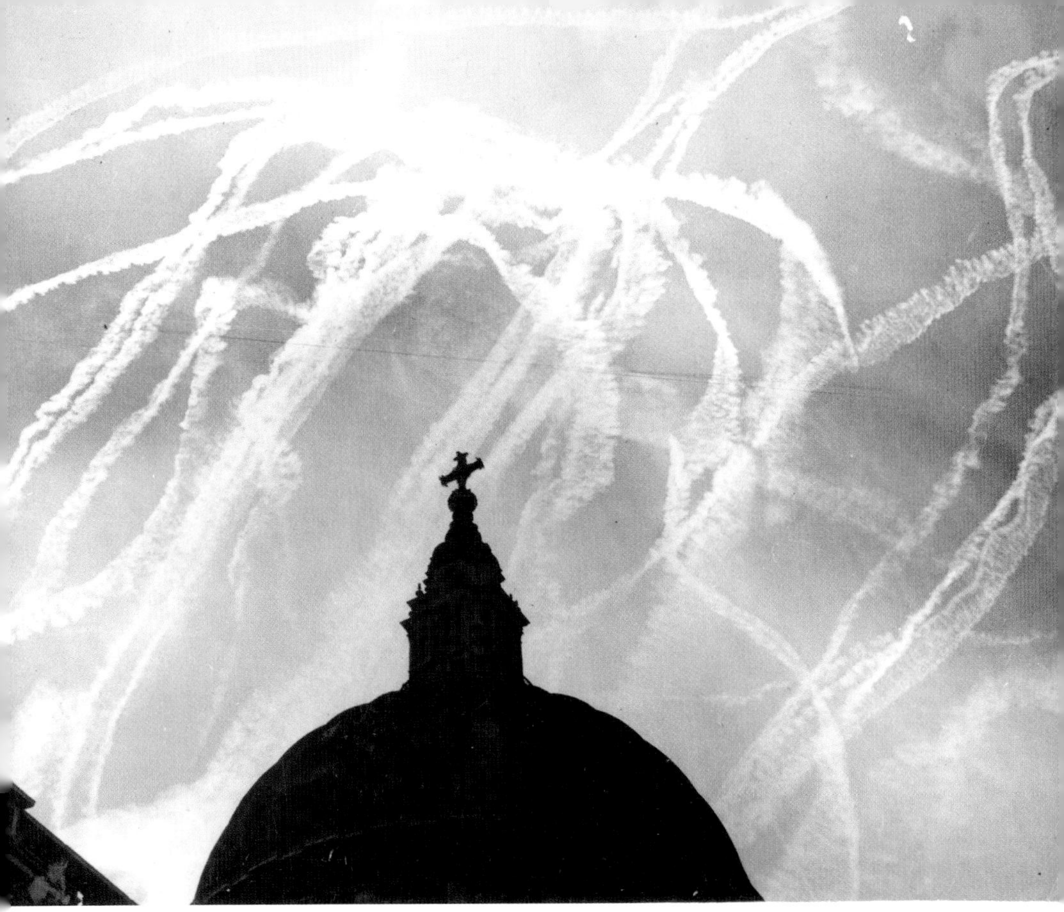

Above: Spitfire and Hurricane Museum, Ramsgate – vapour trails from a dogfight over St Paul's, 1940. (Wikicommons)

Below: Royal Engineers Museum, Chatham – Lieutenant John Rouse Chard VC's pistol as used in the defence of Rorke's Drift, January 1879. (Wikicommons)

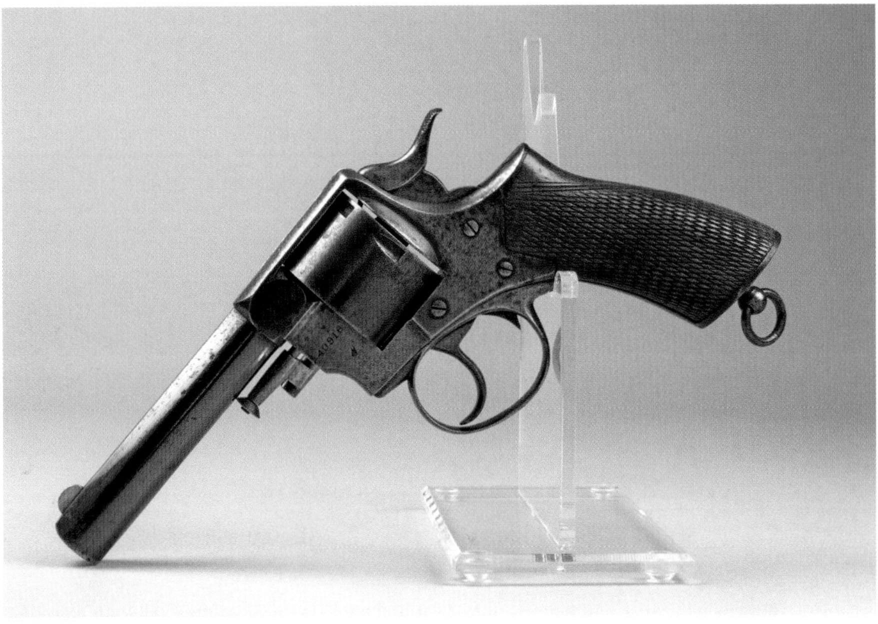

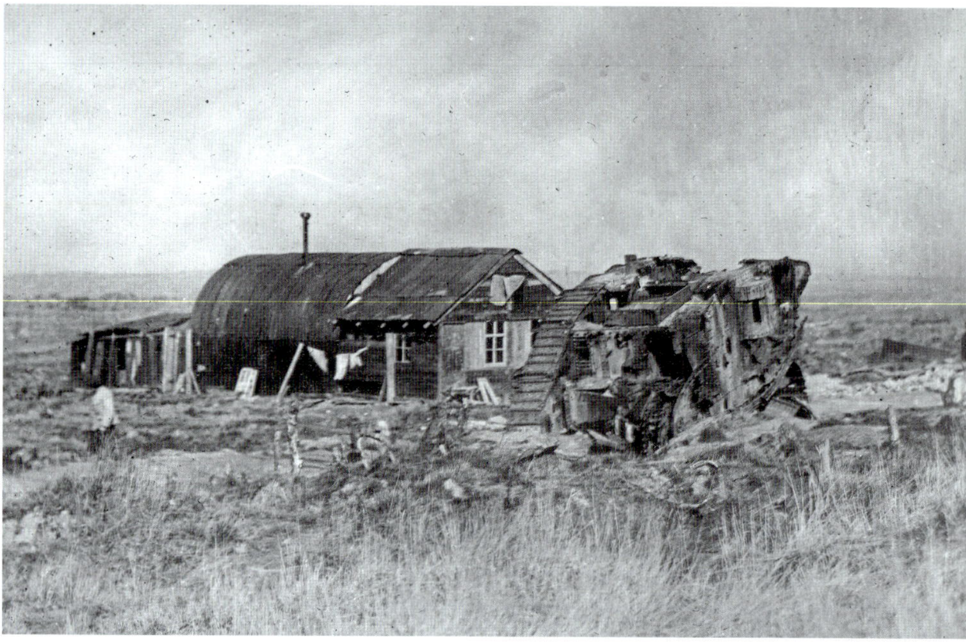

Above: Bovington Tank Museum – Ypres Road 'Tank Cemetery', showing the bare skeleton of a wrecked tank. (Paul Reed)

Below: Norfolk Tank Museum – Stuart Tank. The M3 was a US light model upgraded to the M5 in 1942, no match for the Panzers but light, robust, agile and dependable, with a 37 mm gun as its main armament. Stuarts (or 'Honeys') as we Brits called them, saw extensive service in the Western Desert and in Burma. (Ricardo Tulio Gandelman)

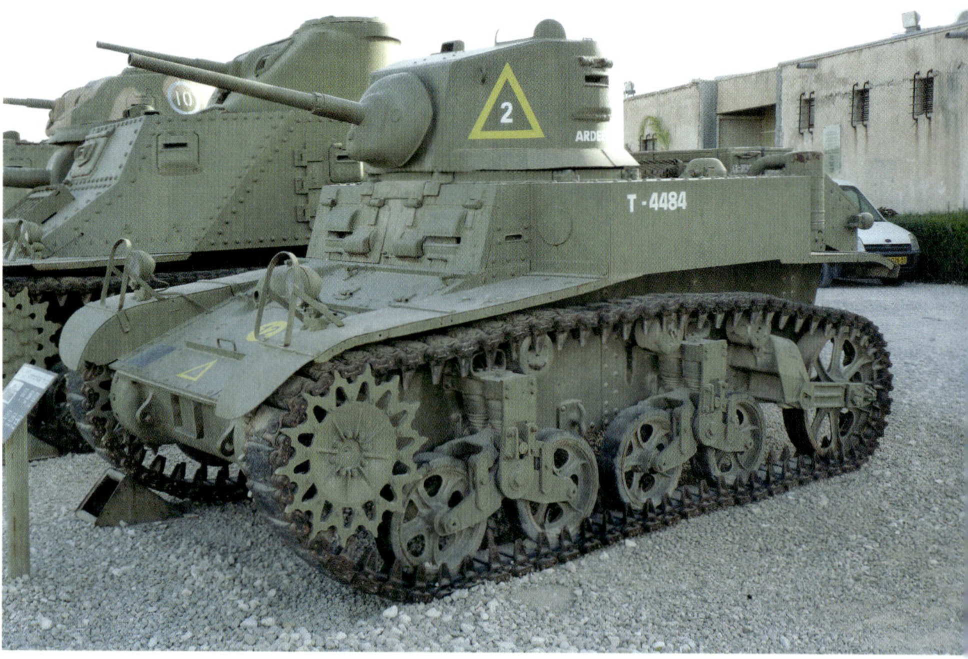

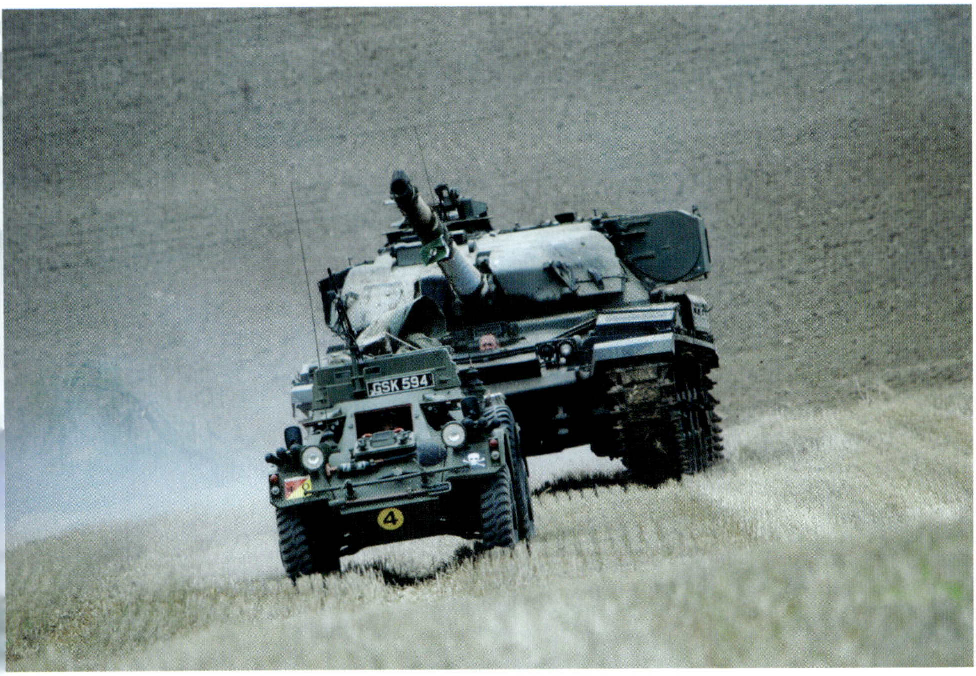

Above: Cold War – a UK Chieftain MBT led by a Ferret scout car; not actually a desert location, just a dry summer in Mid-Northumberland. (Adam Barr)

Right: Home Guardsmen during the Second World War. (Amberley Archive)

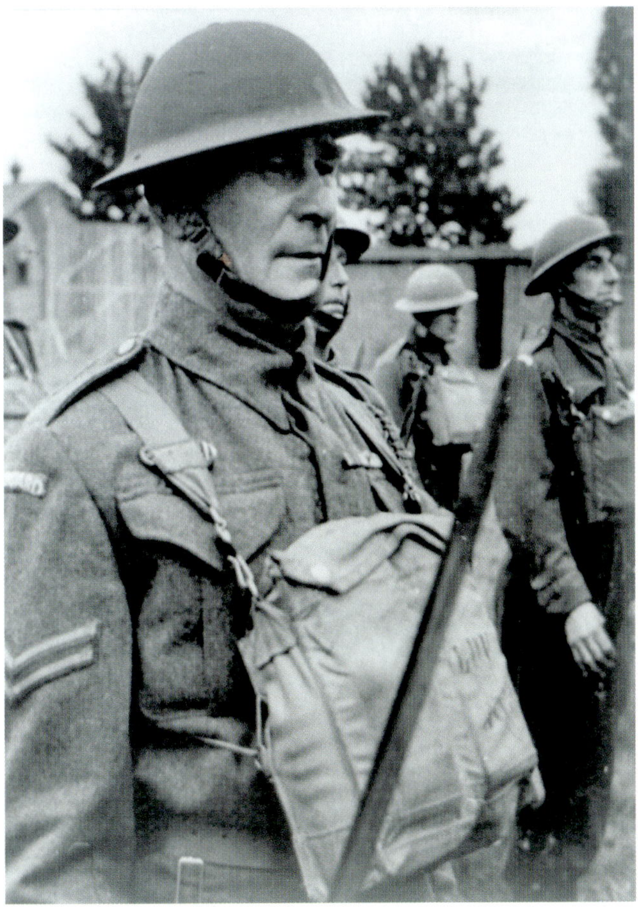

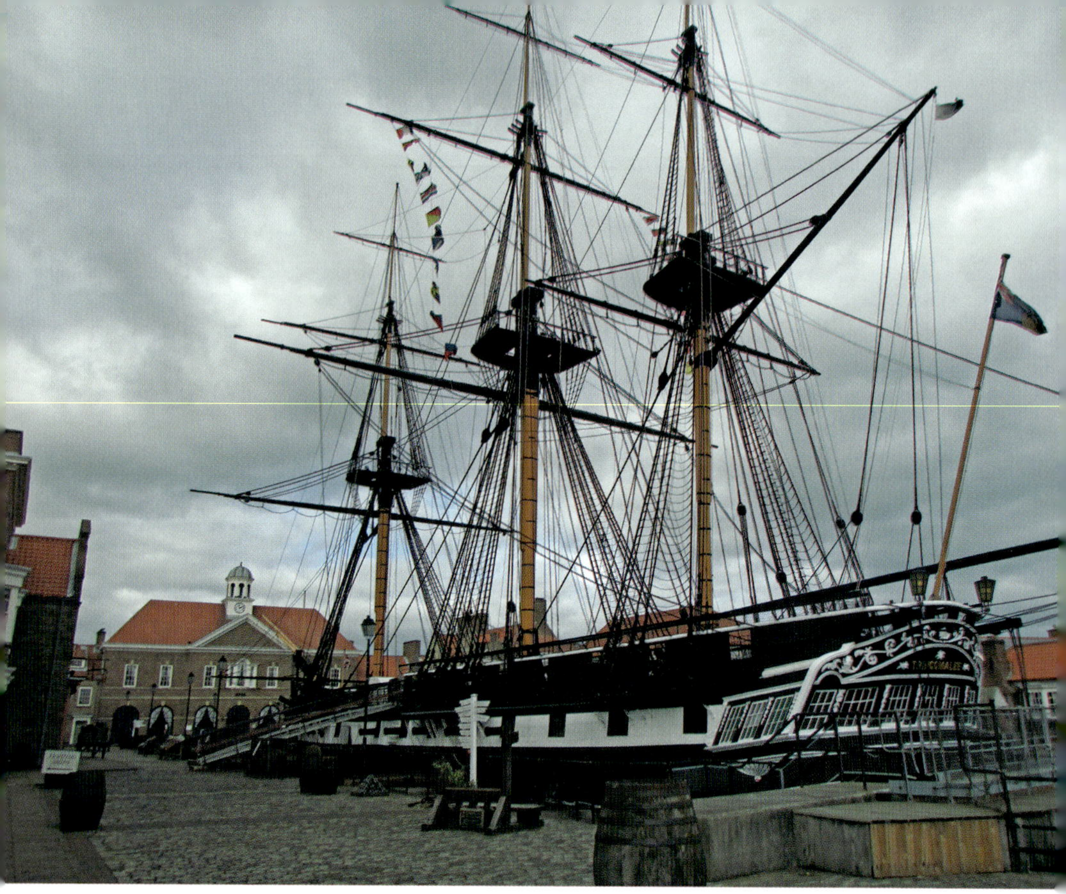

Above and below: National Museum of the Royal Navy, Hartlepool – HMS *Trincomalee*, a late Georgian frigate built from teak in an Indian Yard. (Reading Tom and Martyn Wright)

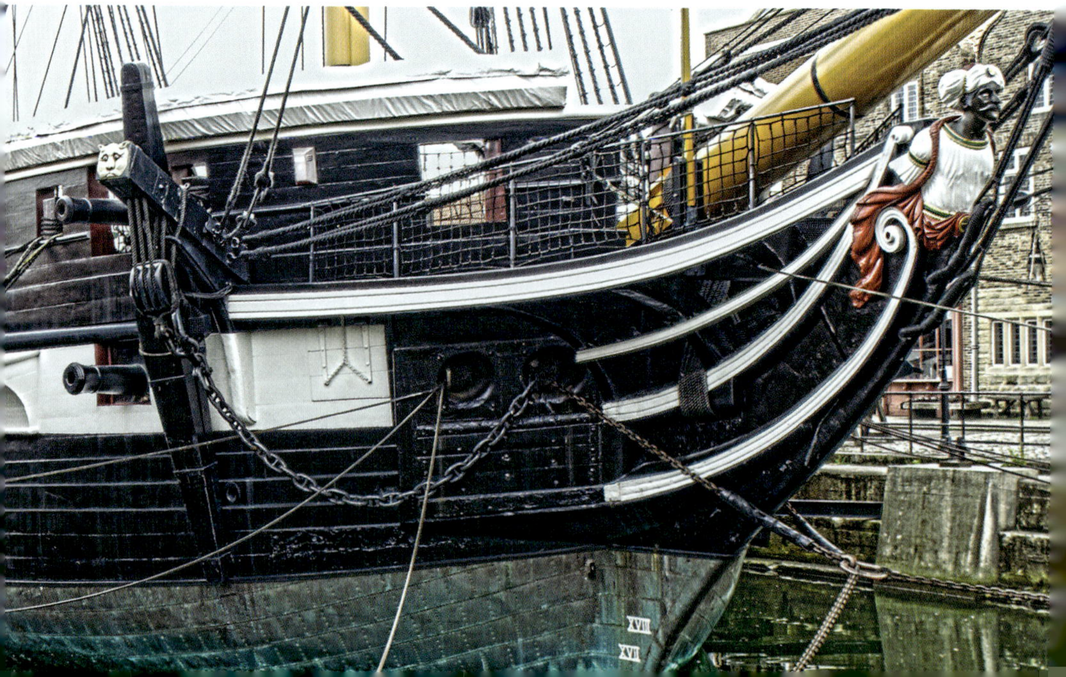

cutting off any escape route. Monmouth marched into Axminster later on 15 June where he gained more recruits, bringing his army total close to 6,000. Among these was the writer Daniel Defoe but, and crucially, support from local landowners failed to materialise. Monmouth might be popular but there was no widespread insurrection.

When, on 19 June, he entered Taunton, he was very enthusiastically received, houses decked out with flowers. His adherents wore symbolic green boughs in their hats. Although the Corporation and the more influential people of Taunton were unwilling to commit, poorer people flocked to his banner. On 20 June, the Duke was proclaimed as 'King' at the Market Cross at Taunton. Alarmed by his progress, King James had offered a reward of £5,000, dead or alive.

Meanwhile, Churchill was closing in, reaching Chard on 19 June. Feversham had secured Bristol, or at least believed he had… Monmouth had continued north, entering Bridgwater on 20 June to a rapturous welcome, taking up residence in the castle. On 21 June, he advanced to Glastonbury, arriving on the 22nd and on to Shepton Mallet by the 23rd. Now, the good weather the rebels had so far enjoyed turned to heavy rain, greatly slowing them and depressing morale. Monmouth's small army spent the night of 24 June at Pensford, where Royalist cavalry under Colonel Oglethorpe launched a sortie to win control of Keynsham, an important crossing point over the River Avon. Inconclusive as this little action was, the continued poor performance of his cavalry rattled the Duke, whose self-confidence, tenuous at best, began to drain away.

Monmouth was planning to seize Bristol. The city was occupied by royalists under Henry Somerset, 1st Duke of Beaufort. This was the decisive moment: had he held his nerve and launched an assault, the campaign could quite possibly have really taken off. Instead, baulking at the challenge, he advanced towards Bath, which refused entry. The city was garrisoned by strong Royalist forces, who, contemptuously and contrary to the rules of war, shot down his messenger. Monmouth then set up his camp at Philips Norton where he was attacked by Feversham's forces. On 27 June, Monmouth's half-brother, Henry FitzRoy, Duke of Grafton (son of Charles II and Barbara Palmer) and serving King James, led a contingent into Philips Norton but was surrounded by the rebels and had to hack through hedges to escape. Churchill came to their rescue and the rebels withdrew. But they had scored a tactical success, seeing off their attackers and inflicting losses.

This may have temporarily buoyed flagging morale, but the writing was still on the wall.

The Duke of Monmouth marched his men through the night to Frome, arriving on 28 June. Morale was badly dented as news of the failure of rebellion in Scotland filtered through, and many deserted. Argyll had successfully landed in Scotland where he was joined by Sir Duncan Campbell, but their attempt soon foundered. Argyll had been captured and executed at Edinburgh by 30 June.

Now heading for Warminster, Monmouth got as far as Trowbridge before being cut off by royalist forces. He then marched back towards Somerset through Shepton Mallet, arriving in Wells on 1 July. Feversham's cautious plan was to contain rebel forces in the southwest until the remainder of his own troops arrived, but the King was insisting on action. Monmouth was pushed back through Shepton Mallet onto the Somerset Levels. He straggled back to Bridgwater on 3 July, where he received a less than enthusiastic welcome. Citizens were wary of seeing their town suffer another sapping siege (they had endured two during the Civil Wars).

The decisive Battle of Sedgemoor was fought on 6 July, the last major pitched battle on open ground between two military forces fought on English soil (the skirmish at Clifton in December 1745 was pretty minor). Feversham camped on the edge of Sedgemoor at the village of Westonzoyland. Monmouth watched the royal army from the tower of the Church of St Mary at Bridgwater before deciding, on local advice and after holding a Council of War, to attack. He led his forces out of Bridgwater at about 22:00 hours, planning a surprise night assault. The rebels were guided quietly out of the town by Richard Godfrey.

As the first rebels attempted to cross a drainage ditch, the Langmoor Rhine, which was about eight feet (2.4 metres) wide, and a mile distant from Feversham's encampment, they were heard by a royalist patrol, and the vital element of surprise was lost. The Royal Army rapidly deployed as flaky Lord Grey of Wark (whose conduct throughout had been tainted by apparent cowardice), advanced with Monmouth's meagre cavalry. The rebels were outflanked and enfiladed, hundreds cut down, many in what is now known as Moor Drove Rhine. Fatalities from Monmouth's army have been estimated as being anywhere between 700 and 2,700, while royalist casualties (fatalities) were under 30 with, say, 70 more wounded.

Monmouth fled, accompanied by Lord Grey. He was hoping to reach the coast at Poole. On arrival at an inn in Woodyates, they parted company. Abandoning his horse, Monmouth continued across country alone but was captured after two days of skulking, sheltering in a ditch dressed in a tattered shepherd's clothes. His rebellion was over; it had lasted just six weeks. Parliament naturally passed an Act of Attainder against Monmouth (stripping him of all titles and declaring him a traitor). Monmouth wrote pleadingly to Charles II's queen, kind-hearted Catherine of Braganza and she generously tried to intercede with her brother-in-law for her stepson's life, but James refused to be swayed. Monmouth was executed on 15 July 1685, on Tower Hill.

As he mounted the scaffold at Tower Hill, Monmouth handed a gold toothpick-case to one of the witnesses, asking him to give it back to his mistress, Lady Henrietta. He tipped the headsman with the words, 'Here are six guineas for you and do not hack me as you did my Lord Russell. I have heard you struck him four or five times; if you strike me twice, I cannot promise you not to stir.' Unfortunately, the executioner took five strokes to accomplish the job and even then, the head was not completely severed and had to be separated with a knife. Monmouth's captured followers suffered dreadfully at the hands of the notorious Judge Jeffreys at the Bloody Assize. Yet within two years, James himself was in exile, betrayed, amongst others, by Churchill.

Sedgemoor Battlefield is a good example of how we fail in our duty to commemorate. Aside from the memorial, there's nothing there, the great ditches have gone, and the modern environs are deadly dull. You need a powerful imagination to put yourself back there but it's well worth the effort. The small Bridgwater Blake Museum is a lovely local collection with a gloriously old-fashioned feel, a must if you're going to the battlefield [56].

A digression into movies as part of commemoration, the Ballachulish Hotel:

John: *Zulu*, nearly the first war movie I ever saw, Haymarket Cinema 1964, clips are still shown in the Royal Armouries Galleries, amazing it's still allowed.

Graham: Especially given that John Prebble, the author, brilliant as he was, leant very pointedly to the left, as did the director and crew as I understand, supposed to be an anti-War/anti-Imperialist blast.

John: Glad to see they cocked that up so royally, probably the most jingoistic movie of all time. Before that I'd just seen all those grand old '50s films, *Sink the Bismarck*, *Battle of the River Plate et al.*, very worthy, very stiff upper lips and pretty dull. First one I saw in the cinema was *Longest Day*, and I loved it, had every American and British actor in you'd ever heard of, some of whom, like Richard Todd, had actually been there.

Graham: He played Major Howard at Pegasus if I remember right?

John: He did, with Peter Lawford as Lovat, I think in real life he was a druggie and a dipso, Lawford that is, but had the good sense to marry a Kennedy so was promoted to presidential pimp. Not necessarily a bad career move.

Graham: I remember *Tobruk* with Rock Hudson and George Peppard, a version of Operation *Agreement*, inevitably led by the Americans with a host of British character actors. Action was great, even if all the tanks were significantly post-war.

John: I was infatuated with *The Blue Max*, loved the aeroplanes and those Uhlan-type uniforms. I'm so sad I can identify most Great War fighters or scouts as they were called then, had all the Airfix models.

Graham: I'm sure none of the other air war films set in WWI could compare, *Aces High* and *Flyboys*, seriously banal by comparison. Now, I gather there's a remake of *Dambusters* on the cards. That would be a travesty, can't imagine anyone but Richard Todd saying, 'I've got letters to write.'

John: Just don't mention the dog...

4

South of England

To borrow from L. P. Hartley: 'The south is a different country, they do things differently there.' That's a northerner's view mind you, but I guess most in the south feel the same way about the north. Still, those outwardly gentler counties have been fought over just as much as our northern hills and the traces are everywhere. From the White Horse of Uffington to the Tower of London, Dover Castle and a great wealth of military museums and monuments. Far too many to profile here. Those I have selected are based on personal choice and nothing more, and the choice is not intended to demean any of the rest.

Gurkha Museum, Winchester

> A Gurkha officer should be comely, sprightly and above all else, confident in his own dress and bearing. He should, where possible, eat a small piece of meat each morning with molasses and beans. He should air himself gracefully when under fire and never place himself in a position of difficulty when being shot at. He should eat his meals comfortably and ahead of his soldiers, for it is he who is more important tactically on the battlefield and therefore he who should be well nourished. His hair should be well groomed and if possible, he should adorn a moustache or similar facial adornment.
>
> When speaking to his soldiers he should appear unnerved and aloof and give direction without in any way involving himself personally in the execution of arduous or un-officer like duties. He should smoke thin panatelas except when in the company of ladies, where he should take only a small gin mixed with lemon tea. He should be an ardent and

> erudite gentleman and woo the ladies both in the formal environment and in the bedroom where he should excel himself beyond the ordinary soldier with his virulent lovemaking prowess. These I say to you are the qualities of an officer that set him apart from the lay person and the common soldier.
>
> Lt Gen Hubert Worthington[1]

In bandit country. There was bother with dacoits,[2] lots of it in fact. On 2 October 1824, the resident magistrate at Saharanpur[3] requested military assistance. In addition to robbing convoys, a large gang of outlaws had seized an old fort at Koonja in the south-eastern part of his district, at least eight hundred of them under two gang-leader brothers, Kulwa and Bhoora – the Jesse and Frank James of the locality. They had instigated a reign of terror, and something had to be done. Frederick Young with two companies drawn from the Sirmoor Battalion was just the man to do it.

When the detachment had marched as far as Saharanpur, they picked up a unit of police and another hundred and fifty Gurkhas with a couple of willing volunteers: Lieutenant Debude, a sapper and young botanist turned adventurer, with Doctor John Forbes Royle.[4] After a forced march of 36 miles in only twelve hours, they arrived in sight of the fort. The dacoits, well aware of their approach, were waiting with an extended skirmish line deployed along the frontage of the village. The fort with its high walls of mud brick occupied the northern part of the settlement.

Greeted by a ragged volley, Young ordered his men to dump their packs and advance in battle order. The bandits got that message pretty quick and scampered back into their redoubt. So far, so good, but the Gurkhas had no big guns or siege equipment. The parapets were manned by sharpshooters armed with long-barrelled matchlocks, old, slow, but deadly enough in the right hands. Young ordered a handy tree cut down, a bit like Kirk Douglas in *The Vikings* – kukris made good pioneer tools. Slung on ropes, the timber beam made an effective ad hoc ram. Frederick Shore, the superintendent from Dehra Dun[5] and Lieutenant Debude guided the battering with Gurkhas swinging the ropes.[6]

As their muskets gave covering fire, the assault team went into action, hammering a wedge into the heavy, iron-bound gates. Only two men could get in at once. Young was first through, next a brace of Gurkhas, then Shore with the rest of the attack group. Straight off, it was hand-to-hand. Henty at his most adventurous could not have outdone the reality. Shore cut down one bandit about to slash Young's neck. An instant later, the superintendent was facing a huge bandit, intent on carving him into very small pieces. Young, opened fire with his double-barrelled Manton pistol. The first chamber misfired but, and that's the advantages of having two, the second one dropped the dacoit, though Shore still took a nasty cut.[7]

Piling in through the breach, Gurkhas were stabbing with their long triangular bayonets and hacking with kukris, spraying bright arterial blood. The rest were, by now, seeking a safer postion elsewhere, leaping from the high walls. One, it is said, begged for his life, stuffing grass into his mouth, a gross humiliation for any Hindu, so profound was his fear. It didn't do him any good.[8] Leaving bayonet squads to flush out any survivors inside the fort, Young took more to beat the dense fields of waving sugar cane, killing another score of dacoits.

Over a hundred and fifty were dead in the wreckage, thirty more in chains. Kulwa the bandit-king was dethroned, dead, though his brother got away, at least long enough to die of his wounds. The Gurkhas lost five men killed and 33 injured. Soon, Kulwa's head was swinging in the breeze in a cage above the gateway of the gaol at Dehra Dun as a warning to others. A pair of small cannon lifted from the fort at Koonja became trophies of 2nd Gurkhas, last seen guarding the barracks in Hong Kong. Young included in his after-action report a comment that none of the many women inside the bandit fortress had been molested.[9]

It should be remembered that in an age when many British recruits were dragged from the dregs of industrial slums or from agricultural recession, for whom going for a soldier was more last resort than vocational choice, the Gurkha was a different breed altogether. Sir Francis Tuker called him honest and incorruptible. Honesty wasn't a characteristic necessarily in evidence with Tommy. A later British commentator, Patrick Davis, who served with the Gurkhas, wrote in 1970: 'In a world where grand and petty larceny are habitual, distortion of truth and self-deceit commonplace, the honesty of the Gurkha seemed unique.'[10]

They also have a constant sense of humour, less of Tommy's habitual grousing. This tends to often be macabre, but military humour often is:

> As the use of paratroops was increasing during the Malaya confrontation in the 1950s, a British colonel asked the leader of a platoon of Gurkha

> if they would be prepared to jump from a C130. Somewhat to the colonel's surprise, the Gurkha sergeant requested a day to talk it over with his men. The next day, the Gurkha duly reported that they would do it, but only over marshy ground with the aircraft flying at no more than 100ft. 'But at a hundred feet the parachutes wouldn't work,' the colonel explained. The Gurkha replied, 'Parachutes? No-one mentioned parachutes!'[11]

I'm with my son-in-law Major Sam Meadows in Winchester to talk about some of his early experiences in Afghanistan. First tour and he was a lieutenant serving with 2 RGR during Herrick 10. Trained as an archaeologist, like Lawrence of Arabia he had made the transition to soldiering. Afghanistan would be his baptism of fire. He kept a diary:

'6th April 2009: As I sat in bed at 22.30, I've finished preparing my kit and equipment for my first operational patrol in Afghanistan which starts tomorrow morning, and I have feelings both of trepidation and great excitement for what lies ahead then, when I will be acting as a rifleman and in nine days' time, when I will be in command of my men with little or no guidance, which in itself, is a slightly daunting thought.

8th April 2009: Another day of familiarization patrols today into Kartolagen Chirmitiri and Mukhtar districts of Lashkar Gar – these areas are narrow urban spaces with a maze of roads and junctions that do not relate very well to the map! As the district centre becomes more peaceful, new people are building in the open areas. The patrol was led by Sergeant Hemanta Limbu and was controlled very well from the start. Similarly, the relationship he had developed with the ANP was excellent and he used his grasp of Urdu to facilitate this: we visited various ANP CPs, where the local people looked generally happy to see us and morale seemed good.

Interestingly, a socio-political problem is currently playing out in Lashkar Gar at the central police CP in the bazaar area of the city. A group known as the Shah Korez militia who 'belong' to the governor of Musa Qala have been staying in the city causing various minor theft and extortion problems. Today, the leader of this militia's brother had been taxed by the ANP at an illegal VCP, so he sent in the militia, and for whatever reason a firefight broke out that resulted in five of the militia being injured and their leader arrested. The ANP training programme paying off here – no casualty but despite the ISAF presence tribal culture still obtains – feudal and dominated by narcotics smugglers and warlords who control the future.

Meanwhile, there was an uneasy truce as Terry had condescended not to attack ISAF during the all-important poppy harvest, this was purely on economic grounds and hostilities would instantly be resumed once the indispensable crop was safely gathered in. In the stillness of the night, the quiet of the dark has been the soldier's constant companion since Trojans watched Greek campfires burning beyond their walls.

12th April 2009: It is now about 00.30 and I've noticed how quiet it is in the location, there is no traffic at night, just a few crickets and the occasional vocal Afghan fighting dogs, so quiet that the strong desert breeze can be heard whistling around the compound and the Afghan national flag ripples with the prevailing wind. The almost full moon casts an incredible amount of light, enhanced by night vision goggles – the night is our friend and the enemy's foe.

Afghanistan is crowded with its own history, a saga of conquests, of rise and fall, rugged crests and sweeping valleys, high desert and fertile alluvial plains, ospreys swirling in the eddying currents. The enemy is largely invisible, old mud-brick forts seeking to dominate timeless settlements crowd hilltops and look over the teeming maze of compounds. But this is the enemy's country, and he really knows no other. He is at home here and his fathers fought the Russians over the same ground and ultimately beat them. Contacts were heard over the radio – '3 km to our west ... the HMG firing back at the enemy.' The Gurkhas' only casualty during this phase was a severe injury caused by 'blue on blue' – one soldier inadvertently shot another at close range with a 9 mm pistol, one of those daft accidents that so dent morale.

23rd–15th April 2009: Operation Pishet Sataal. We were tasked to escort the Shar Karez Militia, a group of essentially bandits and mercenaries to a desert laager south of Musa Quala. This militia had been formed with President Karzai's permission by Mullah Salamand, loyal to Mohammed Nabi, a small man with a fantastic beard and gentle manner who had been a senior mujahedeen commander against the Russians. The militia was being deployed to protect the provincial governor, an ex-Taliban with a big price on his head. The journey covered 100 km of tarmac and 70 more over desert terrain. I had to check out over a hundred culverts for IEDs.

As soon at the convoy got as far as Gereshk, the mercenaries reverted to type and began pillaging, echoes of P.C. Wren as the vehicles moved off tarmac onto sand; one broke down and had to be secured by a defensive cordon, wagon laager style. No baddies were seen but everyone knew they were watching. That night in the desert was amazing, the stars clearer than I'd ever seen and the threat from the enemy made it all very exciting. On the way back, as we passed a jumble of local traffic one man aged 25–35 stared me down and at that moment, I knew he was the enemy. I was staring into his eyes and there was nothing I could do …'

It's a cliché, but from Troy to Helmand, the infantryman's role has in so many ways changed so very little. Helmand was a war that the great-grandfathers of Sam's Gurkha soldiers would have recognised in an instant. I've spent a lot of time in the wonderful Gurkha Museum researching other books and it's a great experience, staff and volunteers are in many cases retired Gurkhas, unfailingly polite and courteous.

Royal Green Jackets (Rifles) Museum, Winchester

In 2015, I was speaking in a local library on the battle of Waterloo, the bicentenary. We tour guides who had survived the epic chaos of the grand re-enactment on the battlefield had come to appreciate how Napoleon must have felt by late evening. In the library and just before I was due to speak a couple who attended most of my local talks brought in their three times great grandfather's relics from the fight in which he'd served (alongside Ensign Ewart as it happens). They had his medal, his papers and joy of all joys, his 1796 heavy cavalry pattern sword. Not often you see one in pristine nick with full provenance.

Why is this weapon so popular today? Well, that's more to do with infantry, riflemen in fact, than cavalry. Bernard Cornwell revolutionised popular historical fiction when his Lieutenant Richard Sharpe crashed onto the shelves and on-screen with Sean Bean in the title role. He carries a heavy cavalry sword which is regularly drawn to deal out death to deserving Frenchmen. He customises the blade to make it less cumbersome and confers upon it a totemic status it never really had during its years of service as, frankly, it's a heavy cleaver that's awkward to use.

In Sharpe's day (had he existed), rifles on the battlefield were something of an innovation. The idea of rifling the barrel of an otherwise smooth-bore musket wasn't new. The problem was they were laborious to load, increasingly so as the barrel fouled with powder residue after each round discharged. Nonetheless, they were far more accurate over much greater distances, you'd be unlucky to be hit by a Brown Bess say much over 80 yards. During the American Revolution battles British redcoats suffered grievously from the accurate fire of American riflemen deploying the elegant and highly lethal Pennsylvania rifle, so ably wielded by other fictional heroes such as Hawkeye and Chingachgook.

Sir John Moore[12] was one of those who took notice and was instrumental in the forming of the 95th Rifles (later the Rifle Brigade), raised by Colonel Coote Manningham in January 1800. These men looked very different, they thought and manoeuvred very differently, and their Baker Rifles[13] changed battlefield dynamics. Their creation was timely, as Britain was about to take France on across the Iberian Peninsula (1808–1814). As part of the soon to be famous Light Division under 'Black Bob' Craufurd, the Rifles would fight in every significant battle, earning a formidable reputation – 'those Green Fellows' or 'grasshoppers' as the French would call them.[14]

One of the regiment's more colourful characters was an Irishman, Thomas Plunkett (1785–1839), who took part in the Battles of Cacabelos during Moore's epic and dreadful retreat to Corunna in the harsh Galician winter of 1809. The Rifles were forming a rearguard with superior French forces pressing hard. Plunkett, seeing the gorgeously uniformed French commander some distance off – safely he'd assume out of effective range, dashed forward to try his luck. He aimed over crossed ankles from a sitting position and squeezed off a round. The distance was likely 300 yards (275 metres), a phenomenal range at the time, but General de Brigade Auguste-Marie-Francois Colbert fell dead from his saddle. As the stricken officer's ADC, Latour Maubourg dashed up to help, Plunkett dropped him as well, in case anyone thought his first shot was just good luck. He then had to scarper pretty smartly back to British lines pursued by vengeful cavalry, but the French attack was effectively stalled. The rifle had arrived.[15]

The Western Desert, 1942. 'Woodcock' and 'Snipe' were a brace of key actions that took place in the closing days of October during the opening phase, Operation *Lightfoot*, of the 2nd battle of El Alamein, which Montgomery intended would finally end the pattern of exhausting advance and retreat of the Desert War. The battle was immensely hard fought, a desperate attritional struggle that ended with the rout of Rommel's forces.

Monty's orders to General Lumsden, commanding 10 Corps, provided that 7th Motor Brigade should put in a night attack. 10 Corps artillery, with support from that of 30 Corps, would lay down a barrage. Kidney Ridge was so named as it is said to resemble the shape of a kidney bean. This was about all that was agreed, there were differing opinions as to its exact location or even whether it was a ridge at all. It appeared to lie at the extreme right flank of Wimberley's Division. Roughly a mile north-west and the same distance south were two enemy positions labelled 'Woodcock' and 'Snipe'.

It was decided these should be eliminated, but there was uncertainty as to exactly where they might be found. The plan for 26 October was that 2nd KRRC was to aim for 'Woodcock' whilst 2nd Rifle Brigade marched on 'Snipe'. Once the infantry had attained their objectives, leaving their start lines at 23.00 hrs, then at dawn on the 27th, 2nd Armoured brigade would deploy north of the former position and 24th Armoured Brigade would move to the south of the latter.

During the afternoon of 26 October, the corps and divisional commanders arrived to settle on the attack. 'At this conference numbers of officers were gathered on the "sheltered" side of a Churchill tank when a passing sergeant of the Royal Corps of Signals let out the most weird and realistic noise of AP shot travelling at speed. Everybody wondered what on earth had happened and then burst out laughing!'[16]

In a fog of dust-blanketed dark, 2nd KRRC advanced from their start lines along Moon track, though without any moonlight. The dispute as to what lay where hindered any effective barrage and the battalion, which also ran into opposition, progressed slowly, to the extent that by dawn they were still south of their objective. The battalion commander judged the ground there untenable and therefore withdrew to a more suitable position, about 1,000 yards north of the ridge and the same distance short of 'Woodcock'.[17]

To the south of 2nd KRRC, 2nd Rifle Brigade encountered similar difficulties, the pall of dust obscured start line and objective alike, compounded by the fact the guns were pounding away to the north. Their commanding officer, Lieutenant-Colonel V.B. 'Vic' Turner, decided to march toward the sound of the guns and lightning flashes of exploding rounds, swinging around and continuing for two relatively uneventful miles. The motor platoons advanced on foot, shielded by a screen of Universal Carriers.[18] Having reached what they assumed was their objective, they dug in.

> The following morning there was a certain amount of controversy about our position, the Jocks having placed 'Stirling' and 'Aberdeen' in places with which we did not agree. I was sent over to one of these battalions on our left to find out (a) where they thought they were, and (b) if they could assist us in any way. The Jocks were right in their map-reading.[19]

Attaining the objective was the signal for the support units to move forward, 16 x 6-pdrs of 'S' company and 11 more from 239 Battery 76th Anti-Tank regiment RA. The medical team was left at the start. Bad ground disabled several vehicles and only 13 of the battalion guns and 6 of the RA got through, bringing vital ammunition for the MGs and much needed water.[20] In such confusion, the artillery officer from 2nd RHA was mislaid, a vital presence, one that would soon be sorely missed. A posse of carriers had investigated rising ground southwest of the battalion. Here they blundered into an Italian leaguer, with a German unit clearly in sight. Undisturbed by these intruders, the Axis columns began a leisurely drive westward at dawn, only to be raked by British guns; 14 'kills' were claimed.[21]

> I went in as a Bren gunner with a foot platoon, loaded with spare mags and hand grenades. There was no number two on the Bren. The other lads were carrying spare mags. Our objective was about 2,000 to 3,000 yards ahead. The Bren carriers were forward of us. We were on our start line at about 23.00 hrs: there wasn't much moon, and we stood waiting. As we waited for the word to advance, the thoughts of how and what's going to happen went through my mind: we were all the same, nerves taut, parched lips and throat. Then suddenly we were off, with the artillery giving covering fire and all hell was let loose.[22]

With battalion HQ sited in an old German dugout and in the centre of an oval-shaped perimeter, 'A' company with 10 guns had the northeast quadrant, 'C' company with four of the 6-pounders held the west and 'B' with half a dozen guns the southeast. Meanwhile, 24th Armoured Brigade had been slow in getting underway. Uncertainty compounded by exhaustion and the swirling pall of cloying dust meant lead units were not motoring forward until 07.30 hrs. Unhelpfully, their first contact was with Turner's men, whom they commenced shelling! It appears that the British tanks may have mistaken some hulks near the riflemen's trenches as 'live' enemy armour.

Having been appraised of their error both they and the Rifles were soon hotly engaged as two dozen or so Axis tanks were observed forming up on the higher ground south-west. 47th Royal Tanks were the lead formation and swiftly saw their number reduced as the battle heated up. Six vehicles were knocked out in rapid succession. Despite gallant attempts, the British armour could make no headway and were withdrawn behind higher ground in rear, half a mile distant from and east of the infantry. Both Italian and German tanks attacked, over a dozen of the former and twice as many of the latter, who were seeking out 24th Armoured brigade. These were engaged and eight knocked out.[23] The riflemen were holding on but reduced to a quartet of guns per company area and low on ammunition.

As the Axis tanks withdrew, their guns deluged the exposed position. Then the enemy tanks tried again, nine Italian vehicles sweeping in from the south. Sergeant Calistan commanded the only gun which could bear upon these swift moving targets, and he found himself with the battalion commander as loader with Lieutenant Toms as no.1. Calistan coolly held his fire, directed by Turner, and destroyed six of the enemy in quick succession. Less than a handful of live shells remained. Toms dashed to his jeep and returned loaded up, with bullets humming around like angry bees as the three surviving Italians stormed ahead, regardless of their losses. The jeep was hit and burst into flames whilst the two officers battled to save its vital cargo. Turner was seriously wounded and prevailed upon to take cover. Calistan then calmly disposed of the surviving enemy. He was the one man unwounded of the three.[24]

> About five in the evening another counterattack came in. by the sound there wasn't so much fire coming from our anti-tank guns. Whether they had been knocked out or were just short of ammunition, we didn't know. We found out alter it was a combination of both. Again, the remaining anti-tank guns did a wonderful job. Round about this time I glanced behind me, and to the right rear was a Panzer III. I told the lads to keep their heads down. This was our lot. Out front there was quite a movement of vehicles, but our thoughts were on the enemy tank at our rear.

> We knew that there was a 6-pdr anti-tank gun to our right and facing the front, but there was no firing from it. Presuming the crew were knocked out or were out of ammo, if one of us were to crawl to it, perhaps we could have a go at the panzer. We did a John Wayne, raising a tin hat above our slit trench. We did this and the tank gave it a burst from his MG. it would have been suicide for any of us to get out and have a go. All of a sudden there was an explosion over our position. We all thought that his was the tank firing at us with his 75. Deeper we crouched in our position. Then there was a pall of smoke and the smell of burning rubber. We looked over the edge of our trench ... the panzer was burning,[25]

Whilst the riflemen's battle raged, the exact whereabouts of their comrades in 2nd KRRC remained unclear and it was late morning before the Queen's Bays made contact. 2nd Armoured had moved forward that morning, initially believing opposition to be weak. They were soon disabused. Axis forces were gearing up for a counterstroke and 21st Panzer were about to be committed. Point 33, which was the true location of 'Woodcock', remained in German hands with 2nd Armoured strung out in a south-westerly line falling just short of the east flank of Kidney Ridge. 'Kills' claimed included a dozen German, two Italian tanks and a quartet of 88s. By 13.30 hrs the Queen's Bays were ordered against Point 33, beefed up by a trio of heavier Churchill tanks.[26] This attack failed, one of the Churchills was disabled by enemy fire, the others by mechanical difficulties, and the survivors withdrew. 10th Hussars were engaged in a sharp fight, claiming three enemy tanks destroyed. Then for a couple of hours, relative calm descended.

> After reporting to HQ, a visit was paid to one of companies of the 'Woodcock' battalion with whom we had a troop; in addition, the 'Deacons'[27] had been put into action to their left rear. Having suffered some casualties, they were withdrawn, having achieved nothing except capturing a few Boche who had remained hidden underground.[28]

Rommel was not idle. Having brought 21st Panzer into the ring, he planned a significant blow towards Kidney Ridge supported by the remnant of 15th Panzer, Littorio and elements of Ariete. In the north, 90th Light Division with cover from JU87s was to strike at the Australians. It was not until 16.00 hrs that the attack developed, forming up had again been hindered by the persistent attentions of

Desert Air Force. As the Axis armour rolled forwards, attempting to pass south of 'Woodcock', Fisher's tanks delivered a murderous fire. They then engaged 10th Hussars, obligingly offering the riflemen a most convenient target for their 6-pdrs.

> Over the ridge they trundled, a tight bunch of a dozen first coming into my view and showing black against the skyline, but too far away from us to be able to do much about it. They stopped at about 3,000 yards and two or three K4s [PzKw Mark IV] began to shell the infantry lying around the tanks. All the while we watched them closely through binoculars and, holding our fire, hoped that they would grow bolder and move in closer. This they refused to do, but one of two K3s skirmishing around the others suddenly came within reach of Shermans on our right and in a couple of minutes one of them was smoking furiously and the crew racing for cover, while the other tanks lumbered off up the ridge out of range.[29]

Between them the two British units likely accounted for a score of enemy tanks, though the mass advancing from the south continued to rain shells on 2nd Rifles battered positions. The counterstroke had been seen off and Rommel was on the point of despair: 'It was later learnt that a determined attack had been launched by enemy tanks on the "Snipe" battalion. A very fine show was put up by the A/T guns of that unit and thirty-seven tanks were destroyed. Part of one of our batteries had the honour of taking part in this action but had a minor role as all the plum positions were occupied by infantry guns.'[30]

Having fought magnificently, the riflemen rather hoped they might now be relieved. That was the intention, but neither carriers nor men appeared. The wounded had been evacuated and remaining survivors later withdrew by companies, and by 23.15 hrs battalion HQ had pulled back. This was fortuitous, as Allied shelling had stirred up a hornet's nest in the adjacent Axis leaguer with the occupants, thus rudely evicted. seeking a new home in the 'Snipe' position. Even if the riflemen's claimed tally of 76 enemy armoured vehicles destroyed veered towards the optimistic, a later, objective, assessment still counted 37, a formidable score.

In 1958, the Ox and Bucks Light Infantry amalgamated with the King's Royal Rifle Corps and the Rifle Brigade to form the Green

Jackets. Eight years after that the Royal Green Jackets were formed by the further amalgamation of the three separate regiments forming the Brigade: 1st Green Jackets (43rd and 52nd), 2nd Green Jackets (KRRC) and the Rifle Brigade. Finally, in November 2005, the Green Jackets underwent yet another coming together when they, with the Light Infantry, Devonshire and Dorset Light Infantry, Royal Gloucestershire, Berkshire and Wiltshire Light Infantry, were stirred into a single brew – The Rifles.

Today the Green Jackets Museum is located within Peninsula Barracks Winchester, next door to the Gurkha Museum. These imposing barrack piles have a long association with the Rifles. In the beginning was William the Conqueror and yet another castle, refreshed by Henry III then duffed up (like so many) by Cromwell during the Civil Wars and slighted in 1651. Charles II devised a neat riposte when he decided to commission Wren to throw up a vast palatial complex on the site, an English Versailles. No doubt the former Lord Protector would be spinning in his grave. But the Merry Monarch died, the money ran out and the half-built shell was left to rot, aside from providing a handy POW camp for prisoners from both the Seven Years War and the American Revolution.

During the Napoleonic Wars, the site was developed as a full-time barracks and this role continued and was indeed extended during Victoria's long reign. By 1856 there was a rail link to the capital and two years after, both KRRC and the Rifle Brigade were stationed there, a long association already begun and formalised in 1872. It was the Prince of Wales, future King Edward VII, who laid the foundation stone for a new and rather grand complex, opened in 1904 and built on the epic scale, true to Wren's original vision. Inside the excellent museum, formally opened by her Late Majesty Elizabeth II in 1989, there's a first-class Waterloo section, unveiled in time for the bicentenary.

D-Day Museum, Portsmouth

> All of Amsterdam, all of Holland, in fact the entire western coast of Europe all the way down to Spain, are talking about the invasion day and night, debating, making bets and – hoping.
>
> Anne Frank, diary entry, 22 May 1944

> Soldiers, Sailors and Airmen of the Allied Expeditionary Force; you are about to embark upon the Great Crusade, toward which we have

> striven these many months. The eyes of the world are upon you. The hopes and prayers of liberty-loving people everywhere march with you. In company with our brave Allies and brothers-in-arms on other Fronts, you will bring about the destruction of the German war machine, the elimination of Nazi tyranny over the oppressed peoples of Europe, and security for ourselves in a free world.
>
> Eisenhower's Order of the Day

On a visit to Normandy as a guide on a brilliant summer day in June 2018, I'd got the group to the wide expanse of Utah Beach, a long sweep of perfect sand, staked out along the tide line to harvest the oyster crop. Aside from the squat, huddled remains of German blockhouses abandoned in the dunes, there was nothing to suggest this tranquil, gilded shore had ever been a battlefield. As we scaled the concrete roofs, a pair of charioteers propelling racing carts along the beach dashed by, like some scene from an altogether earlier age. And that's the trick with Normandy, trying to conjure up the awfulness of battle, which wasted so many towns, villages and farms, which brought the full, crushing power of industrial war to this beautiful landscape.

As Montgomery pointed out, the Britisher was not, in the main, the member of a professional *corps d'élite* as their adversaries in the *Wehrmacht* might claim to be but essentially a citizen under arms. In the main, these young soldiers were a conscript army, dragged from occupations and professions that had nothing to do with warfighting. The experience would be challenging, infinitely tedious, often disagreeable and occasionally terrifying. The task facing Tommy in spring 1944 was to break into Hitler's Fortress Europe, *Festung Europa*, perhaps one of the most daunting challenges any army could possibly face. Fritz had the job of stopping him.

And these young men (Tommy included not just Brits but Free French, Poles, Czechs, Belgians, Danes, Dutch and Norwegians), our fathers and grandfathers, did it, they won. The world post-1945 may not quite have been Churchill's 'broad, sunlit uplands' but democracy and humanity survived. The freedoms and liberties we take for granted today are their legacy. In my own *D-Day – the British Beaches*[31] I focused on the experiences of ordinary men and women who were caught up in these momentous events rather than on the grand strategy. There are many excellent books which do, particularly in my opinion the great John Keegan's *Six Armies in Normandy*[32] and David Howarth's *Dawn of D-Day.*[33].

Over the last forty years I've spent many hundreds of hours with tour groups of all ages in Normandy, the British, Canadian, and American beaches, Pegasus Bridge, Merville Battery et al. So, I've walked, crept, and crawled through Atlantic wall defences throughout Normandy and Brittany, handled and fired most of the small arms and made a career out of talking to veterans.

D-Day was the single greatest military undertaking of all time, and a superbly convoluted deception plan: *Fortitude*, a vast and complex naval armada, *Neptune*, and the amphibious assault, *Overlord*. The job was to breach the mighty walls of Hitler's Fortress Europe and begin the liberation of Germany's victims, France, Belgium and Holland, and then push on through the Siegfried line and into the heart of the Reich, which would be inexorably squeezed from both sides. Stalin would get his second front and the Russian offensive in the east, *Bagration*, would be astonishingly successful. If, as Churchill so memorably described, El Alamein was the end of the beginning, then D-Day was the beginning of the end.

Failure would have been disastrous – it would have taken the Allies years to rebuild their strength sufficiently for another attempt. Hitler could have swung his Normandy garrison eastwards to hold the line there. He could have continued building up his Atlantic Wall defences to the point they would have been truly impregnable. The war might have ended in an armistice with incalculable consequences for mankind. So, the opening airborne assaults on the night of 5 June and the invasion of the beaches next morning could be said to be the decisive hours of World War Two and perhaps of the twentieth century.

It is eighty years since D-day and fewer and fewer veterans survive. I did speak to the son of one; his father had been a beach master on Gold and had never spoken about his experiences. Only when the old man was clearly nearing the end of his long life could he be persuaded to speak of his experiences on 6 June 1944. But all he would say was, 'Well son, it was all very noisy, and I got the very distinct impression we weren't at all welcome.'

As concise histories go, that's pretty hard to beat.

Remy Douin was thirteen and he cycled along the emptied lanes that spring with his artist father Robert. Theirs was an artistic lineage and just because it was wartime that didn't stop Douin *pere* from sporting his artist's beard and pointed moustache nor from wearing a loud, very un-warlike like velvet suit, cravat and elegant wide brimmed fedora. Robert had a badly damaged arm from the last war, but he kept up his vocation of teacher, church artist and restorer. This meant he could travel along the coast inspecting ancient churches pretty much without comment – and church towers commanded an excellent view. What he sketched, cleverly disguised, was the dangerous and forbidden mass of the Atlantic Wall. Each sketch, like a giant mosaic, was a piece in the jigsaw he was building.

He had joined the fledgling resistance at the very outset in November 1940, in his native Caen. Two year later, he was a fully-fledged activist and leader of a local *riseau*. He used his cover and *Cyrano* bravado as disguise in plain sight. It was very dangerous work. A year before D-Day and he knew he was a marked man, the *Geheime Staatspolitzei* and their collaborators were on his tail. Still, though he could have been spirited away to England together with his family, he chose to stay, teaching fine art classes in *l'Ecole des Beaux-Arts* in Caen. The work, his greatest creation, was paramount. Robert Duin was a very brave man.

Remy knew what his father was about; as the pair rode their bicycles though the lovely Norman spring, using the turns of their pedals to calculate distances between bunkers, the net inexorably tightened. At last, he sent Remy south to stay with relatives. On 16 March he walked to the station to meet a contact. Nobody turned up, not until the Milice kicked in his door next day. Robert with many others was dragged to the prison in Caen where he was interrogated and systematically tortured. His left arm, the one he'd damaged last time round, was broken and he was beaten mercilessly before being thrown in the cooler.

When he was at last dumped in an ordinary cell, it was crammed with captured resistors. He managed to confer with another prisoner, Bernard Duval, a teenage carpenter who'd suffered similar torture – ironically, he'd been the one who fitted the heavy timber door to Cell 27, the one he was now in! Robert managed to find some scraps of parcel paper and laboriously crafted letters to his wife and son using a series of innocuous looking dots. These were got out weekly via the laundry basket. On 6 June in the morning, Robert was among 75–80 other resistors who were taken out of their cells and shot in batches against the walls of a blood-slicked courtyard. His body was never found.[34] Remy returned to become a civil servant and businessman, Bernard Duval also survived.

The D-Day Story tells the story of D-Day and the Battle of Normandy. It holds over 10,000 items – preserving, researching, and acquiring objects to share with the public through exhibitions, workshops, and other activities. The museum is fully accredited, meeting nationally agreed standards for UK museums' (https://theddaystory.com/discover/about-us/).

This is a rather bland description from the museum website, which fails to do justice to the place or hint at the enthusiasm and assiduity which has gone into establishing the collection. The D-Day Museum on the sea front in Portsmouth is, literally, a gateway to the Normandy campaign and it's really the best place to begin a battlefield tour.

One of my personal favourites exhibits is an inscribed shell from a 25-pounder that commemorates the service of 321 Infantry Brigade, which, as an element of 50 (Northumbrian) Division, came ashore on Gold Beach. This brigade comprised three regular battalions: 1st Hampshires, 1st Dorsets and 2nd Devons. Originally designated as part of the Malta garrison, they had already seen hard fighting in Sicily and Italy and would see more through northwest Europe once the Battle for Normandy had been won.

In Normandy:

John: Do you wish you'd been here, in the war, a part of all this, I'll confess me freely I do, I never served but hope I'd have had the balls to do something like this.

Graham: This was way beyond anything my generation experienced, the Falklands was as close as anybody I knew ever got while we, the Light Infantry, were on standby only in case we were needed, which would mean something had gone horribly wrong. For the rest of us it was mainly Op *Banner* and Northern Ireland, neither glamour nor glory to be had there.

John: When I bring tour groups here, as you know I do quite often, it's hard for them, looking around at such fabulous scenery and coastline, to understand just how messed up the whole place became and, of course, virtually all of the bocage has now gone.

Graham: Our generation has been spared all that because of this, what happened here in '44. It was the war to end all wars. My lot spent most of our time waiting for WWIII on the plain of North Germany, equipping, training and preparing for it and it never happened, Mutually Assured Destruction saw to that.

John: Until now and Ukraine …

Graham: Until now.

Churchill's War Rooms

I'm old enough to remember Sir Winston Churchill's televised state funeral, in grainy black and white, from 1965. I was of a generation brought up and accustomed to revere the great man, for us perhaps the greatest Briton of all. This image has been challenged by revisionist thinking that recasts Winston as a bullying imperialist. He probably was, and shared his generation's prejudices, but he still saved us from Hitler.

In 1898 Young Winston took part in what is often regarded as the final great cavalry charge in British Military history and like the even more celebrated Charge of the Light Brigade at Balaklava in October 1854, it was not entirely intentional. Churchill was a lieutenant in the Queen's 4th Hussars but rode with four squadrons of the 21st Lancers on the baking morning of 1 September when Lord Kitchener sent his cavalry to reconnoitre Mahdist positions around the dun-coloured smear of Omdurman, across the mighty Nile from Khartoum. Nothing seemed to be moving, but then, over the dry plain, what initially appeared to be an ad hoc thorn bush defence, or zareba. It wasn't, the distant mass began to heave and shift, sprouting the defiant banners of the Khalifa's vast, sprawling army. There would be battle.

Next day battle was indeed joined, and Gordon was duly avenged; Jihadi fervour proved no match for cool deployment of breechloading guns, bolt-action repeating rifles with regular rolling volleys. The Mahdist army was effectively smashed, nearly anyhow. Not much sport for the *arme blanche*, essentially in reserve while firepower did the job. As the broken enemy recoiled, spilling back towards Khartoum, Colonel Martin with the 21st was sent forward to report on the rout and cut up any stragglers who might offer themselves. A thin line of skirmishers offered a spluttering, inconsequential fire from extreme range, say half a mile (800 metres) or so distant. Martin gave the order to wheel right into line and move forward, walk building to a canter. He commanded 320 troopers and one future prime minister.

Yet, things were not what they seemed. As the lancers careered over a shallow ridge leading down into a dried-up watercourse, they were suddenly confronted by a massed phalanx of spearmen, perhaps ten times their number. This wasn't good and the enemy showed no signs of retreating. Men and mounts crashed in a furious, sprawling mêlée. Churchill saw

> ...horses spouting blood, men bleeding from terrible wounds, fish-hook spears stuck right through them, men gasping, crying, expiring. One nearby lieutenant had been wounded in the shoulder and leg, his hand almost severed by a sword strike, and a sergeant's face was cut to pieces ... the whole of his nose, cheeks and lips flapped red bubbles.

Only with extreme difficulty, the lancers manged to extricate themselves and reform. He wrote more soberly to his cousin the Duke of Marlborough on 29 September:

> Now about the war: - Perhaps you will read my letters in the *Morning Post*. They contain the best of my impressions put down as carefully, from the point of view of style, as I can. The military operations, though short, have been interesting. The Battle was a wonderful spectacle. Then I had the good luck to ride through the charge unhurt, indeed untouched – which very few can say. I used a pistol and did not draw my sword. I had no difficulties and felt confident that I should get through if, neither my horse fell, nor I was shot – for I must tell you the ground was execrable and there was the wildest shooting in all directions. Neither of these things happened and such of the enemy as approached me or attacked me I shot; three I think I killed. It is difficult to miss at under a foot's range. The whole thing was a matter of seconds – for as you may have gathered – we burst through their line and formed up the other side. The loss was most severe, one officer and 21 men killed, nine officers and 66 men wounded and 119 horses out of only 320. Such a proportion and such a loss has been sustained by no regiment since the Light Brigade-forty years ago.[35]

For shooting Mahdists, Churchill had purchased a state-of-the-art semi-automatic pistol, the Mauser Construktion 96 or C96 (1896 being the year of design). This was the first pistol to be capable of firing in bursts and it came with a wooden holster that could be attached as a shoulder stock to convert the weapon into a short carbine. The box magazine held ten high velocity 7.63 x 25 mm bullets. Lawrence of Arabia also liked the Mauser. Even though it was never adopted as a service pistol by the German Army, it remained an 'iconic' firearm, wielded by *The Men from Uncle* in my youth and later by Han Solo. Churchill's pistol is held by the IWM in the Cabinet War Rooms below Whitehall.

I recall watching Gary Oldman's remarkable portrait of Churchill in the movie *Darkest Hour*[36] and his brilliantly nuanced climactic delivery of 'We shall fight on the beaches …' which galvanised Parliament on 4 June 1940 and saved Britain, possibly, from a shameful compromise or ignominious capitulation. The scene drew an immense cheer from the cinema audience. Any detractors, if indeed there were any, had to the good sense to stay quiet.

Quite a lot of scenes from the film take place in the urgent claustrophobia of the War Rooms beneath Whitehall. After a long-neglected dormancy these were opened to the public, under the aegis of the IWM, by then PM Margaret Thatcher (a great admirer of her wartime predecessor), on 4 April 1984. This was a providential delivery. Prior to the outbreak of war in 1939, government had dreaded the effect of large-scale, strategic bombing of civilian areas. The cataclysm which engulfed the Basque town of Guernica on 26 April 1937 was a dreadful harbinger; Hitler's Luftwaffe had demonstrated with callous might just what horrors aerial bombardment could unleash.[37]

A year earlier, the Air Ministry was convinced German bombers could kill 200,000 Londoners per week and turn the city into toxic rubble.[38] So prevailing wisdom indicated that government offices should be dispersed through the provinces. In May 1938, Sir Hastings Ismay, Deputy-Secretary to the Imperial Defence Committee, ordered a survey of government buildings in Whitehall to identify a suitable location for an emergency underground command centre. The site chosen was the basement of the New Public Offices building on the corner of Horse Guards Row with Great George Street (it now accommodates HM Treasury).

Work was immediately put in hand. This basement area was fitted with comms and broadcasting gear, soundproofing, ventilation, protected by massive reinforcement. Army, Navy and Air Force agreed on the notion of a Central War Room to accommodate the joint chiefs of staff (not necessarily eliminating inter-service rivalries!). Clearly, the War Cabinet needed full access to its soldier-servants and in the spring of 1939, as tensions fizzed throughout Europe and nobody really believed that 'Peace in our Time' would endure, it was decided that the War Cabinet would be set up within the Central War Room. The whole setup went live on 27 August, only a few days before German forces surged across the Polish border.

These War Rooms became the hub of government, the Map Room was manned 24/7 by officers drawn from all three services, jointly charged with creating the daily strategic summary. While Chamberlain remained at the head of the table, he visited the Cabinet Room on only one occasion. It was Churchill who, when he came there in May 1940, decided this was to be the nexus of his decision making. In total, 115 Cabinet meetings took place within the underground complex over five years, the last being on 28 March 1945 as the final German onslaught, their V-2s, spluttered out and victory was clearly in sight. Fears of the effectiveness of Luftwaffe attacks at the height of the Blitz led to further ferro-concrete protection being added, 'The Slab' as it was known, over 1.5 metres thick.

As war continued and spread over several fronts, the whole underground township grew like topsy. Rooms for typists and switchboard operators, staff dormitories, quarters for senior officers and politicians. One major innovation was the setting up of a Transatlantic Telephone Room with Winston's own combined office and sleeping quarters. The PM could speak securely to US President Roosevelt, and this personal relationship was a vital contribution to final Allied success. Churchill delivered four wartime broadcasts from the War Rooms, but he didn't like sleeping there, preferring 10 Downing Street or the No. 10 Annexe (a newly refurbished apartment in the New Public offices, located directly over the Cabinet War Rooms).

These War Rooms were the epicentre, the beating heart of a vast imperial effort that stretched across the globe and a war that was fought out over huge expanses of territory in each hemisphere. Perhaps 55 million people died. In Marshal Zhukov's memorable words, 'nobody was counting.' Then there was peace, and this bustling subterranean complex emptied. It had only one purpose, that purpose was past, and the world moves on. Cobwebs gathered instead of generals.

No more high command, just the dull old Ministry of Works. Yet, as early as spring 1948, there was growing interest in the possibility of some form of public access, a gaggle of journalists guided by Ismay did get the tour on 17 March that year. Thereafter, you could get a look around by appointment. By the 1970s, this clamour had swelled exponentially. The place was now under the care of the Department for the Environment. Only 5,000 visitors a year got lucky, and the War Rooms were looking distinctly moth-eaten, furnishings and documents began to look more Miss Haversham than Winston Churchill.

As early as 1974, the IWM was canvassed but declined on the grounds of expense and the Museum only finally agreed to take the site on eight years later, once the trustees had received sufficient assurances of government financial support. Margaret Thatcher provided much of the impetus; the restoration was something of an electoral boost as well, though by the time the doors were opened, the Conservatives had already won a landslide victory.[39] The Department for the Environment remained in charge for another five years before IWM finally took control.

In 2003, there was a significant expansion of the historic site when Churchill's suite was refurbished and added to the tour. Two years after that, another substantive makeover led to the addition of the Churchill Museum, 9,100 square feet (850 square metres) of additional space dedicated to the great man's life and achievement. Most recently, a decade ago, a new external entrance with an impressive faceted bronze vestibule was put in place, described as 'appropriately martial and bulldog like'. Surely Winston would have approved.[40]

In London SW1:

Graham: Churchill was of his generation of course, yes he did take a patronising view of people from India or other parts of the empire, reprehensible in today's terms but that's just the way it was.

John: In my day, I suppose we were very much the other way, the Empire, Commonwealth was good, everything British was naturally superior and we'd won the war, Churchill opened the way to 'broad sunlit uplands' and all that.

Graham: Just remember, in your day you were posh if you had a telly.

John: Last year I took Henry [grandson no.1] to the Discovery Museum and there is a large display case full of fifties TVs, tiny screens. I asked him what he thought they were, and he looked pityingly at me, as the young are wont to do, and replied with a hint of exasperation, 'microwaves *of course*.'

'Firepower' Royal Artillery Museum

The magnificent Nine. A few years ago, I went to see the remake of *the Magnificent Seven*. The original, one of those movies which does deserve the much-overworked term 'iconic', was the first film I ever saw. You don't expect too much of remakes, most are pretty dire but this one isn't. It's more politically correct; the bad guys are grasping capitalists rather than misunderstood Mexican bandits (I'll bet Donald Trump didn't like it) but it works. The film is entirely fictional, depending on your view of private enterprise, but we do have our own real-life version. 2022 was the fiftieth anniversary of the battle at Mirbat in Oman.

On 19 July 1972, nine members of Britain's Special Air Service (SAS) were the backbone of a British Army Training Team (BATT), stationed in the small, *Beau Geste* port of Mirbat in Southern Oman. This was one of those messy little conflicts that officially never really happened. Communist-backed guerrillas, known locally as the Adoo (not their official title, which was the usual extended Monty Python bombast – The Popular Front for the Liberation of the Occupied Arabian Gulf, later renamed the Popular Front for the Liberation of Oman and the Arabian Gulf) were infiltrating from South Yemen.

They, the enemy, wished to capture the small port but would have to take out the BATT position, barricaded in a stout compound building. This was forward of the town proper, where a series of shelving ridges led to the settlement. It's not certain how many attackers there were but rather a lot; anywhere from 300 to 500. The morning was overcast with low cloud, so the defenders couldn't hope for air support.

The BATT men weren't alone. There was a weak Omani platoon and some local militia, all of whom could be best described as doubtful. The Adoo advanced at first light, the locals guarding the forward slopes didn't stay to argue. The SAS with a single I-Corps specialist manned the parapet *Fort Sahara* style and opened up on the attackers with standard SLR rifles, a single Browning .50 calibre machine gun and one light mortar. Captain Mike Kealy ordered the supporting Pakistani and Omani personnel to join in the defence – they weren't that enthusiastic.

The odds were unfortunate, and the defenders lacked any really heavy firepower. Some 800 metres from the BATT house, a group of Omani Special forces occupied a classic mud-walled fort bolstered by a 25-pounder. By this point, the Special Forces hadn't done anything special, in fact they were keeping their heads down and not at home to visitors. Sergeant Talaiasi Labalaba decided to do something about that and sprinted over the fire-swept gap to reach the gun. Normally, a 25-pounder needs a six-man crew, but he worked single-handed, banging off rounds at the encroaching enemy. Despite severe wounds, he kept fighting.

Trooper Sekonaia Takavesi volunteered to make the dash. With the SAS laying down covering fire he reached his injured comrade, and they kept up the good work, unaided by the Omanis in the fort. One man did come out, and the rest began firing. The Adoo were virtually within feet of the emplacement. Despite his wounds, Labalaba attempted to reach a 60mm mortar but was hit again, this time mortally. Takavesi was also wounded. Kealy and Trooper Tobin were the next to brave the gauntlet. Takavesi was hit again and Tobin badly. By now, after several hours, the skies had cleared, and airstrikes pounded the Adoo, who by 12.30 hours had fallen back, leaving probably a hundred dead. Trooper Tobin did not survive.[41]

At present, the battered 25-pounder stands in the Royal Artillery Museum.[42] Sergeant Labalaba was awarded a posthumous Mention in Dispatches (MID). In the circumstances this was pretty miserly and many feel, rightly, he should have received the Victoria Cross. But this was a battle that never occurred in a war that never happened. Britain's involvement was clandestine and such engagements, however heroic, were an embarrassment. Three years passed before any medals were handed out. At the present time, the museum itself languishes in limbo awaiting a new home. Hopefully this will be found and opened soon.

John: Should Sergeant Labalaba have got a VC?

Graham: Of course, but it wasn't going to happen as, officially the fight never happened and the war never existed. That's politics.

John: Did that bother you, the indifference of politicians or their downright shiftiness?

Graham: Can't say it did, we'd no time for politicos and most soldiers never think of the 'bigger' picture. We had to allow killers like Adams and McGuiness off the hook as the price of peace in Northern Ireland. We didn't have to like it; nobody would but that was just the price that had to be paid. Men who had the blood of our comrades and countless innocents on their hands had to be welcomed into the fold simply so the killing could stop. There wasn't an alternative, it would just have gone on and on.

Dover Castle (Operation Dynamo)

Bad King John hasn't had a favourable press and there's good reason for this. He lost most of the Angevin Empire his father had created and fell out with his own great lords. They compelled him in 1215 to sign Magna Carta. He'd not the least intention of keeping to his word and civil war inevitably ensued. Within a year, the baronial faction was fast losing ground. To bolster their flagging cause, they offered

Prince Louis of France the throne if he'd back them. He did. That May he landed a substantial force in Kent and John retreated rapidly. He was so disliked that many towns were prepared to recognise the Frenchman as a liberator.

There's always a 'but' and this but was Hubert de Burgh, who held out at Dover. And he had plenty of form: in 1204/1205 he had led an epic defence of Chinon. He had around 140 men at arms and plenty of auxiliaries behind Dover's walls and he wasn't frightened of Frenchmen. In July, Louis got going, launching a serious bombardment against a stout timber barbican defending the main gate. Stone-throwing engines kept up a steady barrage, miners were at work trying to dig under the outer timber palisade. Crossbowmen crowded into a tall siege tower to snipe at defenders.

Finally, the French launched an infantry assault and took the barbican. Next up was the main gate, built of course in stone. Again, Louis' miners went to work and actually brought down one of the twin towers. He sent his infantry in again, but this time the defenders, who had thrown up improvised barricades, flung them back. Alarmed by his heavy losses, the Frenchman settled down to starve the garrison out. The whole place was closely blockaded, and Louis promised a grim reckoning for any survivors. De Burgh wasn't impressed and continued to hold out. This stalemate lasted until mid-October and a truce. Though frustrated at Dover, the French had still netted significant gains in the southeast. The cease-fire came into effect on 14 October and John did the most helpful thing he could have done by dying four days later.

Louis called on de Burgh to throw in the towel, who was he fighting for now? De Burgh decided to fight for the infant King Henry. Louis left a token force to mask Dover and moved his field army. De Burgh didn't waste the opportunity, packing in provisions and repairing the damage. He had the aid of an effective 5th column of guerrillas led by the enigmatic 'Willikin of the Weald', a kind of Robin Hood type who attacked French outposts and communications while English ships at sea played havoc with Louis' re-supply.

The Frenchman wasn't done though. In May 1217, a year on, he came back but with more men plus more and bigger kit. A huge trebuchet 'Malvoisin' ('bad neighbour') was deployed. But the English irregulars beat up his camp, stealing supplies and cutting throats. The besiegers, under threat from land and sea, were close to becoming the besieged. Worse, that serious bruiser William the Marshal, the boy-king's champion, trounced the French at Lincoln. Louis packed up his baggage and departed. This time he wouldn't be back. Defeat on land was followed by disaster at sea. Fittingly it was de Burgh's own flagship which led the charge. Dover was safe.

The great brooding bulk of the fortress has watched over our southern coast ever since, and what was perhaps its greatest hour was yet to come and hitherto almost impossible to have foreseen. In 1939, it was a very different kind of war, but the threat was just as real. Germany invaded Poland so Britain and France, as Poland's allies, declared war on Germany. This didn't save the Poles, who were swiftly overrun and brutally crushed. *Their* nightmare was only just beginning. In the west there was a weird hiatus, the Phoney War, as British troops deployed to support the French. Belgium, with a kind of desperate optimism, stayed neutral, imagining this might save them this time. That hadn't worked for them in 1914 and it wouldn't in 1940. Hitler had no interest in declared neutrality.

On 10 May 1940, Blitzkrieg was unleashed as Panzer divisions deployed from the leafy depths of the Ardennes Forest to blast across the Meuse at Sedan. The French were wholly unprepared, neither their strategy nor morale was fit for purpose. Disaster unfurled like a banner, the Germans raced for the sea, allied armies were cut in half, and nobody had any idea what to do about it. Initiative was in the hands of the Wehrmacht and stayed with them. Three tank corps were headed for the Channel, and by 21 May had effectively trapped the BEF, the Belgians and three French corps into a shrinking enclave along the coast of Northern France with their backs to the sea. Lord Gort, commanding British forces rightfully concluded that evacuation was his only hope of salvation – it was a pretty slender hope.

Soon, over 300,000 soldiers were marooned on the beaches with the panzers closing in. A miracle was needed and two such occurred. On 23 May, Von Rundstedt, commanding Army Group A, halted his tanks, worn down by the very speed of their penetration and the job of finishing off what remained of Allied forces was left to Herman's Luftwaffe. Göring would gain a reputation for promising what he couldn't deliver and the beleaguered BEF gained a form of respite, not that anyone on the beach at Dunkirk under screaming Stukas would have appreciated his good fortune. Those three days when the German armour paused and drew breath permitted the Allies to form a defensive cordon of sorts. The British fought from 29–31 May to secured Dunkirk, whilst the French battled to delay seven Axis divisions (three of which were armoured) at Lille.

A catastrophe loomed. If the remnant of the BEF was destroyed or captured, then Britain could not hope to continue the war. The spectre of a European continent dominated by the Swastika and ruled by the jackboot loomed very large indeed. A second miracle was needed. That came in the form of Vice-Admiral Sir Bertram Ramsay (1883–1945), one of Britain's greatest war heroes, yet you don't hear much about him – and what's he got to do with Dover or Dunkirk?

Born into a distinguished family, Ramsay was always destined for a career in the Royal Nany, joining as a cadet in 1898. By 1915, he was in command of a monitor.[43] He served in the vital work of the Dover Patrol and then on 9 May 1918 took part in the 2nd Ostend Raid where his courage and leadership ensured a Mention in Dispatches.[44] By 1938, he had retired from the service but allowed Winston Churchill with his typical persuasiveness to entice hm back to the colours. He was appointed Commander-in-Chief Dover, responsible for guarding the south coast and keeping the Channel clear, an onerous responsibility. Not half as burdensome as the mission now entrusted to him, to get whatever could be got of the BEF off the beach at Dunkirk, a Sisyphean task which appeared to hold very little prospect of even partial success.

Dynamo was the code name for the evacuation. Ramsay and his staff planned the rescue from their HQ deep in the bowels of the great fortress. It was said that the name came from the dynamo room powering the subterranean labyrinth of tunnels, first excavated during the Napoleonic Wars, a nice touch but probably apocryphal. What is certain is that Ramsay got his instructions on 19 May and the plan had to be operational by the 26th.

Only 7,669 personnel were lifted on that first day and prospects looked grim. One feature still standing from the bombed-out shell of the harbour, its great ominous clouds of black smoke billowing like the devil's shroud, was the Mole (a long pier or breakwater). Patiently, in a long snaking queue, men blanketed the Mole while 39 RN, four Canadian and three French destroyers shuttled back and forth.

Providential as the Mole might be, it could never be enough. The nature of shallow, sloping beaches meant big ships couldn't get in close enough. Enterprising engineers constructed impromptu moles from abandoned trucks and those legendary flotillas of small, civilian craft responded to the call, a bizarre miscellany of yachts, ferries, commercial vessels of every shape and size, the whole eclectic fleet helping to deliver the miracle. By the eighth day, 338,226 allied servicemen had been recovered. The beach was empty, and the final German blow fell on air. Hitler had just lost the war, even though, as Churchill, trenchantly as ever, pointed out, 'Wars are not won by evacuations.' The BEF had sustained 68,000 losses, mostly POWs, during the course of the Battle for France; all of its vehicles and guns were abandoned, but Britain still had an army. Against all odds *Dynamo* had succeeded.

By 1943, the tunnels of Dover Castle were serving as a combined HQ for all three services and Admiral Ramsay would go on to have a very active role in planning Operations *Torch*, the Allied landings in North Africa 1942, together with *Husky*, the amphibious attack on Sicily in 1943. Both succeeded. His biggest challenge would come with planning for the greatest invasion of all, Operation *Neptune* – the D-Day landings. This was the final twist to *Dynamo*, this time the Allies were going back, undertaking the boldest and most daring campaign of the war, possibly of all time, and it was Ramsay who shouldered this immense burden. Again, he succeeded. As Correlli Barnet noted, '*Neptune* constituted a never surpassed masterpiece of planning'. It's very hard to disagree.

Admiral Sir Bertram Ramsay did not survive the war. He was killed in a plane crash on 2 January 1945. His achievement was immense.

Spitfire and Hurricane Memorial Museum, Ramsgate

As an aircraft builder in the early thirties, R.J. Mitchell became aware of how German aircraft design, particularly of single-seat fighters, was racing ahead of British design. The mainstay of the RAF was the Hawker Hurricane, a handy plane and a stable gun platform, easy to fly but outmatched by the monocoque ME109. Mitchell was already a sick man and was told he could perhaps prolong his life only by avoiding the stress of work. He chose to sacrifice himself, he worked 24/7 to leave the RAF, Britain and for that matter the free world, his brilliant legacy. He died before the war but lived long enough to see the Supermarine Spitfire fly.[45]

The Luftwaffe had experimented on Guernica. Both Göring and 'Bomber' Harris drew the wrong conclusions that strategic air power alone could bring an enemy to his knees. Göring's other mistake was the overall concept of the Luftwaffe. It was a superb *tactical* air force but unsuited to a *strategic* role. The Battle of Britain itself was one of attrition, whoever's nerve broke first would lose. It was a damned close-run thing and really ended in stalemate, with both sides fought out. Losses in terms of numbers were almost even but any German shot down alive over England would be captured, whilst RAF crew, if they lived and were whole, might fly again.

Having won the Battle of France, the Luftwaffe re-grouped into three *Luftflotten* (Air Fleets). *Luftflotte* 2, commanded by *Generalfeldmarschall* Albert Kesselring ('Smiling Albert'), was to be responsible for the bombing of southeast England and the London area. *Luftflotte* 3, under *Generalfeldmarschall* Hugo Sperrle, targeted the West Country, Wales, the Midlands, and northwest England. *Luftflotte* 5, led by *Generaloberst* Hans-Jürgen Stumpff from his lair in Norway, looked towards the north of England and Scotland. As the fighting raged, command responsibility shifted, with *Luftflotte* 3 taking more responsibility for the night-time Blitz attacks while the main daylight operations fell to *Luftflotte* 2.[46]

Early German predictions suggested it would take a mere four days to incapacitate RAF Fighter Command in southern England. This would be followed by a further four-week offensive during which twin-engined bombers and long-range fighters would destroy all military installations throughout the country, and British capacity to produce more planes. The campaign would commence with attacks on aerodromes near the beaches, gradually creeping inland to deplete that magic circle of sector airfields protecting the capital.

Subsequent adjustments gave the Luftwaffe five weeks, running from 8 August to 15 September, to win hegemony over English skies. To achieve this objective, which would signal an effective end to the war, Fighter Command had to be wasted, either on the ground or in the air. In achieving this, the Luftwaffe needed a clear win; not a draw or a pyrrhic victory that would leave it too weak to support any subsequent amphibious invasion. In short 'they' had to knock far more of 'us' down than we did of them. There was only one alternative strategy, which would be to repeat Guernica on a much larger scale, devastating British cities. For the moment, Hitler baulked at this.[47]

Overall, the Battle of Britain can be divided into four fairly distinct phases: 10 July–11 August, *Kanalkampf* ('The Channel battles'); 12–23 August, *Adlerangriff* ('Eagle Attack'), the grandly dubbed onslaught against the coastal airfields; 24 August–6 September, hitting the aerodromes generally; and finally, 7 September onwards, the raids switching to targeting towns and cities.

One of those 'up' on 10 July was Ira Jones, hero of the previous war with 40 kills. At this point the Welshman was a rear echelon wing commander based at a training field. Some 63 Ju 88s were approaching the West Country and radar warning came too late to scramble fighters. 'Grandpa Tiger' Jones wasn't one to be deterred by the odds, or the fact his antiquated Henley target-tug was totally unarmed. He'd knocked down plenty Boche in the last show and wasn't going to let a fresh opportunity pass.

Jones got onto the tail of one Ju 88 – 'I felt the old joy of action coursing through my body.' His only weapon was a Very pistol that fired a pretty multi-coloured cartridge. He came in front of the German pilot (who must truly have wondered what the hell was going on), banked and fired off his single flare. The bomber was lit up like a Christmas tree. It worked, the Ju 88 promptly sheared off and ran for it.

The Luftwaffe preferred to fly in a loose section of two (nicknamed the *rotte*); a leader (*rottenführer*), followed at about 200 metres by his wingman (*rottenhund* or *katschmarkes*), who also flew slightly higher and was trained to stay with his boss. Two of these pairs formed a four-man flight (*schwarm*). Each flight in the squadron (*staffel*) flew at staggered heights with, say, 200 metres between them, making the whole formation trickier to spot at a distance and facilitating flexibility. By using a tight cross-over turn, the *schwarm* could quickly change direction. Slower and heavier Bf 110s adopted similar tactics.

The RAF during the inter-war years had planned to meet enemy bomber offensives, fighters had been ruled out. So the tactical imperative had been the ability to concentrate and deliver a hefty punch; these 'fighting area tactics' focused on a series of flying manoeuvres aimed at maximizing firepower. Thus, fighters were expected to fly in tight, V-shaped flights ('vics') of three, with four vics in tight formation. By 1940, this was outmoded and the Luftwaffe's more flexible approach clearly superior.

Well-armed but lacking agility, Bf 110s proved vulnerable to nimbler RAF fighters. This meant the bulk of bomber nannies would have to be made up of ME 109s. Though the need to protect vulnerable bombers was paramount Göring, who knew all about an ace's hunting instincts, was keen to free up fighters for 'free hunts' (*freie jagd*). Fleet-winged ME 109s would range ahead of the bombers like beaters flushing game. The closer the fighters were tied to the bombers' apron strings, the harder it was for them to manoeuvre and losses in dogfights climbed accordingly.

RAF fighter controllers with their finely tuned antennae were often able to detect these skirmishers and move squadrons around to avoid them. Dowding's plan was always to conserve his fighters in order to inflict maximum damage on Boche bombers. Extending their hunters' lures, the Luftwaffe resorted to adding small formations of bombers as tastier bait, covering them with packs of ME 109s.

At the age of 21, Pilot Officer John Beard flew his Hurricane in the defence of London. We have the classic image of RAF flyers chilling in deck chairs, fully kitted out and tense for the call when the cloudless blue sky of an English summer begins to vibrate with the thrumming of bomber engines. That afternoon, Beard's squadron was scrambled and raced to their planes, deafened by the roar of propellers as they climbed to 15,000 feet:

> Minutes went by. Green fields and roads were now beneath us. I scanned the sky and the horizon for the first glimpse of the Germans. A new vector came through on the R.T. [radio telephone] and we swung round with the sun behind us. Swift on the heels of this I heard Yellow flight leader call through the earphones. I looked quickly toward Yellow's position, and there they were!
>
> It was really a terrific sight and quite beautiful. First, they seemed just a cloud of light as the sun caught the many glistening chromium parts of their engines, their windshields, and the spin of their airscrew discs. Then, as our squadron hurtled nearer, the details stood out. I could see the bright-yellow noses of Messerschmitt fighters sandwiching the bombers and could even pick out some of the types. The sky seemed full of them, packed in layers thousands of feet deep. They came on steadily, wavering up and down along the horizon. 'Oh, golly,' I thought, 'golly, golly…[48]

Orders came through the earphones and the Hurricanes circled to begin their attack before screaming down like maniacal banshees.

Beard kept both hands on the stick: 'You have to steady a fighter just as you have to steady a rifle before you fire it.' The afternoon sky was full of aircraft, a maelstrom. Beard focused on his target, a Heinkel, swelling in his vision until locked in the red dot of his sights. He wondered why the bomber didn't move, his finger on the trigger. Eight Brownings hammered away, two bursts each of as many seconds, the acrid stench of cordite filling the cockpit.

His rush took him above the quarry. He saw rounds hitting, a red glow blossoming inside the canopy, bursting out into streaks of flame. The bomber staggered and began its final fall, detritus spinning off. There were more. He climbed and saw another flock of bombers circled by shepherding ME 109s. As he raced in, he detected three fast flights of Spitfires coming up, breaking into a fleur-de-lys attack, carving a bloody swathe through the invaders. Eight of the enemy went down.

The sprawling mêlée had spread out over London. He could see the streets and parks below, laid out like a familiar jigsaw: 'I had an instant's glimpse of the Round Pond, where I sailed boats as a child.' He spotted a Dornier hotly chased by a Hurricane. The British fighter in turn stalked by two ME 109s. The Hurricane pilot hadn't seen them, but neither had they spotted Beard. He charged head on into the teeth of one, shooting him to pieces. He aimed at the other, but the German executed a smart Immelmann half turn and got clear. Beard had no time to celebrate as another brace were now on his tail. He managed a fairly smart climb himself, tracer streaking underneath. He was low on fuel and ammo, happily as it seemed, so were they and broke off.

RAF Manston has been in the thick of it for quite a while, beginning during the Great War then continuing through the Battle of Britain all the way to Operation *Market Garden*. As a location, the years have not diminished the impact of the place, and the Memorial Museum certainly has the right 'feel'. Perhaps the two prime exhibits are a Mk XVI Spitfire, a late model, but this one saw action in 1945, together with a Hurricane IILF751. You can have a go on a very lively Spitfire simulator, great fun of course, but it does give you an idea of just how tricky piloting such a thoroughbred fighter plane would have been – not for the fainthearted.

Royal Engineers Museum, Chatham

Fredrick James Mulqueen DSO MC, a Canadian by birth and civil engineer by profession, served firstly with 172 and latterly 182 Tunnelling companies on the Western Front. In the summer of 1914, he was working in South America on the engineering staff of the Sao Paulo Tramway, Light and Power Company. Pre-war recession diminished the opportunities, and he transferred his skills to the Anglo-Brazilian Iron and Steel Company. Their order book was equally reduced, and, on 4 July 1914 he embarked for Liverpool on the SS *Darbo*.

On 15 February, Mulqueen dined with an acquaintance who had returned from war work in the Far East, an engineer who spoke of the increasing need for skilled professionals at the front. A transfer to one of the new tunnelling companies was discussed, but in the short term led nowhere. By March he had orders for France as a gunner and landed at Le Havre on the 27th. Barely a week later, he was sent up to the front via Bailleul. It was in the course of his baptism with the guns that he decided to switch to engineering: 'On my return to the battery, I reported what I had done, and I was looked upon as quite insane. To join a tunnelling company was considered sheer suicide.' By 22 April, he had orders to join 172 Company RE.[49]

On that same day he met and was briefed by Norton-Griffiths himself, already something of a titan.[50] It was a dark time. The Germans had used gas for the first time in the Salient and opened a gap in the line; the Major had encountered 'the remnants of a Zouave battalion some distance behind the lines with gas in their lungs and fear in their eyes'. The war had entered a newer, even more terrible phase. Sappers were already burrowing beneath Hill 60 in the sector, but Mulqueen was sent to company HQ in Dickebush. He was to relieve a Lieutenant Hickling in the line. The British position, dug in amidst the ruins of St. Eloi, was compromised by German control of the 'Mound' overlooking the shattered huddle: 'A heading had been started from the village toward the Mound and was about halfway there at that time. Our orders were to charge a mine in the heading, which was to be fired in the event of an adverse situation developing.'[51]

That evening, Frederick moved up towards the front line. 'In those days there were no communication trenches or, if there were, we didn't use them. The night was dark, and the road was muddy. I had thought the vocabulary of the ranker major was complete, but I can assure you that, on that night, I was to learn it was far from it!'[52] The night sky was lit by the fitful glare of arcing Very lights and enlivened by the zip of spent rounds coming close. Rumour insisted German snipers had infiltrated and lurked in cover, ready to pick off the unwary.

> Eventually we arrived and reported to Lieutenant Hickling, whom I found to be a first-rate fellow. The place was alive with men, shoulder to shoulder in a position which consisted mostly of barricaded buildings, half-demolished and shallow trenches. Hickling showed me the shaft and heading and the mine which he had laid but which was not completely tamped. The electric leads from the detonators had been carried to the surface and run behind a stout wall to a cellar to the north. I have forgotten the exact details of the charge or its distance from the shaft, but my recollection is that the former was more than ample and the latter far from it! The exploder was ready and waiting but it was not connected.

The relief completed, Mulqueen was now very much on his own. He was a civil rather than a mining engineer,

> I had never been in a mine! The depth of the shaft was probably 14 feet and the gallery leading from it was exceedingly small, possibly 3′6″ x 2′0″. It was much smaller than that which was later made standard. The sergeant and I explored it, and following Hickling's instructions we outlined to the men the amount of additional tamping to be placed. Sandbags filled with clay were used for this purpose ... there was room for only one person at the 'face' at any time. The sandbags were passed from hand to hand along the length of the gallery and securely packed adjacent to the charge where the tamping was completed, a check was again made of the leads, exploder etc.[53]

Mulqueen was under the immediate orders of the infantry commander in his sector, a lieutenant-colonel of 4th Battalion Northumberland Fusiliers. It was the infantry officer who would give the order to fire the mine. Meantime, the Fusiliers were crammed into an isolated post, the infantry's strength concentrated in the forward positions.

German gunners were active. Desultory shells came over but there seemed no fear of imminent assault...

> The term whizz-bang is an extraordinarily effective description of the type of high velocity shell. In the case of a round fired from a howitzer, one

> could hear the explosion of the gun some seconds before the shell arrived. Not so the whizz-bang; one hears a whizz and then the bang almost simultaneously. We tunnellers were well established in dugouts adjoining the shafts. Fusiliers were ready and determined, machine-guns well sited to enfilade any attacks on the battalion front. In daylight the position appeared even more precarious; the Germans occupied the other half of the village. The Mound dominated the whole position and the Germans had captured this after exploding a mine of their own on 14th March.[54]

Since the first use of gas, much talk revolved around its lethal effects.

> There was a great deal of discussion on the use of gas by the Germans and the best way of combating it. It was agreed that it could only be used when wind conditions were favourable and, at such times, extra sentries should be posted. The Germans were known to have used masks and it was understood we would eventually get them. Pending their receipt there was little one could do ... one unique suggestion was that one should urinate on one's handkerchief and place it over one's nose. It was stated that urine (ammonia) neutralized the chlorine. I can say I never tried it!

Meanwhile, the business of digging continued: 'I spent the following day in the line with Hickling and he explained his plans to protect the post underground. Already difficulties were being encountered in that the disposition of the soil proved a problem. The clay was filled into sandbags underground, hoisted up the shafts by winches and the sandbags piled outside the so-called dugouts which had been built over the shaft-heads.'[55]

For the infantry most materials were in short supply, but the engineers fared rather better, drawing on precious stocks of timber and cannibalizing the wrecked buildings around.

> The infantry lived in makeshift dugouts, but Hickling had designed a more rugged affair to cover our shafts and to house the men who were off shift. He scrounged timbers from the destroyed houses and used them as uprights to support wooden joists (also from the houses) on top of which he placed corrugated iron sheets and protected the top and sides with layers of sandbags containing the earth brought up from below. These dugouts were relatively safe from shrapnel. In any event one had a sense of security in them and psychologically, they helped the men.

... Hickling had decided that the dimensions of the shaft and heading of the first mine were wrong and should be enlarged. It is probable that the subject had been considered at Company HQ and with the other tunnelling companies. In any event, standards were agreed upon; the shaft was 6′0″ x 4′0″ and the galleries 3′6″ x 2′6″. The ground was sandy clay and was wet, therefore at the bottom of the shaft a sump was dug and a double acting, hand-operated pump installed over it. The pump was worked by an infantry fatigue party and, as the men usually hated the job, the pump received some pretty rough treatment. It was also necessary to supply fresh air to the working face and this was delivered through a three- or two-inch airline connected to a Rootes hand-operated blower placed at the top of the shaft. This blower was also operated by an infantry fatigue party, and it was not a rare occurrence for the officer on duty to fins both the pump and the blower idle!

... Hickling had been feeling his way downwards seeing if he could probe deeper before hitting the water table. The engineers were beginning to master the practice of this branch of their profession. Geologists, hydrologists and other experts were co-opted, and a new science was developing, embodying such eclectic elements as crater formulae, explosives, mining theory, listening instruments and mine rescue. At 35 feet the diggers encountered drier clay and the original charge was lifted from the first, shallow mine and deposited deeper. The first shaft was converted into a listening post to detect counter-mining operations.

... The mine in this heading had been placed directly under the junction of the roads leading to Ypres, Voormezeele, Wytschaete and Hollebeke. The road junction consisted of an area of heavy pave, i.e. oblong granite blocks approximately 10″ x 4″ x 6″. Based on our subsequent experience, had we fired the mine, the pave would have been thrown in all directions and would probably have destroyed not only the Germans but also many of our troops including the officer who exploded the mine! Some of us were born under a lucky star!'

... Under Hickling's able administration our section commenced a series of galleries which not only protected the St. Eloi post but subsequently forced the Hun into defensive action close to his own lines. A ready use was found for the sand-bagged soil. It was built into an extension of the trench system both to left and right. It was customary to stack the sandbags by day and leave the more exposed work of building trenches till night-time. Mulqueen came across a party working blithely by day. They were convinced their area was in dead ground and couldn't be overlooked by German marksmen. Frederick decided to cautiously

> investigate. No sooner had he peeped out than a sniper round passed clean through his hat (this was prior to the issue of 'tin' helmets), a lucky star indeed!
>
> ... Tunnellers were not very popular with the infantry, largely because the spoil from underground was easily spotted by the Germans and drew fire. Also water, pumped from diggings, tended to flood the trenches.[56]

On 6 May, Mulqueen endured what he laconically describes as a rather one-sided talk with the infantry CO after the latter had descended rather precipitously into the run-off. Wisely, he decided to seek a solution. Behind the main front-line trench ran an older choked irrigation drain or channel. If this could be cleaned out it could provide the necessary additional drainage to absorb the waste from the diggings. That night he went out with a small working party. Despite darkness and diligence, the enemy detected them and began throwing grenades. One 'sank into the muddy water between the three of us and exploded about two feet from my head.' Miraculously, he emerged largely unscathed, though both his companions suffered wounds.[57]

Despite his apparent escape, Mulqueen suffered headaches and dizziness. This was doubly unfortunate as he was at this point responsible for the positioning of a 'collar' over a newly dug shaft. This was a tricky business. The collar itself was fabricated in rear and from timber. It than had to be carefully and exactly located over the head of the shaft. Planks lining the opening were suspended from the collar, so correct siting was of the essence. With his headache nearly blinding him, Frederick had been obliged to leave the fine detail to his sappers, whose efforts were less than outstanding. The unfortunate result was a' hell of a mess', and Hickling, proud architect, 'as mad a hornet'.[58] Obviously, his injuries from the grenade blast were more severe and he was obliged to report sick. Here he first heard of the sinking of the *Lusitania* engendering 'a further loathing of the Huns if that was possible.'[59]

An initial period of four weeks rest was extended by another six. He did not return to St. Eloi until 24 July. His comrades had not been idle. The underground labyrinth had progressed prodigiously, and new drainage carried away the excess water. His luck held when another German rifle grenade missed by a whisker, but the shock caused an unexpected nervous reaction, legacy of his previous close shave. The work continued:

> The respective shafts and underground workings at St. Eloi were lettered in accordance with a general working pattern in vogue on the front.

> Headings were driven from our front line towards the German trenches at varying depths of 15–35 feet. As those headings approached their objectives, they encountered German defensive operations. Alternatively, in some places, they forestalled German offensive workings. On some other fronts, shafts were sunk in the support trenches and were less vulnerable to enemy raids, attacks etc. these were connected by lateral galleries running beneath or slightly in front of the front-line trench system. In August 1915, we had not managed to build any laterals at St. Eloi, as the Hun did not give us time to develop such refinements![60]

Listening was the key. The key to what was becoming a deadly subterranean game of cat and mouse, of hunter and hunted, one where the roles could be reversed in an instant, the price of inattention – death or entombment. Listening

> ... was accomplished by a carefully controlled system. While there were several different kinds of listening instrument, the first with which we were supplied and, in my opinion the best was the geophone, constructed on the principle of the ordinary medical stethoscope. It consisted of a circular wooden disc, about three inches in diameter and an inch thick, within which was enclosed a flat mass of mercury held in place by two mica discs. Two small nipples led to air spaces on the outside of the upper and lower mica discs respectively. Rubber tubes connected the nipples to the ears. To obtain direction of sound, two instruments and a compass were used.[61]

This business of listening was highly complex and subtle. Sound ranges differed in changing soil conditions. Operators had to become adept in filtering out extraneous noise, work which became the province of highly specialized elites,

> ... whose acumen was uncanny. I remember in one of our headings on the Vimy Ridge which was in chalk, we traced the German operator from over 90 feet away to within 10 feet of our position. In the early days, listeners were trained 'under fire'. It was a dangerous undertaking. If the neophyte's judgment failed him, he and his comrades were apt to find themselves buried under tons of clay with little hope of rescue. Not all volunteers were temperamentally suitable. The successful listener must be a cool personality. Often, he will find himself alone in a dark gallery. He must not allow his imagination to run away with him. On

the other hand, he must not overlook even the slightest sound indicative of any activity.[62]

> ... The moist gallery, 35 feet below ground, half-lit with a flickering candle, apparently remote in its very stillness and yet, was it remote? Was the dull sound which the geophone picked up the burst of a distant shell or was it clay carelessly dropped in a nearby German gallery. The two geophones were shifted along the floor until the sound reaching each ear was uniform. The compass was set midway between the two instruments. Had sounds been heard previously from that general direction? Were the sounds coming from a higher or lower level? The instruments were shifted up and down the face of the gallery, up and down the timbers. Moments of indecision, one must not become 'windy'![63]
>
> ... Yes, the Germans were working forward to the left and slightly higher but how much higher and how far away? The sergeant took the instruments. What did he think? Should we let the Hun come closer or should we blow? A hasty check of the distance from our trenches, and from the German – a hasty review of previous listening records; how quickly was he working forwards – were we too distant from both and should we concentrate on the destruction of his gallery by firing a camouflet [an early sapper's version of a shaped charge].

Only the officer could make the final, vital decision. Once made, a process unfolds. The 100 lb camouflet had to be located within a bored-out chamber, some 16 feet in length. Normally, the sergeant worked the auger; silence is, in every sense, golden.

> The mining officer must advise the infantry of his decision and its probable effect. He must have the explosives and canisters brought forward from the magazine. He must warn the shifts working in nearby galleries. He must assemble the electric leads, detonators, and exploders. Possibly some two or three hours after the decision was made, the mine, completely tamped would be ready to fire![64]

As the underground war intensified and mining operations ballooned, the choice of suitable explosives was modified: 'From the day when I joined the Tunnelling Companies in April 1915 to the day we fired our last mine in December 1916, we thought and dreamed explosives and their effect. Black powder was the first, followed by wet gun cotton, then blastine and, finally, ammonal; of these ammonal was by far the most satisfactory.'[65]

Rebecca Blackburn has curated the museum and archive for seven years (at the time of writing). She has a background in regimental museums but sees the RE collections as being far wider in scope. The Engineers have fought battles, but they've also devised engineering solutions for problems spread over the globe and in every form of terrain. They build structures, create bridges, conduct sieges, dig trenches and redoubts. As with Chard at Rorke's Drift or Gordon in Khartoum, they deploy engineering responses to tactical situations. Their experience embraces cartography and a wide range of skills with adaptability as the keynote. At the end of the day, they are soldiers and, as with such exhibits as a V-2 missile, we have to appreciate that death and destruction are an integral part.

For Rebecca, one object she finds particularly expressive is a saw-backed Russian pioneer's sword from the Crimea, a prize from the desperate trenchscapes around Sebastopol. Like its British equivalent, this is both tool and weapon, something which so ably symbolises the role of the military engineer. It still resonates. I do get this – my eldest daughter has served in the Royal Engineers and her Afghan experience mirrors this tradition.

At the museum:

John: the exhibit that did it for me is Chard's revolver, no surprises I suppose, and I always wanted a Webley-Pryse.

Graham: *Zulu* has a lot to answer for. Bizarrely perhaps the one that really intrigued me is the V2, I know it's just here on loan, but I found it remarkable, never seen one up close, the technology is astonishing.

John: One day you should go to the V2 launch site 'La Coupole' near St. Omer, bloody astonishing, looks like the set from a Bond movie, they were miles ahead of us.

Graham: But it was all a bit of a white elephant, unimaginably huge amounts of resources and their best brains tied up in a project that aside from its terror effect mad no difference at all?

John: You could say that the Germans' technical brilliance and engineering excellence ultimately worked against them, they over-designed and over-specified, outstripped their capacity to maintain production and delivery. The Mark V (Panther) and Mark VI (Tiger) were brilliant but too complex and rushing them into service just exposed all the teething problems and lack of reliability.

Graham: Classic alternative was the Russian T34, basic, even crude, but superb, built to fight across the worst of Russian terrain and one size fits all. The panther I think was designed to be the German equivalent but, despite being a first-rate design, missed the whole point about simplicity. KISS.

John: 'Keep it simple, stupid.'

Bletchley Park, near Milton Keynes

Graham and I went to Bletchley Park in summer 2019, those halcyon days before pandemic shut us all down. For our post-war generation reared on *Lion* and *Victor*, Bletchley is part of our cultural narrative. Wildy eccentric boffins in Harris tweed jackets, invariably smoking pipes, thwarting dastardly blond Nazi types in black leather greatcoats. The reality was far less stereotyped and completely astonishing.

The place goes back a long way, mentioned in the Domesday Book, a house was constructed on the estate, a rural pile in 1711. This was demolished later that century and the architect Sir Samuel Lipscomb Seckham bult a new property in 1877. Six years on and Sir Herbert Samuel Leon acquired both the estate (some 600 acres – 235 hectares), and expanded the dwelling into an altogether grander, eclectic mismatch of styles to create what has been described as a 'maudlin and monstrous pile'. It's hard to disagree, it wouldn't be architectural merit that led to fame.

For that we can thank then then head of the Secret Intelligence Service (SIS/MI6) in 1938, Admiral Sir Hugh Sinclair. He acquired the house itself with 58 acres (23 hectares). The estate was up for redevelopment as housing, to feed the pre-war boom for southern suburbia. He paid £6.000, a very substantial sum at the time[66]. The Admiral saw its potential as a base for Government Code and Cyphers School (GCandCS). Thank God for his farsightedness and his deep wallet (though it's likely he did have access to SIS reserve funds!).

Aesthetically challenged as it was. Bletchley possessed several key advantages – easy access both from London and Cambridge in a quiet rural location not likely to attract the attentions of the Luftwaffe. A first wave of furious construction followed as a rash of timber huts, rather less pretentious and more workaday than the house itself, with communication networks, were installed so that by August 1939, GCandCS could open its top-secret doors. Bletchley would gather together a unique and divergent set of people whose combined skills, if they didn't win the war for the Allies, certainly prevented Britain from losing it.

From the outset GCandCS was more than a gathering of experts. It would pioneer the concept of close, inter-service liaison and employ an almost production line method. The need for such a tightly structured approach was recognised as early as 1940 when the Phoney War suddenly became anything but. Gordon Welchman, a Cambridge mathematician, following earlier work by Alastair Denniston (head since 1919) divined (he claimed in his book) that successful codebreaking proceeded by stages: collection of data, breaking the codes, evaluation, and dissemination. Key to this was the need to ensure the enemy, once his codes were fractured, should never get wind of the fact; he should continue in the belief his comms were safely encrypted. It's not possible here, in what is really only a brief introduction, to provide a detailed history. There are many and the museum's own guidebook gives a very accessible and extremely well-presented account.[67]

Ultra, Britain's best kept secret by 1941, was born of Enigma and this was the brainchild of a German inventor, Arthur Scherbius, whose objective had been to design a machine that could both encipher and decipher automatically. The concept was not a novel one. The machines all worked on a rotating disc principle and Scherbius was not the first to attempt a mass-produced version. His design came onto the market in 1923 and was adopted by the German Army five years later. Its capabilities were sufficient to enthuse the Navy and later the Luftwaffe. Its value was in encrypting secret messaging that was vulnerable to interception, primarily radio traffic. Enigma's unique selling point, its 'USP', was its reflector disc, which empowered the machine to both crypt and decrypt. Whilst all such machines achieve the same function, with Enigma there was no need to move from an encipherment setting to decipherment mode – albeit this was also a potential weakness. Powered by dry cell batteries, it was also lightweight and easily portable. In appearance it seemed as innocuous as a contemporary portable typewriter. Its potential in the field of modern warfare was immense.

Because of its ability to multiply possible encryptions to such an infinite degree, it was thought utterly impregnable to 'cracking' by even the most gifted of cryptanalysts. The Germans believed that hundreds or even thousands of mathematicians could labour for generations without success. This belief persisted even once the codes had been thoroughly penetrated. It wasn't just codebreakers at Bletchley who first began to fracture the Enigma; the cryptanalysts of the Polish army achieved miracles by means of pure applied mathematics with some 'mechanical aids'. The latter, which were to assume increasing importance at Bletchley, were electromechanical devices that tested the solutions of encrypts much faster than could be achieved by pure manpower, 'bombes' as they were known.

In July 1939, as the threat of war loomed, British and French Intelligence officers attended their Polish counterparts in Warsaw where they were presented with a facsimile of the Enigma the ingenious Poles had constructed. However, they'd been beaten by the Enigma designers who'd now added two additional discs, thus multiplying the complexity. Bletchley continued the Polish model with the recruitment of civilian academics (in fact it was nominally a branch of the Foreign Office if funded through SIS). The most outstanding of these was Alan Turing, whose Olympian intellect stood out even in such gifted company.

An element of operator laziness and, in the case of Luftwaffe personnel, inexperience, greatly facilitated the cryptanalysts' work, as did the use of some partial guessing or 'cribs' as they were known. The great intellectual powerhouse that was Hut Six, where operators put in thousands of hours of painstaking work, saw them achieve miracles. Firstly, the Luftwaffe ciphers were broken and then others. Some of the naval codes and that used by the Gestapo were never cracked. *Shark*, the Atlantic U-boat key, remained inviolate all through the murderous months of February–November 1942, when the Battle of the Atlantic hung in the balance and thousands of seaman with hundreds of thousands of tons of merchant shipping succumbed to U-boat attacks.

One of the limitations with Ultra, as the Enigma intelligence was designated, was that officers in the field never saw the original decrypts. These were frequently unintelligible and needed to be translated and effectively interpreted, so that which was passed on was in an edited form. In the early days, this fine art of interpretation was left to linguists rather than Ios, with the risk that an experienced IO might have read the decrypt in a subtly different way. As Ultra was Britain's most closely guarded secret, perceived as the trump card that could tip the finely wrought scales between survival and defeat, entry to the circle of initiates was restricted. The source of the intelligence had to be concealed so that the Germans would not suspect Enigma had been broken. Field commanders were generally told that information had been gleaned from well-placed agents or detailed reconnaissance.

Bletchley's legend is inextricably linked to that of the man regarded as its presiding genius (out of a star-studded cast), Alan Mathison Turing (1912–1954). His paternal family had been servants of the Raj, he was born in Maida Vale though his parents did travel to and from India during his childhood, when he was boarded with a retired army couple at Baston Lodge, Upper Maze Hill, St. Leonards-on-Sea (now with blue plaque). Even at primary school, his outstanding abilities were recognised. From the age of 13 he boarded at Sherborne School, though not all of his teachers appreciated what they saw as a rather narrow focus. His first day coincided with the General Strike of 1926, so no trains for the 60-mile (97-kilometre) journey. Undeterred, he set off and covered the distance on his bike. History would show that Alan Turing didn't give up.

While at Sherborne he met and befriended Christopher Morcom and the two became very close. Some have claimed that Morcom was Turing's 'first love'. Whether this was physical or not we can't say as their relationship, undoubtedly intense, was cut short by Morcom's early death from bovine TB in 1930. Alan remained in correspondence with his dead friend's mother for years afterward. From 1931–1934, he studied mathematics at King's College Cambridge, graduating with 1st class honours. During the mid-thirties, he continued his studies at Princeton, working towards a PhD. He also began to work on building his 'Universal Machine' (now the 'Turing Universal Machine'). His concept was that this machine could perform the tasks of any other computation machine – computing anything that's computable.

By September 1938, he was working part-time in academia but also for GCandCS. On 4 September, the day after war was declared, he presented himself for service at Bletchley; the ground was about to shift. Historian Asa Briggs who also worked at GSandCS later wrote: 'You needed exceptional talent; you needed genius at Bletchley and Turing's was that genius.'[68]

Creating the bombe was the first of five major cryptanalytical advances that Turing made during his time at Bletchley. The other four were: deducing the indicator procedure used by the German navy; devising a statistical procedure, branded *Banburismus* for making significantly more efficient use of the bombes; initiating a procedure labelled *Turingery* for working out the cam settings of the wheels from the Lorenz SZ 40/42(*Tunny*) cipher machine and, later in the war, developing a portable, secure voice scrambler at Hanslope Park, codenamed *Delilah*. Collectively, these utterly transformed the science and practice of cryptology, Turing, with other colleagues such as Tommy Flowers[69] helped defeat Hitler and changed history, pretty much forever.

Seven years after the war ended, Turing was charged, in rather convoluted circumstances, with gross indecency, alongside his alleged lover 19-year-old Arnold Murray. On the advice of both his solicitor and his brother, Turing pleaded guilty and was convicted. This could have spelt gaol, but he was permitted to opt for 'probation' – a banal term covering a form of physical and mental torture by drug therapies, badged as 'chemical castration', a draconian and grossly invasive treatment with significant, permanent side effects. The conviction was enough to cost him his security clearance, losing his position at GCHQ, and he was barred from entry to the USA, though he retained his academic appointments. He was found dead at home on 8 June 1954, from cyanide poisoning, via a half -eaten apple found by his bead. The coroner's verdict was one of suicide, though this has subsequently been questioned and some commentators are convinced he may have ingested the toxin accidentally. Conspiracy theories inevitably abound. *Snow White* was his favourite movie, featuring of course a poisoned apple, but he often did take fruit to bed with him and regularly left these unfinished. What is certain is that his treatment was cruel and inhuman. His rehabilitation has taken a long time, though now he has gained the recognition his achievement always merited, and he was introduced to popular audiences by a movie biography, *The Imitation Game*, with a stellar central performance from Benedict Cumberbatch.[70] Alan Turing is now a national hero and symbol of the evils of intolerance.

David Kenyon is an archaeologist and military historian. He is, even though it's a much-overworked term, passionate about Bletchley and with good reason; the displays, ambience and sheer scale of the place are unparalleled. It matters for a variety of reasons. Obviously, it was a key location in Britain's war effort. The work undertaken by the cryptographers affected the course of history and certainly helped very considerably to avoid an Allied defeat in the early years, as well as greatly facilitating the long road to final victory. That's just half the story; work done there, the technology developed, has changed us and changed the world we live in. You could say computer science was born here, the I-phone you carry is the direct descendant of Colossus, incredible genius and dedication distilled into today's everyday objects.

Codes also fascinate David, who was creating his own when aged ten! He sees his work at Bletchley as bringing the whole story more and more into the public eye. Turing died as a convicted sex offender but has since been recognised and rehabilitated as not only a beacon of sheer brilliance but as a man who was grossly abused by a brutally censorious legal system. We also must remember the Official Secrets Act, which bound all who signed to a perpetual silence. As David points out, there is no actual limitation period; papers may be released after certain set periods, but this isn't automatic. Many who worked at Bletchley took their experiences, undisclosed, to the grave.

As I observed earlier, asking a curator to name their favourite object from a collection is daft in many ways, as David points out, it's the whole rather than individual parts which make up the story. He does mention a bundle of hand-written decrypts from 6 June 1944, a date of some significance. German messages being sent in the wake of the Normandy Landings were on Admiral Ramsay's desk at Dover within three hours, probably rather sooner than they were on Hitler's.

Bletchley Park is unique, both in its critical importance and its very being. The Bletchley Park Trust manage the place magnificently and the heritage offer is, in our view, absolutely second to none. Not only is the story so compellingly presented, the wartime atmosphere and the sheer scale of the operation come very vividly to life. We can't wait to go back.

Coffee shop in Milton Keynes:

Graham: Bletchley Park must be one of the most remarkable places I've been to, if for no other reason, the sheer scale and complexity of it.

John: I can't think so many men and women of such towering genius can ever have been collected in one place or who were able to exercise such a profound effect on history. And they could never talk about it.

Graham: They wouldn't have dreamed of it, look at Tommy Flowers, always worked for the GPO, and though he was given £1,000 and an

MBE never sought any special recognition, just did his bit and went back to his civvy job and pension after the war. Nonetheless, once knowledge of Colossus became public knowledge, he certainly did talk about it and some might say was prone to exaggeration where his own role was concerned!

John: And Turing, of course, some gratitude he got.

Graham: It can't help but affect us, all of the soldiers, sailors, airmen, all of the support service, the merchant marine, were dependent, for theirs and Britain's survival on a disparate crew of mathematical and other geniuses, they never even knew about.

John: And the finding of the Naval Enigma, HMS *Bulldog* and all that...

Graham: Her captain was a Northumbrian, wasn't he?

John: He was, 9 May 1941, recovered a naval enigma machine plus codebooks from *U-110*, a game-changer for Bletchley.

Graham: Now, we all thought it was the Americans?

John: So, Hollywood would have you believe – but this was months before Pearl Harbor!

Bovington Tank Museum

On 16 September 1916, German soldiers on the Somme defending Flers-Courcellete, probably thought they'd seen the very worst war had to offer. In the mellow dawn of a late summer's morning, they saw a vision of hell. Armoured monsters ground inexorably towards them, seemingly impervious to small arms, chewing up barbed wire and skipping over trenches, the demons of industrial war, orcs on tracks. The Germans ran, can't say I blame 'em. In this case the donkeys got it right and saw the potential for these lumbering, rhomboid leviathans. It had been on 20 January that year that William Foster and Co of Lincoln, in peacetime manufacturers of agricultural machinery, demonstrated their prototype.

The Admiralty Landships Committee was impressed as the ungainly, belching monster easily crossed a trench line and made sauerkraut of barbed wire. Could this be the answer to the terrible deadlock of the Western Front could it finally be broken and the path to victory unlocked? It had been realized since the war bogged down into messy attrition that the advantage lay heavily with the defenders, well dug in, machine guns sighted to provide murderous crossfire, trench-mortars and artillery in support.

To a generation of generals brought up in the spirit of the offensive, this was a baffling conundrum. The plain fact was that flesh and blood, the latter already poured out in torrents, could not hope to prevail. All the traditional values of dash, élan and courage just added numbers to the casualty list.

Despite this impressive debut, Kitchener wasn't impressed; happily, Haig was. He was a man in need of a breakthrough and his Somme offensive, which opened on 1 July, would be the biggest and bloodiest in history. For once he could rely on Lloyd George, then minister of munitions, to let slip the cogs of war and issue contracts for 150 tanks.

As befitted the naval origins of the concept, the vehicles were to be manned by 'crews' – eight men in total, one officer and seven ORs, drivers, gunners, and gears-men. All was top secret, so those who volunteered didn't quite know what they were volunteering for. Riding to war in an armoured dragon might seem a less depressing prospect than footslogging over no-man's-land in the teeth of withering fire. It wasn't. The heat and exhaust fumes inside the overcrowded hull were atrocious, exacerbated by the extremely rough ride. Noise levels were so bad it was nearly impossible to communicate and as there was no tactical radio, the officer had to dismount to liaise with his accompanying infantry. The beasts were thirsty and unreliable; they could make no more than 4 mph and needed to refuel every 25 miles.

Bizarrely, there were both 'male' and 'female' tanks, though successful breeding seemed unlikely. Males carried two 6-pounders in the side turrets or sponsons and three Hotchkiss MGs to provide hosing fire. Females carried four Vickers MGs and a lone Hotchkiss. They were all petrol-engined with tracks being driven through a complex set of three gearboxes, the 16-litre, 6-cylinder Daimler engine was under-powered to shift the 28-ton hull.

Could these new-fangled monsters change the course of the battle? Any early hopes of breakthrough had soon died, as had many thousands. By mid-September, Haig badly needed a result. One Tommy who saw the leviathans come rumbling up from the skeletal remains of Delville Wood was Private Arnold Ridley, destined to be far more famous as his fictional character, Private Godfrey. Of 50 tanks deployed, less than half made it to the start line and seven more promptly broke down. There was no breakthrough, but the tank proved its worth and would go on to spearhead the break-in battles of 1918.

Fray Bentos – the real *Fury*. I'm sure many readers will have watched the gung-ho US movie *Fury* where Brad Pitt, following in the footsteps of Errol Flynn and commanding a disabled Sherman, stands off a whole SS Battalion in Germany.[69] That never happened, at least not in Germany in WWII. However, it did happen in the previous war, in Flanders, and it was a British tank, a Mark IV Tank F41, otherwise known as 'Fray Bentos'. Captain Donald Richardson, commanding the nine-strong crew, came from a line of grocers and named the metal warhorse.

On 22 August 1917, F 41 was part of a troop sent in against a fortified German outpost. Linear trenches were gone in favour of a deeper grid system fronted by a belt of solid bunkers, bristling with machine guns. The ground was a nightmare, the heavy, sucking mud of Passchendaele, as the whole battle would come to be known, dragged down men, beasts and machines. It was especially hard going for tanks. Richardson and Lieutenant Hill, his 2I/C, were both wounded by bullet fragments, Hill driving. Fray Bentos slid into a crater. This was very bad news as the vehicle was immobilized and its heavy guns were tilted at impossible angles. Private Brady clambered out to free the beams slung on the hull to provide traction in just this scenario, a good design feature but one which made the tanker a prime target for every MG in the district.

Brady was killed. Richardson couldn't send another man out, so their AFV was now a steel bunker, its armour ½ inch thick. Mobility was the key to survival; they'd lost theirs and shell on frantic shell plus a blizzard of Maxim rounds and small arms ricocheted off the hull. The armour plate held up, but the incessant battering caused red hot splinters to spit around inside the hull. Injuries mounted; private Budd was crushed when they attempted to get the tank moving and it pivoted horribly on the trapped gun barrel. Inside, temperatures soared. Being stuck and partly submerged in the ditch had its advantages. German gunners consistently missed, and the British began to fight back, using Lewis guns, rifles and the officers' Webley pistols. Although the main Allied attack had ground to a blood-soaked halt in the terrible slime, Tank F41 was forming its very own mini-salient into enemy lines. To counterattack, the Germans would have to knock out Fray Bentos.

They failed; each rush was seen off with piles of German dead building around the besieged hull. The fight went on all day and into that first night. Under cover of darkness a storming party got close enough for at least one to get a grenade in, it was thrown back, and the intruder shot down. Time and again, on the second day, repeated rushes came on and every time they were thrown back with more corpses joining the already impressive heap. Inside, one of the wounded died, more were wounded, some badly; food and water ran out. They were reduced to drinking filthy sludge from radiators. From their position they were, between rushes, able to enfilade the main German counterattacks, effectively holding back the whole show.

For a second long night, exhausted, thirsty, hungry, and bleeding, they held on; ammunition was by now running very low. Next day brought no relief and somehow they fought on until every round was used up. After 72 hours of constant battle, on the third night, Richardson decided to break out; there was no thought of surrender. Astonishingly, the seven survivors, in various states of disrepair, all made it back to Allied lines. Both Richardson and Hill won the MC, and every other man was also decorated.

Fray Bentos II, also commanded by Richardson, fought at Cambrai later that year. The image in the plate section, unpublished elsewhere, is from the archive of Paul Reed and part of the picture-postcard industry that developed as early as 1919 when the first battlefield tourists began to arrive in Ypres. The landscape was like nothing anyone had imagined, and they all wanted to take postcards of it home to show people. The level of devastation was staggering. Tanks were like strange monoliths on the landscape, and they appear in many such images from the time.

The face of battle had just shifted. And 'Bovvie' is where you really get it. The Tank Museum, Bovington, is a world-class collection, nobody does it better. We're regular attendees there for their annual 'Tankfest', serious heavy metal. Kipling, vicarious *beau sabreur* of the Age of Empire, who had sacrificed a son on the insatiable altar of industrial war, came to view the graveyard, a vast sea of twisted abandoned leviathans, in 1923. This gave him the idea of establishing a museum, but the project only reached fruition 24 years later, after the next round. A series of distinguished curators, including George Forty and David Fletcher, undertook a steady expansion.

Current galleries cover the rise of the tank and eclipse of the cavalry, the Great War role and into World War II, true heyday of heavy armour, then bringing the story up to the present day. Human history blends seamlessly with technical progress. There aren't many tanks that aren't shown at Bovvie, my own favourites, the Stuart, Chaffee, Conqueror and Chieftain, are all there.

Graham and I do have a thing about armour, even if he was light infantry. For a while until the reality of costs hit home, Graham was the owner both of a Fox Armoured Car and Ferret Scout Car.[71] The former had the inestimable boon of both power steering and a pre-selector gearbox, which took a bit of mastering but was smooth enough. The Ferret being slightly older, lacked PAS and you really felt it. Both had a prodigious thirst for fuel. I'm going back a decade here so the cost of running a classic armoured vehicle at 2024 prices will be prohibitive. The proprietor and staff at our local garage used to line up to salute and lay down a red carpet, by the time we got back the needle had slid alarmingly. Domestic pressures began to mount.

I remember interviewing DLI veterans (Italian Campaign), who'd expressed an infantryman's envy of the perceived greater security their armoured brethren enjoyed in the steel womb of their Shermans. Those who, as additional pathfinders, enjoyed brief sojourns inside the armoured carapace were quickly disabused. They found the heat, noise, claustrophobia and frequent stark terror utterly unnerving. I began to get that as we trundled along the quiet lanes of Northumberland, alarming groups of cyclists into shocked awe. This was Northumberland in winter where the relentless cold of being atop the turret was a constant discomfort, but to be fair, nobody was shooting at us.

At Bovington:

John: What about armour then, do tanks have a future when your five-thousand-pound missile can knock out a five-million-dollar tank with such relative ease?

Graham: No point asking a light infantryman. Comrade Putin certainly thought so, sending in his armoured columns at the start of the fight.

John: In part, that comes under the generic heading of just plain daft.

Graham: Well, when you begin by defining a war as just a 'special military operation', you may have miscalculated before you started. You don't have to be Rommel to know launching a major offensive at the end of an eastern winter isn't clever. Worse if you assume like, say, when fighting Georgia, your opponents will simply give in to shock and awe. Not much of either when you're stuck fast in the clarts.

John: doesn't mean tanks are fully obsolete?

Graham: God, no. In open country with proper support and adequate air cover, tanks can still do the business. Best, as ever, if they bypass urban centres, avoid getting bogged down in street fighting, look how NATO's upgrading its inventory. Loads more fun with tanks to be had yet...

5

East of England

War is cruelty. There is no use trying to reform it. The crueller it is, the sooner it will be over.

W. T. Sherman (The State of Georgia would not necessarily agree.)

Peace is but a cessation of hostilities in a war that is never-ending.

Thucydides

At the Beresford Arms, Whalton, Northumberland:

John: Why did you decide to join the British Army in the first place, and did your initial experience conform to your expectations, if not, how did it differ?

Graham: I suppose, despite having gained a law degree, I wanted to do something different, excitement if you like, it certainly wasn't about serving Queen and Country per se, though I was proud to do so.

John: Do you think northerners are different to southerners in their approaches to both service and conflict generally?

Graham: Well, as you know, the Rifles recruits heavily in our region but mainly squaddies and NCOs, officers come from a wider spread. There are strong traditions of service in this region, part of that's an economic response as recruiting always is, but there is a pride in serving. I believe that's less so in the south and there's more anti-military sentiment amongst the left urban elite.

John: What did the notion of service mean to you at that time and why did you choose the light infantry role?

Graham: It's hard to define. I suspect it was just the idea of adventure, not some boring office job, hard to say at that age, you don't necessarily analyse these things at the time. Light infantry, well that just appealed and we had the example of the DLI to look up to.

John What were and are the main differences between officer cadets and ORs?

Graham: Well, most officer cadets came from middle class or upper middle class backgrounds, most, if not all had been to public schools. A good many also descended from military families, a conscious career choice. ORs were different, more impoverished backgrounds, economic choice, or necessity but quite a few had fathers and brothers who served. And for a good few, especially those with NCO potential, it was a conscious career decision.

John: How would you define leadership? And how do you define morale and its importance?

Graham: Bloody hard question and I'm not sure it can be easily defined. You need natural authority and self-confidence, got to be born with those I suppose but the training hardens and sharpens, and the army is immersive. It gives you the tools and expects you to use them. Morale is bloody vital and that's a core part of an officer's job. Being banged up in crappy police barracks in NI for weeks is debilitating and sapping. That's when you make or break as a platoon commander. Being on active service is easier oddly enough, as there is the adrenalin and excitement of action or certainly the possibility of action, it's the slow attrition of boredom that's harder to deal with.

John: How do you see the role of HM Armed Forces now as opposed to when you joined?

Graham: We were waiting to fight the Russians in a grand industrial conflagration in North Germany. Even Op. Banner was a kind of sideshow. All eyes were focused on the North German Plain and NATO vs Warsaw Pact. The wars today's soldiers have faced are completely different. Counter Insurgency wars like Iraq and Afghanistan feel more like Kipling than *Dr Strangelove*.

John: What do you perceive as the primary challenges in the 21st century?

Graham: A different world or, until recently I'd have said it was. I'd have said like everybody else, that the age of major interstate wars was over. Well, Vlad the Invader surprised us all and changed everything. Turned the clock back. Otherwise, I'd have said the main challenges this century were the rise of ISIS and its affiliates/successors, unrest in uncontrolled swelling of impoverished urban populations, wars of failed states like Syria. But now the rule book's out of the window altogether.

I've always liked East Anglia, the sweep of Norfolk coastline and the quiet intimacy of Suffolk. It's very different in so many ways from Northumberland, less stark, but in some ways almost more remote and, for me, defined by its castles. Orford's lean angular tower brooding over a flat and largely featureless coast reminds me of Charlton Heston's donjon in *The Warlord* (1965, directed by Franklin J. Schaffner). For whatever reason, it's not where I'd necessarily expect to find a museum dedicated to the Cold War, though of course the region has its strong wartime and post-war connections to RAF/USAAF bases.

Bentwaters Cold War Museum

Some years ago, I took a group of year 8 pupils on a guided trip to visit the Great War practice trenches at Silloans on the Otterburn MoD Ranges. In the afternoon, I took them to the fringe of high ground looking east down over Holystone and the peaceable reaches of Coquetdale. In the opposing slopes of a gentle V-shaped valley, squadrons of skeletal tanks, Russian T 55s (I think), on one side face off against Chieftains on the other, frozen at the point of contact, their mute barrels upraised as a final salute to the Gods of War. *Avē Imperātor, moritūrī tē salūtant.*[1] Here they are, the hollowed out sentinels of the Cold War, deployed in a final berserker charge, a mini battle of Prokhorovka[2] in the Northumbrian Fells, half shot up, eerie and inert. The kids loved the spectacle but none of them had ever heard of the Cold War.

The Cold War for God's sake. We grew up with it. Omnipresent. Entire focus of the West's defence strategies. For half a century Britain maintained her armed forces at wartime levels to meet the Beast from the East. The Beast never did come though, and then, overnight it seemed, the Berlin Wall came down as the whole Soviet 'Empire' imploded, and it was Peace in our Time. Well until February 2022 anyway, when, to all our surprises the Cold War sprang back into being and suddenly warmed up. As I write, the spectre of Armageddon hangs over us once again.

Everyone thought that Germany, dismembered by the victorious allies would be the battleground. According to the terms of the victors' deal, the Potsdam Agreement:[3]

- Germany west of the Oder-Neisse line was to be divided into four zones, USSR, UK, USA, France.
- The Soviets subsequently refused to allow post-war reconstruction (Marshall Plan) in their sector. Stalin wanted an emasculated rump of Germany left impotent.
- Stalin also hoped to squeeze the UK out and imagined the USA would retire, opening the way to extend the Eastern Bloc across the whole of Germany and pave the way for a satellite, communist-controlled state. That didn't work.
- The East became a ruthlessly oppressive police state with its dire mechanisms, *Stasiland*. Huge differences between Federal Republic of Germany (FDR) and German Democratic Republic (GDR) arose, massive economic growth in the west during the fifties, total stagnation in the east, *the Valley of Unknowing*.
- In 1948 Stalin instigated the Berlin Blockade, which led to the Berlin Airlift, a PR disaster of the first magnitude for the Soviets, the communists' share of the vote in FDR slashed, the blockade lifted in May 1949.
- Before the frontier was closed, some 3.5 million people escaped to the West from East Germany. The Berlin Wall with all the pathos, drama and spy-fiction writer's heaven that surrounded it became the very symbol of the Divide, its eventual if unexpected tumbling signalling the ostensible demise of the Cold War.

An early manifestation of the conflict here in the UK was a form of early warning system or 'ROTOR' – and this resulted in the construction of a chain of radar stations studding the entire east coast of mainland Britain.[4] Where the threat was perceived to be greatest, these were placed underground and that at Troywood in Fife (1951)[5] is the largest survivor, now restored and open as a museum. These bunkers were massive affairs. An initial excavation was sunk to a depth of 150 feet (45m), and a bed or lining of aggregate was topped by a reinforced concrete structure, some 10 feet (3m) thick, the reinforcement in the form of one-inch (25mm) tungsten steel rods. An outer skin or casing of brick was revetted with netting soaked in pitch. Spoil from the excavation was used to form a mound, laced with additional concrete blocks, to provide 'burster caps'. From the surface the landscape remained innocuous, a typical Fife farm steading.

Despite this enormous outlay in sweat, ingenuity and taxpayers' cash, the site was run down in the late fifties but enjoyed a new lease of life after 1958, when it was re-designated as a civil defence bunker, suitable for use as a Regional Seat of Government (RGS). This new role endured for a decade until the demise of Civil Defence, when it was again re-designated and re-fitted as the Central Government HQ for Scotland in the event of a nuclear conflict (RGHQ). Preparing the site as an emergency location for the Scottish administration necessitated a considerable re-build, and the place was due another makeover in 1992, but the dwindling threat from the Soviet bloc overtook the rebuilding and it was officially mothballed in 1993. On 1 April 1994 it opened again, this time not as a top-secret establishment but as a visitor attraction. Perhaps part of the essence of heritage, yesterday's threat is today's museum.

The North Atlantic Treaty Organisation (NATO) formally came into being on 4 April 1949 as a planned response to perceived Russian aggression. The confrontation had begun when allied and Russian forces joined hands on the Elbe in May 1945. No sooner was the Second World War over than the foundations and battle lines of the next were being laid down. The Soviet riposte came on 14 May 1955 with the formation of the Warsaw Pact, with Northern Germany being viewed as the likely battleground.

Both sides, at the outset, believed an armed confrontation was inevitable and this dark shadow hung over the world for nearly two generations. Sea power was expressed not only by surface fleets but in the form, sinister and deadly, of the nuclear-powered and nuclear-armed submarine, carrying its frightful payload of ballistic missiles (SLBMs). Britain had entered the first rank of nuclear powers, albeit as a poor relation and with the deterrent initially focussing on a force of medium range 'V' bombers.[6] A shift to sea-power came during the sixties.

Organised and large-scale anti-war movements are, as far as we can surmise, a peculiarly twentieth-century phenomenon, beginning with groups such as the Peace Pledge Union (PPU) after the Great War. The nuclear threat produced a reaction in the form of the Campaign for Nuclear Disarmament (CND). Unilateral disarmament was an early mantra of the more radical left, the idea, arguably dangerously naïve, was that if the west disarmed, Stalin and his successors wouldn't trouble us. Those behind the Iron Curtain had a more realistic view. When I was young, we had several Hungarian families living in the area. They'd got out of Hungary in the wake of the Russian retaliation in 1956 and had no illusions.

In the nineties, once the threat had evaporated or seemed to have done, I recall seeing in barns across North Northumberland an impressive array of Soviet era armour and ordnance, a T72 where I might have anticipated a combine harvester and the odd 155mm howitzer. A lot of this was pulled together by a local very enthusiastic enthusiast, who was intending to create a museum to the War 'that never was'. Perhaps it's as well that's still on a back burner somewhere.

RAF Bentwaters came into being during 1942 when the Allied bomber offensive against the Axis was beginning to gather its fearful, near biblical momentum, poised to rain fire on Hitler's sacred Fatherland. Coventry and the Blitz were about to be avenged with accrued interest. This site, near Woodbridge in Suffolk, apparently takes its name from a pair of artisans' cottages sited on what would become the main runway. Anyone my age will remember Godfrey's Cottage, similarly imperilled, in *Dad's Army*. The aerodrome, operated by Bomber Command, opened its runways in April 1944, in time for the D-Day landings. Before the end of that year, it was transferred to 11 Group RAF Fighter Command.

From early May 1945, just before the end of the war in Europe, USAAF fighters based there were flying shotgun for Bomber raids over Germany, as Hitler's 'Thousand Year Reich' ran into the sand after a dozen. The Allies erstwhile ally, Soviet Russia, became the new foe. By March 1951, the base was formally leased to the USAAF, home to 81st Fighter Interceptor Group. They'd be there for 40 years, playing a vital role in the projected defence of Western Europe. They set out with F-86 A Sabre Jets, then F4D Phantoms and finally A-10A Thunderbolts. On 23 March 1993, the last mission took off from Bentwaters. Within another three months the place had been de-commissioned.

And now it's a museum, a Cold War Museum[7] – entirely fitting given the provenance. Bentwaters' curatorial team have fully restored the War Operations Room, Battle Cabin, BT Telephone Exchange, decontamination showers and airlock. There are exhibition spaces highlighting its wartime role, unit histories, especially that of 81st FW, plus space dedicated to its twin-base complex at RAF Woodbridge, which features a display on Special Operations/Rescue Squadrons based there and the Aggressor Squadrons stationed at Bentwaters.

There are nods to other linked sites – USAF Cobra Mist on Orfordness, together with 2164 Communications Squadron.

On 24 February 2022, Russia invaded Ukraine, disingenuously badged as a 'special military operation'. This was the start of a major European war, essentially the first since 1945 (I exclude the Balkan Conflicts of the 1990s, as being civil strife). All those old Cold War demons burst out of their closets and started jumping up and down again. At the time of writing, this new chapter is very far from being completed and the aftershocks will reverberate for a very long time. (See below, *The Way Ahead*.)

Norfolk Tank Museum

I first got excited about tanks watching a 1960s Hollywood blockbuster, *Battle of the Bulge*.[8] Today's armour historians would scream the place down, as the movie admittedly bore only a passing resemblance to the bitter battle in the Ardennes during that deep winter of 1944/1945. Robert Shaw, a British actor, did a good job of portraying a ruthless Nazi tank general, based I suspect on Joachim Peiper. Robert Ryan, Henry Fonda, Charles Bronson, and Telly Savalas took him on and won – no surprises there. Most of the action was filmed in the far more agreeable weather conditions in Spain and Shaw's King Tigers were in fact US M47 Pattons. Robert Ryan's M4 Shermans were in reality M24 Chaffees. At age 12, I didn't care.

German armour and superior tactics would bring dazzling results when they bounded out from the Ardennes in May 1940 and burst across the Meuse at Sedan. Late 1940 and the launch of General Wavell's offensive Operation *Compass* would initiate a new era of armoured conflict. The Western Desert appeared to open up a vast canvas over which tanks could race, sleek as frigates and potent as ships of the line: pure warfare where the featureless desert formed an inland sea, unencumbered by towns, rivers, forest or marsh, and with a tiny, nomadic and largely unconsidered civilian population.

Reality inevitably proved to be somewhat different. In 1939, numerous British armoured units had only recently converted from cavalry, either regular or yeomanry[9] regiments with proud traditions and social cachet. These tended to rather look down upon the rude mechanicals of the RTR who, in turn resented the amateurs' condescension. Infantry disliked all alike. Relationships between infantry, armour and artillery were frequently based on ignorance, disdain and suspicion. Infantry in the desert felt exposed without armoured support and resented the ability of their mechanized comrades to withdraw from the front of an evening and their ability to stock up on luxuries. They did not understand the need for tanks to re-fuel and replenish and the intricacies of maintenance in trying desert conditions, dust and harsh surfaces taking a fearful toll on moving parts.

British armour often proved to be inferior to its German opponent and the reasons for these deficiencies were partly tactical and partly technological. The inter-war years had seen a rivalry between the 'tank' school, led by J. F. C. ('Boney') Fuller with Basil Liddell Hart, and the 'cavalry' school, who chose to ignore telling lessons from the Great War. General Sir Percy Hobart[10] was a passionate advocate of the former and championed to the full the Fuller/Liddell Hart concept of armoured warfare.

This envisaged tanks as an independent, striking arm not dependent on a tail of infantry and guns but playing a cavalry role in firstly pinning enemy forces by a frontal feint, then executing a series of flanking movements to strike the foe from side and rear. To exactly what extent these theories contributed to *Blitzkrieg* is not our present concern, but Guderian's exploits in Poland (though not strategically decisive) and the crushing blow delivered in France in 1940 served to reinforce this view. Thus, British doctrine was based essentially on a cavalry charge 'Balaklavering' – where tank fought tank. What was not fully considered was the crucial role to be played by anti-tank guns.

In 1938, General Percy ('Hobo') Hobart had been appointed to lead what would become 7th Armoured Division.[11] He lasted only a year before being dismissed and disappearing into temporary obscurity.[12] His theory influenced his successors, and the success of Operation *Compass* in dismembering Il Duce's army in Libya appeared to fully vindicate the 'all-tank' approach. How a tank might best deliver fire was also a subject of debate. Cautious counsel preferred static fire from a 'hull-down' position, where the vehicle takes advantage of natural cover or folds in the ground to expose only the turret. Such fire will be, by definition, more accurate than where the tank fires on the move. Officers in Britain's Royal Tank Regiment were great exponents of the latter school and trained hard, but movement must inhibit accuracy and if the tank manoeuvres to deliver fire side-on, akin to the naval broadside, it is consequentially more exposed to AT guns.

Within British concepts of armoured warfare lay a dichotomy as to the role of the tank and thus the creation of two distinct types. First of these was the 'I' or Infantry Tank whose role, as its name implies, was to advance in support of the footsloggers. The early versions during the Great War had lumbered over desolate trenchscapes at ponderously slow speed, unable to exploit a breakthrough when one occurred. That role was better accomplished by light or 'cruiser' tanks. Both the recently converted cavalry formations with their long traditions and RTR itself preferred the notion of speed and dash.

Passing through the front line to wreak havoc in the enemy rear echelons, was a function performed in the First war by light Whippet tanks and armoured cars, which, though not decisive, had filled the role more spectacularly than cavalry. In the armoured regiment some three or four tanks would form a troop, four of these plus an HQ component of three formed a squadron and three squadrons a regiment. Behind the AFVs was an 'A' or logistics echelon, with the 'B' echelon for those who for whatever reason comprised the LOOBs relegated to leaguer.

Protection, mobility, and firepower are the three principal concerns of the tank designer, and every variant thus embodies elements of compromise. A tank which moves fast cannot carry a greater weight of armour, a heavier tank is likely to be slower. Attempts were made to improve the armament on the Crusader Mark II and Valentine Mark II by replacing the obsolete 2-pdr with a 6-pdr. This conferred limited benefits. The Crusader III suffered from the same lack of adequate armour and mechanical reliability. Valentine IX, whilst gaining a 6-pounder, lost its 7.92 mm machine gun, a significant omission. It was not until the introduction of Sherman Marks II and III (fitted with twin diesel motors) that the Allies acquired a tank that could hope to match its Axis opponents.

Shermans boasted strong tracks and heavier deflective sloping armour. The 75 mm M3 main armament was a distinct improvement on its predecessors, including the 75 mm M2 mounted in the sponson of the Grant tank. A further crucial disadvantage of Lee/Grants[13] was the severe restriction imposed on the tanks seeking a hull-down position. Mounting its main gun in the sponson inevitably meant the turret remained fully exposed. Better M 48 HE shells gave the Sherman more effective killing power against German armour.

British Churchill tanks were other welcome newcomers to the Desert War. The Mark I was still lumbered with the 2-pounder, though the Mark III was up-gunned with a 6-pounder. The Churchill proved rugged and reliable in service and its cross-country capability was remarkable. Its armour was effective against the German 5-cm Pak (*Panzerabwehrkanone*) anti-tank weapon but ineffective against the dreaded 8.8-cm.

If German armour was initially superior in design, then German tactics were equally superior in execution. The élan of the cavalry was not an element in German tactical thinking. Rommel did not expect his panzers to engage in Homeric duels with their enemies. Killing Allied tanks was primarily the function of anti-tank guns, leaving the armour free to punch through and wreak havoc. When the panzers did take on Allied tanks, their superior armour and armament told. British cruisers might sweep forward in a gallant but doomed charge, but well-sited anti-tank weapons exacted a fearful toll. The issue decided before the panzers discharged their first round. Indeed, the German armour might be nothing more than bait to tempt British opponents into pelting towards a well concealed gun line.

Bltizkrieg as a concept had always depended not primarily upon tanks but on an all-arms offensive, avoiding dispersal and concentrating armoured forces into a decisive fist, avoiding any tendency to spread tanks around in penny packets. In such dispositions, speed was not the first imperative and tank commanders' urges to rush ahead should be curtailed to maintain cohesion. To retain this high level of overall mobility, the ability to recover and repair damaged vehicles was paramount. Thus, there was an incentive to seek to hold ground won at the end of the day rather than retire to leaguer. Radio communication was an essential tool. There could be no victory without it.

With an all-arms concept, firepower and aerial bombardment could and should properly be employed to unleash a hail of destruction, pulverising enemy positions prior to any advance. Towed anti-tank guns accompanied armoured columns to be used in the offensive as well as in the purely defensive role. Reconnaissance was another vital ingredient. Enemy positions had to be located and then, after softening up, assessed to identify if the bombardment had produced sufficient cracks to allow an advance to proceed. If the commander could not be satisfied that the targets were sufficiently reduced or damage, then he should not hazard his tanks.

Deutsche Afrika Korps (DAK) employed two main workhorses in its Panzer divisions. The first was the *PzKwIII*, designed in the mid-thirties, with a five-man crew, and armed initially with a

3.7-cm gun. This relatively light tank, around 15 tons (15.2 tonnes), underwent a constant series of improvements and modifications mainly involving more and better armour and heavier armament. By the time of the Desert War many were armed with the vastly better long 5cm KwK L/60 (model J); models N and O were fitted with the low-velocity 7.5cm KwK L/24.[14] Second was the *PzKwIV*, heavier at some 20 tons. It was originally intended as a support tank with the *PzKwIII* being the main workhorse.

Gradually the Mark IV grew into this role and continued in use throughout the war after production of the Mark III was abandoned in 1943. Early models were armed with the short-barrelled 7.5-cm KwK L/24 until this was replaced by the long-barrelled 7.5-cm KwK L/43. Like the Mark III, successive variants had increasingly improved armour.[15] The long-barrelled 7.5-cm became the preferred gun for all panzers during 1942; HE rounds were becoming more common than AP and hollow-charge ammunition for this gun was issued from the summer of 1942.[16]

In the autumn of 1942, General Bernard Law Montgomery attacked Rommel's mixed German and Italian forces. By now the pendulum of Desert Warfare had swung tellingly in the Allies favour. This second and ultimately decisive Battle of El Alamein was hard, bitter and bloody. Monty's armour struggled to penetrate the crust of Axis defenders but by the end, the line began to fracture, then splinter. British armoured cars sprinted through the cracks. More sedate, the tanks of 22nd Armoured brigade were also moving forward, at last in open ground. The infantry was still unimpressed at the sight: 'They flew pennants. I did not know these little flags had any meaning. I thought they were typical of the bullshit bravado of tank mobs.'[17]

Some, however, witnessing the beginnings of the developing tank battle between 7th Armoured Division and Italian XX Corps, were caught by the imagery of warships, majestic on the ocean of the sands: 'The tanks – there seemed thousands of them – looked like a whole fleet of little boats on the sea. Miles away, to the left of me, there was a great dark cloud where there was a great "stonk"[18] going on. There were little shiny specks dodging about. I could see flashes where there were guns firing.'[19] The fight for North Africa was very far from over, but this would be the final swing of the pendulum.

I'm a big fan of this museum, armour does it for me and it's conveniently located a couple of miles from Long Stratton on the A140, Norwich to Ipswich. The collection includes the Cold War. Most of the armoured fighting vehicles are either runners or on their way. Steve MacHaye, the wonderfully enthusiastic curator, loves all his vehicles (a big nod towards the Centurion)[20] but he has a great affection for Deborah II, a rebuilt Mark IV tank (as featured on *Guy Martin's WWI Tank*). Fittingly so, as his grandfather served in one during the great tank battle at Cambrai in late 1917.

As to Centurions, the collection includes an AVRE[21] Mk13 Gun Tank and an ARV, whose mighty Meteor engines can be heard seductively roaring during an annual 'Armourfest' held in August. There's a Chieftain (love those) and Saladin Armoured Scout Car. For those seeking further stimulation, you can experience an off-road driving course in a Hagglunds BV206, All Terrain Vehicle; depends on how much fun you can handle.

In conversation with Stephen at the Tank Museum:

John: Love to hear more about the Mark IV project. I've frequently taken tour groups to Cambrai.

Stephen: Where do I start? My grandfather died before I had a chance to talk to him, but my Nanna recalled his stories: he told her how the Germans would set fire to early tanks with the petrol tanks on the tanks' roofs. The conditions in the tanks, and especially the conditions in the trenches on the Western Font, was horrendous. The soldiers were often knee deep in water and suffering from Trench Foot amongst all the other miseries of warfare. These stories ignited my passion for military history and tanks, particularly the tanks of the Great War.

North One TV approached me and the Norfolk Tank Museum in 2017 with the idea of building a working replica Mk IV tank to celebrate the battle of Cambrai. The Battle was the first ever large-scale use of tanks. There were 375 tanks, plus supply vehicles, that attacked over a 10-mile front; one tank per 50 yards.

The plan was to build the tank with the help of JCB and Chastead UK; if we were going to build a replica then it had to be as accurate as feasibly possible. JCB reproduced the hull, Chastead the tracks and small machined components, and the Norfolk Tank Museum assembled the drive train and constructed the tank as a whole from all the parts. On 9 November 2017, after six months' incredibly hard work, Deborah II as she is now known, was loaded onto a lorry destined for Cambrai to commemorate all those that sadly lost their lives a 100 years ago. Deborah II is now on permanent display here at the Norfolk Tank Museum for everyone to enjoy.

John: What inspires you and your fellow trustees and volunteers to start and manage the collection?

Stephen: The collection began many years ago as my private collection, inspired by the stories my grandmother told me and my love of history and engineering. The first real piece in the collection was my grandfather's webbing belt and holster from WWII, which I found in an old barn at my childhood family farm. He had been in the Home Guard and had helped with the anti-aircraft guns on the point at Canvey Island. Following on from that first piece I started collecting deactivated firearms, uniforms, and personal kit, then progressing to restoring my first Land Rover and a German Auto Union Munga. [The name*Munga* comes from the phrase *Mehrzweck Universal Geländewagen mit Allradantrieb* – 'multi-purpose universal off-road car with all-wheel drive.'] When I turned 25, I bought my first bit of 'armour', a Ferret Mk 2/3, which I eventually swapped for a Saladin Armoured car two years later. Everything else really just followed there and in 1997 I bought my first Main Battle Tank, a Centurion Armoured Royal Engineers Assault Vehicle.

In 2009, I decided to open the MacHaye Collection to the public. We started with open days a few times a year, progressing to two Sundays a month, then weekends, and finally to four days a week April to October. We close to the public in the winter to carry out restoration In 2018 we registered the Collection as a charity and we are now in the process of becoming an Accredited Museum.

John: Why do you see this aspect of preserving the UK's military history as important? Needless to say, I'm already a believer.

Stephen: Firstly, I believe all aspects of history need to be preserved: how can we learn from the past if it's no longer available to learn from? For me military history is special, it's a combination of my own passion for the subject, especially for the last 150 years with the explosion in technology. With rifles we have moved from the musket to the Bolt Action Rifle, to the machine gun. From tanks that could travel at 4 mph through to the Challenger 2 at 45 mph and being able to destroy a target 2.5 miles away. And, of course, the soldiers.

Where would we be without them and the ultimate sacrifice some of them have made? If we don't remember them, then all was in vain.

John: How do people react to the collection?

Stephen: We get a very positive reaction at the museum. Some people ask me, 'Why on earth did you start the collection?' and the favourite, 'Where do you get a tank from?' and I explain it's just a passion and a love for engineering and history.

John: What would be your dream acquisition?

Stephen: I am very lucky, because I think I probably already have that. For many years I had dreamed about the very first tanks and what it must have been like to drive them. In those early days the crews were learning every day how to get the best out of the tanks and so were the designers. The very first Mk 1 tank appeared on 15 September 1916; in less than six months, by May 1917 we had the Mk IV with better armaments and a whole new fuel system that made the tank so much safer for the crew. My grandfather served in those very early tanks in WW1 it's such a shame that he died when I was only six and I never had a chance to ask him the millions of questions I have! If I could have another dream acquisition it would be ... a Challenger 1 or a famous WWII tank like a Churchill or a Sherman, have you got one?

Dad's Army Museum, Thetford

'They don't like it up 'em you know.'[22]

Corporal Jones[23] became a national icon along with Mannering, Wilson, Pike et al. *Dad's Army* does contain a great deal of truth: the muddling amateurishness, chronic shortages of weapons and equipment, Heath Robinson hardware and a wide divergence of personal backgrounds all strike a factual chord. But the Home Guard remains affectionately risible because it was never tested. Those who fought alongside worker's militias in Spain had witnessed a very different reality. In the event of an actual German invasion, the volunteers of 1940 would have been expected to fight and almost certainly they would have. A memorable scene from the TV series features Mannering's ill-assorted heroes manning a makeshift barricade, doling out their few shotgun cartridges and awaiting German tanks. Obviously, these never came; had they done so, the results would have been swift, brutal and anything but comic.[24]

If one tours the north-east coast and inland, evidence or warlike intent abounds: Northumbrian beaches are studded with tank traps, concrete and steel gun emplacements, pillboxes[25] and some superbly restored batteries, particularly Blyth[26] and Hartlepool.[27] In quiet Northumbrian lanes and in most unexpected places one comes across further emplacements, seemingly in the most random pattern. These are, in fact, traces of the various 'stop' lines set up as a defence in depth against invasion. It was anticipated the Home Guard would be charged with resisting the initial onslaught on the beaches, then holding a series of fallback positions to blunt the overall attack until a successful counterattack could be launched. Most would have died or been captured.

Sergeant John Williams, serving for most of the war with 9 DLI, did a stint with the Home Guard: 'We were sent to guard Vaux Brewery, which we did and, in the process, consumed a good deal of Vaux Ale! The Major, can't mind his name, was an old-type DLI bloke. Did some training with the 3-inch mortar – well, that was love at first sight!'[28] The LDV legally came into being on 17 May 1940. At the outset, this process did resemble 1914 and Kitchener's volunteers, in that there was a near-absolute dearth of weapons, uniforms, materiel and officers. Many who had seen service in the last, and indeed previous, conflicts found themselves back in command of troops, except their current companies were little more than an enthusiastic but untrained mob. Under the provisions of its initial statutorily defined role, the LDV was not paramilitary but an 'armed police constabulary'.

Their primary function was to observe German troop movements in the event of an invasion rather than resist, whilst maintaining a security presence at key strategic locations. This was rather too passive for the martial ardour of volunteers and commanders, who preferred a more proactive stance. Churchill, now Prime Minister, endorsed this, and by late July 1940 had instructed Eden to rebrand the volunteers as the Home Guard. The latter's resistance was based entirely upon the pragmatic argument that a million LDV armbands had been purchased and were now to be wasted.

Thomas Myers served as an NCO with 6 DLI but spent time as a Home Guard instructor; he was not overly impressed:

> The Home Guard I saw never had a regular instructor with them, they were very inexperienced; most were overage and very unfit. I did my best but half of them were generally laughing at me and usually just

> said, 'Come on, we're going to the pub.' Well, that was usually the end of training for that day! Some of them had been in the last war but this one was very different; you see we'd learnt a lot from the Desert and from Sicily; it was no longer a matter of just digging trenches. They didn't have many weapons other than rifles and whilst they were fine for guarding bridges and the like, you can't expect men like that to go running round the countryside rounding up paratroopers, can you?[29]

Not wishing to be excluded, women, initially denied a role in the Home Guard took to forming their own units such as the Amazon Defence Corps.[30] Those ladies of a less overtly martial inclination might have preferred membership of the Women's Home Defence,[31] which, though not official, had branches all over the country. Women were latterly admitted to the ranks of the Home Guard, though strictly in non-military supporting roles. Once the first terror of invasion had passed and as the Allies began to gain ground, the Home Guard units continued in a domestic security role, permitting regulars to take up active service overseas.

Once the initial rush, in clouds of confusion and frustration, had subsided and the Home Guard was a full citizen's domestic reaction force, provision could be introduced for its proper establishment. Organisation came into the hands of Lord-Lieutenants of individual counties together with the senior military area commanders. These then appointed area, zone and group delegates. Those who had seen service were permitted to retain their unit badges in addition to the LDV armband.

After August 1940 the HG units were affiliated to their local county regiments and used their cap badge; thus, all Durham formations wore the DLI bugle. Each community would field a platoon, grouped into companies and battalions in the usual way. County Durham was a region with many men, miners and farmers in particular, in reserved occupations. These now came forward in large numbers to volunteer. Of those who offered to serve, perhaps half had military experience from the Great War. There was no medical test or examination as such, a core level of basic fitness was simply assumed. Durham raised some 26 Home Guard Battalions, one of which was the 1st Durham (Blaydon) Battalion Home Guard. Its catchment area was spread across several small settlements: Byermoor, Marley Hill, Sunniside, Whickham and Swalwell. Some larger factories and businesses even had their own 'in-house' platoons – the Derwenthaugh Coke Works and Vickers Armstrong, just over the Tyne were prime examples. Larger settlements were often able to raise more than one battalion; Gateshead and Sunderland each raised two and Newcastle could boast three.

In 1943, the original zonal structure was altered so that battalions were brigaded in sectors, divided into sub-districts and districts. The Blaydon Battalion thus joined with Chester-le-Street, Lanchester, Consett, Hamsterley and Stanley to form the Consett Sector, its HQ in Hamsterley Hall. Consett Sector, together with Houghton-le-Spring and Weardale plus Sunderland garrison, comprised a sub-district, one of three in the Northumbrian District of Northern Command.

From an assortment of 'civvies' and old items of uniform, some of which had last seen service on the Veldt or North West Frontier, the HG were issued with denim battledress, standard ammunition, boots and forage caps. These gradually gave way to standard wool kit with usual impedimenta and caps carrying the coveted DLI badge. Webbing rather predated the '37 canvas pattern, it even predated the venerable '08 pattern which had sustained Tommy Atkins in the trenches. This leather gear had been volunteer pattern pre-Haldane's reforms and came in a rather jaunty lighter shade. Efforts at darkening the colour proved tricky and varied – the clear favourite being an overnight soaking in a chamber-pot (full). At the outset, the volunteers still wore the HG armband on battledress until distinctive shoulder flashes were provided. Some units, like the Rowland's Gill platoon, echoing the fads of Victorian forbears, also sported the DLI green lanyard. All wore the letters 'DHM' with the battalion number stitched below.

Albert Shippen served with the 4th ITC (Infantry Training Centre) at Brancepeth 1941–1944. He remembered small arms and bayonet training:

> Well, we used the long 17-inch bayonet and we lunged at sacks stuffed with straw. We did bayonet drill where you had your bayonet and the other fellow had a kind of six-foot pole with a boxing glove on one end and a loose ring on the other. He was there to parry, and you had to get your point through the ring; if you weren't smart enough the glove would come round and belt you! I also remember grenade throwing. We had a nasty accident when one fellow simply panicked after he'd pulled the pin and the lever flew off, he dropped the grenade back into the pit and another private and NCO couldn't get out quick enough; both were very badly injured, had to be rushed to hospital.[32]

From knobkerries and pikes to shotguns and rook rifles, the volunteers arrived armed with an antique arsenal of weapons, none of them suitable for fighting the *Wehrmacht*. One early suggestion was a form of pike constructed from a length of pipe fitted with a long sword bayonet, effective if the enemy was obliging or chivalrous enough to permit the volunteer to get within stabbing range. Gradually, a supply of rifles was sourced, some old US P17s from 1917, a reliable weapon though the round was incompatible with that of the Lee Enfield.[33] Some of the US .45 cal. Thompson or 'Tommy' guns[34] were provided, the sort of gun most Englishmen had only witnessed in the hands of James Cagney or George Raft. The lack of a reliable 9mm submachine gun to combat the German MP38/40 was remedied by the introduction of the British Sten, a runt born of wartime expediency.

Improvised rifle ranges appeared like a rash, as usually the units did not train with live rounds. The Rowlands Gill men practised on the riverside by Lockhaugh Viaduct, and local lads would dig out the spent bullets from targets as souvenirs. After 1941, grenades became more readily available in reasonable quantities. Ernest Harvey served as a private in the DLI depot, though at the outset he was too young to join up:

> I was 16 when war broke out, working as a trainee grocer. I'd wanted to join the navy but ended up for a while in the Home Guard. We used to meet in a local hall... Can't remember where I joined, wasn't the police station; you just had to be 16, no medicals. We met twice a week and I had to do some nights fire-watching at the Co-op. We didn't see many raids over Hetton, but Sunderland caught it quite regularly.
>
> Most of those who joined up seemed to be businessmen. Major Sharp, who was in charge, had a shoe shop in Hetton I recall, and there was Norman Land, he was the foreman at the pit. We had no rifles to start with, none at all, and they only arrived in dribs and drabs; it was quite something when you got your rifle mind! We did an exercise in woods once, I remember, and a pair of us got captured, not surprising as we'd spent the whole time catching rabbits. We had no live firing really, only got up to a range at Hetton Colliery every two weeks. The training was quite tough but as we were volunteers and not paid, if you didn't fancy turning up you just sent in your apologies. But most of us looked forward to it, it was very social, you met your friends and by the end we were ready for anything I'd say.[35]

A particular and for a while near-hysterical fear was of enemy parachutists. These were likely to descend in the form either of agents or mass air landings. The art of paratrooper warfare – 'aerial envelopment' (conquest entirely from the air) as one of its Axis champions, General Kurt Student, defined his ultimate ambition – was new. Both the Nazis and Soviets had considered the use of both parachutists and airborne. That is, glider-borne infantry to spearhead attacks by seizing vital installations by *coup de main*. In practice, the Germans achieved some brilliant successes; the capture of the Belgian redoubts an Eban-Emael in 1940 was a stroke of genius, as was the later British descent on Pegasus Bridge, curtain-raiser to the Normandy landings. Larger-scale operations – most notably Student's intended masterpiece, the conquest of Crete by airborne units alone in May 1941 – proved disastrous and though ultimately successful, a Pyrrhic victory, which persuaded Hitler to abandon further attempts.

It was, as much as anything, the fear of paratroops which commended the idea of the volunteers to Anthony Eden and the War Cabinet:

> I want to speak to you tonight about the form of warfare which the Germans have been employing so extensively against Holland and Belgium – namely the dropping of parachute troops behind the main defensive lines... In order to leave nothing to chance, and to supplement from our forces yet untapped the means of defiance already arranged, we are going to ask you to help us in a manner which I know will be welcome to thousands of you. Since the war began the government have received countless inquiries from all over the kingdom from men of all ages who are for one reason or another not engaged in military service, and who wish to do something for the defence of their country. Well, now is your opportunity.[36]

In 1940, in the wake of Dunkirk, this threat appeared very real. Home Guard units thus set up observation posts to watch the skies; shotguns were the handiest of weapons, devastating at close quarters. German paratroops, *Fallschirmjager*,[37] did not jump carrying small arms other than pistols; their weapons were dropped in separate containers, thus individuals were vulnerable as they hit the ground. 'Cromwell' was the code word indicating airborne assault was in the offing; 'Oliver' meant the parachutists were in fact landing. As their silk chutes billowed in mild summer skies, church bells, traditional alarum, would be sounded by the volunteers. In typical British fashion this led, in best

Dad's Army style, to some strenuous disputes as to who should have custody of belfry keys.

As the immediate threat of invasion receded, the Home Guard drew breath and continued. As weapons, uniform and kit began filtering through, a number of motorcoach and transport companies were formed. Their function, in addition to maintaining vehicles, was to sweep up requisitioned transport when need arose; both Northumberland and Durham had Home Guard Transport Columns by 1944. The former county, following yeomanry tradition, also formed a mounted patrol with a miniature marine contingent of half a dozen assorted small boats on Lake Windemere. Neither of these formations was ever tested in action.[38]

Where else would we expect to find a Dad's Army Museum other than in Thetford, which provided so many locations for the series? The team there offer regular tours and Captain Mannering even has his own rather impressive bronze statue. And if that's not sufficient, you can even see Jones's van. This is an odd museum for a past that never quite was – a TV version of reality, like a distorted double image. It's all great fun but it runs a risk of descending into faux-sentiment, offering a distorted view of what actually was, a Potemkin Village effect that, like the original can drown out truth. Still, well worth a visit!

Parfield Home Guard Auxiliers, Woodbridge

> In order to stay behind we needed somewhere to stay; and by sucking up to the Sappers we had already brought into being what might very loosely be called a network of subterranean hideouts in which not only the striking force – Strix and about 15 other idiots – but our far-flung, hand-picked collaborators in the Home Guard, would bide their time before emerging to wreak, in a variety of ill-defined ways, havoc among the invaders.[39]

For those members of the Home Guard who showed a special aptitude there was a secret guerrilla army; highly trained to cause maximum disruption to the invading Germans. Little is known about the Auxiliary Units, even today, but there were three battalions covering Scotland, Northern England, and Southern England and full of the cream of the Home Guard. These men, all hand-picked volunteers, laid down months of supplies of both arms and food, ready to go into hiding the day the invasion started. Their job was to slow the invasion down

> by blowing up arms and fuel dumps, bridges, railway lines and generally causing confusion to the invaders. Most knew if the invasion came, they would be on suicide missions.[40]

Anyone who has read Owen Sheers' atmospheric novel *Resistance* will have an idea of the conditions of active service for these behind-the-lines, last-ditch commandos. The book is both fiction and counter-factual but nonetheless paints a very intriguing picture of what might have been. We must be thankful the reality never occurred. Resistance movements sprang up everywhere the Nazi jackboot prevailed, most famously in France but more concertedly in Poland. There, resistors, the Home Army (AK), did rise in August 1944 and bled dreadfully in the grim attritional fight for the streets of Warsaw. Otto Skorzeny, Hitler's personal gladiator, was ordered, as the end drew near for Germany, to activate a 'Werewolf' network to harass the encroaching allies, plans that happily came to nothing.

> At our most forlorn moment when our army was pouring back from Dunkirk through gates we could never have shut against the invading enemy, Colonel Grand conceived the plan of organising through Britain a closely co-ordinated sabotage and intelligence network among the civilian population who would be left behind in any territories which the German armies might temporarily be able to occupy.[41]

Members of 202nd Home Guard Battalion were trained to act as behind-the-lines saboteurs and assassins in the event of a German invasion. A total of 39 specially constructed underground operational bases – bunkers – were excavated in the region, 27 in Northumberland, the remainder in Durham. This was a highly secret war in waiting; those who took part were thoroughly trained in the arts of killing and mayhem. Recruits were not old, these were younger, very fit men, generally drawn from farmers and gamekeepers; an intimate local knowledge and skill at concealment was essential, as was the willingness to kill. Their role would have been to spread terror amongst both the invader and those who might be tempted into collaboration, including active Mosleyites. No mercy would be extended. Such was the level of secrecy employed that as the northern bunkers were being dug, dummy air raids were laid on to cover the sound of blasting.[42]

A seemingly innocuous publication, *The Countryman's Diary 1939*, covered a whole range of topics not normally associated with field sports – the Auxiliers, mostly trained at the Guerrilla Warfare School at Coleshill in Oxfordshire, were after rather larger game:

> There was an Auxilier Unit just four miles from Rowlands Gill; it was in the woods to the north of Beamish. The man in charge was the colliery engineer, Joe Harker. He recalled that when he was first recruited, he had to go down to Sunningdale, between Bracknell and London, for an initial briefing. Naturally he was told not to talk about the reason for his journey, and when he arrived, he was rather surprised to see that among the other recruits were the three men who had shared his railway compartment all the way from Newcastle.[43]

Command of 202 Battalion and its fellows, 201 (south of the Thames) and 203 (Scotland) Battalions, was exercised by Colin Gubbins, more famous as the founder and paladin of SOE.[44] The clandestine activities of Auxiliers in the northern region have never been fully examined by historians. In charge was 'The Intelligence officer for Northumberland', Anthony Quayle, later famous as an actor: 'He assigned a unit to attack his car which he parked on a remote road; after waiting beside his car for 2½ hours he decided that the men were not going to make it, but when he got back into the car he found two of them in the back seat and then he noticed that two more were under the car and the remainder had him covered from a ditch on the other side of the road ... the men of the Northumberland units were so highly regarded that they were given the job of protecting the Royal Family when they were staying at Balmoral.[45]

The Auxiliers were advised at the outset that their life expectancy, once they were behind enemy lines, would be unlikely to exceed 14 days. Reprisals for their actions against the civilian population were inevitable and would, given the German penchant for brutality, be draconian. The network could only function with the active cooperation of civilians, who would share equal risks in the event of capture: 'For instance a schoolmistress had ripped the treble hammers out of her upright piano to create a hidden space. A chicken farmer dug chalk out of his land to make an underground room in which to use his [radio] set: and a vicar told me no one would find his radio. It was hidden in his pulpit: "Who else but me goes up there?"'[46]

Removal of all traces and destruction of most of the hideouts was an act of policy following the end of the war. The Auxiliaries had been officially stood down in November 1944, by which time their numbers had swelled to around 3,000 and a cloak of secrecy thrown over their operations: 'What is the most frequent question asked by visitors to the Museum of the British Resistance Organisation? Answer: "Why is it still so secret?"'[47]

The assassination of potential collaborators had been finessed to a level where, it is said, each unit had a death-list, people targeted for killing in the event of invasion. This is a long way from Captain Mannering's war. People would live after the war in their communities knowing that they had possessed instructions to eliminate without trial or process – simple murder – numbers of those who might easily be neighbours and who went through the conflict with no inkling of this sword of Damocles hanging over their heads. All in all, this was a very sensitive business.

> I remember as a lad, I think it was 1951, we were scouting in the park, camping and we came across what we clearly saw was a bunker, very well camouflaged amongst the trees. We were really excited as there was a ladder down into the interior but the gamekeeper, a youngish fellow, chased us off, told us to keep away. Next day, would you believe a big concrete mixer turned up and filled the bunker in completely, sealed off the hatch and they dumped soil and vegetation over the site; I couldn't find it now.[48]

On 6 September 1944, as the savagery of the Normandy battle had receded and with Paris now liberated, the HG were clearly no longer needed and partially demobilised, finally disappearing on 1 November. Theirs had not been a glamorous role though always a potentially vital one. The crowning mercy was that they had not been put to the final test. Had this been so, then their history would have been very different. As the threat of invasion faded and finally evaporated, the way was clear for the affectionate legend of Dad's Army to grow. Those secret bunkers of the Auxiliers quietly mouldered in deliberate obscurity, the death-lists discreetly disposed of, but not necessarily forgotten.

Parfield, a former USAAF Bomber base, combines two primary exhibitions: first, 390th Bomb Group Memorial and second, the Museum of the British Resistance Organisation. Both are important but I plump for the Auxiliers, as they stayed so long in the shadows and their reality, had it ever been activated, would have been as far from the cosy fiction of *Dad's Army* as it gets. Had the Germans successfully invaded, our collective memory would be wholly different and by no means entertaining. The Auxiliers would have performed their role, we can be sure of that, and hundreds of families would have had cause not just to mourn their inevitable sacrifice but the loss of potentially thousands of civilian hostages murdered in reprisal. If we want to understand that reality, then we must go to Oradour sur Glane and witness the stark horror of Nazi tyranny and reflect on just how lucky we were here.

IWM Duxford

I first went to Duxford in the nineties accompanied by a squad or more or less enthusiastic children who, girls and boys alike, really loved it. As my eldest child has subsequently pursued a military career, I may have something to answer for. For a start, it's huge, an ex-RAF/USAAF base used in any number of films, perhaps most memorably in Guy Hamilton's brilliantly paced *Battle of Britain* (1969).

Since I first went, the IWM had added a splendid land warfare gallery, beautifully realised on a truly massive scale, but even that's dwarfed by the stellar dome of the USAAF 8th Air Force hangar, a building that commands as much respect as the exhibits. The sort of structure which, when I was young, you only saw in Bond movies. If Blofeld was to emerge from some unseen passage, you'd not be entirely surprised. It tells you all you need to know about the scale that a B-52 within doesn't look that big. It is big, of course, very, and perhaps one of the most impressive aircraft I've ever seen. How in God's name to you get that much metal up in the air and keep it there for as long as it takes to cover a potential range of several thousand miles?

There's plenty of the altogether flimsier machines from the Great War as well. In the opening, pre-credits sequence of the 1966 screen version of Jack D. Hunter's *Blue Max*, George Peppard as the ambitious anti-hero, Bruno Stachel, is slogging through the mud of Flanders, squalor, death and despair around him. A Fokker Triplane flies overhead. He looks up and is entranced by this residual vestige of distant chivalry, some knightly champion way above the filth, his machine painted a vibrant scarlet. No grubby *feldgrau* for him. Bruno knows where the glamour and glory is.[49]

That image has persisted. In the next war and while Tommy was lifted from the beaches of Dunkirk, exhausted, bloodied, and unwashed, Hurricanes and Spitfires knocked down Dorniers, Heinkels and ME109s. They went on doing it over the summer pastures of southern England, saved the day and possibly the world. 'The Few' became legend. Even now, if there is one military aircraft everyone knows the name of, it's the Spitfire. Its creator R.J. Mitchell was, arguably, the first casualty of the Battle of Britain, shortening his life to bring his vision from the drawing board to the air. Seventy-five years on and it's still our finest hour. Decades on I was taking part in a battle re-enactment, Vikings I think we were, going at it hammer and tongs when overhead, as part of a display, flew a Spitfire, the real thing and the deep growl of that Merlin engine was intoxicating. We didn't really want to be Vikings anymore.

Youth and beauty, fighting far above the choked morass, the excitement of violent death in lethal jousts, far nearer Hector and Achilles than Tommy Atkins, led the aces to premiership stardom. In reality it was very far from glorious, a hideous death from bullets, immolation or crashing just round every corner, with the certain knowledge that one day it would come. The survival odds were not encouraging.

Still, it was attractive. The gilded aircraft soaring above a desolate earthbound hell endowed these fliers with god-like qualities, their exploits hungered after by a public raised on the chivalric tradition of Malory or Scott. The war on the ground could never be seen as any kind of crusade, 'the good fight'. but in the air, it might just be different. Even Lloyd George got excited: 'The heavens are their battlefield; they are the cavalry of the clouds. High above the squalor and the mud, so high in the firmament that they are not visible from the earth, they fight out the eternal issue of right and wrong.'

The then Prime Minister goes on to extol the qualities of the airman, not just as an individual warrior pitted against a worthy adversary but possessed of this truly Olympian power to dispense death from the skies. He shoots up enemy trenches and lumbering convoys, swooping like the hawk to devour his lumpen prey, '... every flight is a romance; every report is an epic. They are the knighthood of the war, without fear and without reproach.'

Post Great War literary output resonates with chivalry – *Knight of Germany* (the career of Oswald Boelcke), *the Red Knight of Germany* (von Richthofen) and *Guynemer, Knight of the Air.* Much of this reverential mythology ignores the reality of air combat. It was hard, exhausting, physically demanding, morale- and resilience-sapping and always dangerous. Though the airmen, while they were frequently billeted under wet canvas, did enjoy better conditions than the infantry. The cavalry proper were largely redundant. Even when the war became mobile again in 1918, it was armoured cars and light tanks which led the charge. The long dominance of brilliant horsemen was at an end.

This idea of the ace, the celebrity warrior, persisted after 1918. Even during the war itself, as the size of the vast aerial fleets expanded and the conflict, like some titanic Moloch (as Lloyd George described Haig), witnessed larger and larger armies thrown into the fire, the predominance of the ace declined. Britain and the other combatant nations had entered the struggle in 1914 with a relative handful of primitive aircraft, slow, flimsy, unreliable and not at all suitable as gun platforms. The position a mere four years on was transformed. The big players all fielded thousands of fast fighters, reconnaissance planes and the new four-engined bombers. Strategic air power had arrived.

W.E. Johns served at Gallipoli as an infantryman but was commissioned into the RFC in 1917. He didn't take to the skies over the Western Front till August 1918, piloting a De Havilland DH4 two-seater. He was shot down and his observer killed. Johns spent the rest of the war as a POW. His fictional creation Captain James Bigglesworth – 'Biggles' – did rather better, joining up aged only 17 in 1917. He has a fast, furious, and successful career in the skies which extends to nearly a hundred novels, published between 1932 and 1999, the last some years after the author's death. Biggles embodies the romantic spirit of the ace, quintessentially English of course. Today he reads like the last hurrah of Empire before the word became toxic.[50]

Biggles was a bit too dated for the Second World War. I preferred Frederick E. Smith's classic *633 Squadron*, both the book and the subsequent movie.[51] The hero Grenville, flying Mosquitoes, like Spitfires one of *the* planes of the war, is more angst ridden than Biggles, more taciturn and withdrawn. His crack squadron will be sent on a near suicidal mission to attack a German heavy water plant secure in some remote Norwegian fjord. But not before Grenville has to fly a solo mission, one he hates, but necessary to eliminate any possibility of a captured agent being tortured into spilling the beans. The target is his girlfriend's brother. Enemy flak is tremendous. Despite the fury of the barrage, Grenville the lone crusader continues the unequal fight. When the Gestapo HQ comes into his sights, he gleefully strafes the fleeing torturers. The happy slaughter of these savage bullies is unlikely ever to count as any kind of war crime.

Whilst the flyers are the heroes, so are their planes, the 'Mossie' is an icon in itself. Planes are great fun, the realisation of man's long-held dream to reach the heavens. The first recorded instance of aerial combat occurred in 1913, during the Mexican Civil War when American airmen Phil Rader and Dean Ivan lamb, each hired by opposing factions, blazed away with revolvers from their primitive machines, a wild west moment in the skies. Neither was hit. Once war was declared in August 1914, duelling above the battlefield of that late summer became the latest fashion in aviation.

On 25 August, emerging French ace Roland Garros and Lieutenant de Bernis, flying a Morane Parasol, shot up a Boche plane which escaped by diving into cloud, though one of the two-man crew was injured. A few weeks later, on 7 September, a Russian pilot Pyotr Nesterov, actually downed an enemy aircraft but used his own plane, a Morane, as the weapon in a kamikaze style ramming tactic, taking on an Austrian Albatross. There were no survivors.

The first recorded actual victory was won by a pair of French aviators, Sergeant Joseph Frantz and his gunner Louis Quenault. They were flying a Voisin biplane, which in their case mounted a front-facing 8 mm Hotchkiss machine gun. This was do-able because the engine was a pusher rather than a propeller, rear-mounted behind the cockpit, giving the gunner a clear field of fire. This was not a policy decision; Frantz had acted on his own initiative in fixing up the MG; his comrades in V 25 *Escadrille* ridiculed him for his 'Jules Verne' idea.[52]

On 5 October, Franz was on a routine patrol near the hamlet of Jonchery-sur-Vesle in the Rheims sector. He gave chase to a German two-man Aviatik[9] and the enemy observer lifted a rifle, ready to open fire. Quenault opened up with the Hotchkiss. One round hit the German's fuel tank and the plane, belching smoke, smacked into a swamp. The pilot Wilhelm Schlichting was already dead, drilled by one of the Frenchmen's rounds, his observer Fritz von Zangen was killed on impact. First blood went to Franz, who won the Legion d'honneur and Louis, who got the *Medaille Militaire*. The duel attracted a fair-sized crowd: 'All the French troops on the spot forgot the danger of passing shells and jumped out of the trenches to watch the air fight.'[53]

Major Lanoe Hawker VC, DSO was the British *beau sabreur.* Like his soon-to-be nemesis, he was a hunter, though he'd transferred from the Royal Engineers. Hehad rigged up a .30 calibre Westley Richards stalking rifle on a bespoke mount that acted as an outrigger to clear his propeller. He turned like a goshawk and struck like a serpent. Rifles might be yesterday's weapon in the air, but he used his to deadly effect, Germans fell to his front-mounted Lewis gun, but the single lethal shot claimed a few more. So sure, so deft was his touch, like a Borgia assassin he'd manoeuvre to send that one round through an enemy's brain or his engine – either would be fatal. His policy was one of *a outrance* – all-out aggression. He designed a better clamping mechanism for the Lewis gin in his DH2, and also designed warmer, thigh-length sheepskin flying boots.

The Albatross completely outclassed the British 'pusher' and very quickly made its presence felt. Hawker himself survived several near misses. Like all aces, he became the 'must-kill' target for young bucks in the hungry new *jastas.* The RNAS was beginning to get newer and better planes, the Sopwith Pup and Triplane, but No. 24 Squadron were still stuck with the lumbering DH2s, those who survived anyway.

On 16 November 1916, pursuing an attack on German machines, Hawker, his wingman out of action, found himself up against von Richthofen, a duel of titans. But a very unequal one, the DH2 was an antique. The fight which followed was drawn out, tortuous and nerve shredding. The rising German ace made sure he never offered his adversary a steady target for his one deadly shot. They circled like mongoose and cobra, one strike would be enough, life and death.

As the infantry below, dirty, wet, exhausted, and lice-ridden, watched the show above, the two fliers executed a brilliant series of turns, each seeking the advantage. Hawker was heading for allied lines and safety, von Richthofen was determined he'd never arrive. Hawker's magical touch enabled him to side slip at the apex of the turn, frustrating the greater speed and power of the Albatross. He swooped and dived, turned, and jinked. But you can only dive so far until you run out of airspace.

After perhaps twenty minutes, both planes were very low indeed, Hawker nearly clipped the trees and narrowly dodged chimney stacks and rooftops, von Richthofen on his tail. Pockets of air would lift the DH2 like a magic carpet and give him a few hundred more feet of sky to work with. Von Richthofen was about 60 metres away as Hawker struggled to make the final turn that would swing him about to meet his tormentor face to face and which might just be enough. It wasn't, the nimble Albatross surged forward guns blazing, raking Hawker's plane.

One round took off the back of the Englishman's head and his plane spun down to burst into flames on impact. This was just south of Bapoume, and German soldiers recovered what was left and buried him nearby. Von Richthofen recovered the twisted Lewis gun from Hawker's plane as a trophy and they say he hung it above the entrance to his billet. L.G. Hawker has no known grave and is commemorated on the Arras Memorial. Von Richthofen called him the 'British Boelcke'. High praise indeed.

William Barker was a Canadian infantryman who had transferred to the RFC from the infantry, one of those almost unimaginably courageous soldiers who stood firm against the German gas attack at 2nd Ypres. He first flew as an observer but soon showed his hunter's credentials, bringing down an Albatross on his fourth flight. He went on to train as a pilot and showed exceptional ability even with the ungainly RE8. Moving to the far more deadly Camel, he quickly racked up nine kills. He served both in the UK and on the Italian Front. By 1918, the German Air Force was heavily outnumbered, but the sleek Fokker DVII was giving out a last hurrah. Barker transferred back to the Western front, joining 201 Squadron and flying the new Sopwith Snipe, the RAF's response to the DVII. By 26th October he had knocked down a total of 46 enemy planes. On that day he was ordered home and set off next morning for the flight home. This should have been uneventful.

He had scarcely been in the air for 20 minutes before spotting a sedate Rumpler, temptingly close. Beware the lure of the helpless victim, as many aces had already found to their cost. He blitzed the lumbering two-seater, which soon spiralled down to its death. He kept pace, making sure there'd be no second chances. But it wasn't just going to be a bad day for the Germans he'd just killed. Fire rippled through the air and a tracer round smashed the bone in his right leg. He couldn't get the damaged limb to respond, so lost right rudder. The pain was intense.

As he banked, he saw the size of the opposition. The odds were unfortunate. He was flying into the maw of a full *Jagdgeshwader*, four *jastas* in stepped formation. Three score enemy now confronted him and all flying DVIIs. Despite his wound and despite the odds, he just charged, knocking down the plane that had first fired on him. One down and only 59 to go! He swiftly made that just 58, his second victim exploding in flames.

His adversaries, confident his lone plane wasn't the tip of an unpleasant iceberg, gave him their full and undivided attention, bouncing his aircraft in packs of five with more in reserve above and below. Whatever manoeuvre he attempted, he'd be open to attack. Three hundred bullets riddled the Snipe and he was hit again, this time in the left leg. He could barely operate the rudder at all and was flying by joystick and throttle.

The cockpit was awash with his blood, and he temporarily passed out. the plane went into a spin but providentially, the rush of air brought him round. He'd dropped 6,000 feet and the hungry bees were still on his shot-up tail. Despite the pain and blood loss he was still full of fight and determined he'd ram whichever Boche got nearest. He still had ammunition and scored another DVII before he received his third wound as a round hit his left elbow. He passed out again and the aircraft went into another spin. Somehow, he came round again and got the Snipe into a straight line. He could now only use one of his four limbs. Being so low discouraged his pursuers and in his lucid moments was able to think about how and where he might land.

Where was a handy field, just behind the British lines. How was pretty rough. He skidded down at 90 mph, sheared the wheels and numerous other parts, bounced a few times and ended up with the shattered undercarriage in the air. The Scots from the Highland Light Infantry who pulled him free assumed from his shop-worn state and the amount of blood decorating the wrecked cockpit that they were dragging out a corpse. They weren't. Though in a coma for ten days, he made a full recovery and was given his VC by the king in person. He had brought down four enemy aircraft in forty minutes and survived the most unequal aerial duel of the entire war. A dozen days later and it was finally over.

The war in the air was not glamorous. No war ever is. It was not chivalric or poetic. It was hard, bloody, brutal, lonely, and perilous. The rate of attrition was fierce and relentless. Nonetheless, these fliers, these superlative aces, were heroes in the true sense. There was honour in their quest, and steadfastness, loyalty and, above all, a sublime courage. In the words of the benediction, we should remember them:

> Flying alone! Nothing gives such a sense of mastery over mechanism, mastery indeed over space, time and life itself, as this. A hundred miles,

> north, south, east, west, thirty thousand square miles of unbroken cloud plains! No traveller in the desert, no pioneer to the Poles had ever seen such an expanse of sand or snow. Only the lonely threshers of the sky, hidden from the earth, had gazed on it. Only we who went up into the high places in the shadow of wings![54]

The Parachute Regiment and Airborne Forces Museum

Paratroops were a World War Two innovation. German General Kurt Student had shown what his *Fallschirmjager* could do in 1940 when they brilliantly seized the great Belgian Fortress of Eban Emael.[55] Britain was a long way behind in this novel brand of airborne assault, which was a child of Russian and German cooperation. Nonetheless, the Parachute Regiment commenced recruiting on 22 June 1940 and would swell to a formidable force of 17 Battalions, making up 1st and 7th Airborne Divisions with 2nd Independent Parachute Brigade. Paras, with their distinctive red berets and the aggression of a true elite, fought first in North Africa, then Sicily, Italy, Greece, France and of course disastrously but with extreme gallantry at Arnhem, their 'Bridge too Far'.

In the run-up to D-Day, Allied planners were hoping for a masterful *coup de main* ahead of the amphibious landings. At 02.00 on 6 June 6th Airborne Division, coming in by glider, would attempt to seize the vital bridges over the Orne River and canal, a task allotted to D Company of the Oxford and Buckinghamshire Light Infantry. Other elements of 5th Para Brigade would take and hold ground around Ranville.

To 3rd Para Brigade fell the tasks of knocking out the Merville Battery and destroying the Dives River crossings at Varaville, Robehomme, Bures and Troarn. Sorting out the battery was paramount as it was feared the big guns there could command the length of Sword Beach and wreak havoc among the landing craft. With the Dives crossings gone, the Germans would struggle to get reinforcements in from the east, the paras, if successful, would effectively seal off that flank.

The most easterly beach, Sword, would be attacked at 07.25, the same time as men would go ashore at Gold Beach further west, the Canadians going for Juno between the two British beaches. The 3rd British Division had last seen action on rather different beaches at Dunkirk, hopefully this time it would be different and much of their time in the intervening four years had been spent in getting ready.

Lead elements coming ashore first would be 8th Infantry Brigade Group with armour and DD tanks. Lack of naval gunfire wouldn't be a limiting factor as it had so tragically been at Dieppe; two battleships, a monitor, five cruisers and 13 destroyers were there to provide their enormous combined barrages.

On the division's eastern flank, the Orne River and Canal created a formidable wet gap, and it was intended the paras backed by the Special Service Brigade would have secured those two vital crossings. 'Queen' Beach was divided into White and Red Sectors and would be clobbered by the attacking brigade. Time was very much of the essence; if the Germans got their tanks, 21st Panzer and 12th SS Panzer, moving, then the British could be in for a very rough ride indeed. Getting inland and doing it quickly was the key to consolidation.

The infantry had to relieve the paras at Pegasus Bridge while No. 4 Commando moved east on Ouistreham. No 41 Royal Marine Commando would head west to link with the Canadians coming ashore on Juno Beach (see below). Once Pegasus was secured, 1st Special Service Brigade would advance eastwards from the crossings with 185th Infantry Brigade following – their objective was Caen itself. The troops landing on Sword Beach would only have achieved their objectives if they took the city of Caen on 6 June. This was asking a very great deal.

General Richard ('Windy') Gale, CO of British 6th Airborne Division, was tasked to seize the crossings of the Orne River and Canal at Benouville, eliminate the dangerous Merville Battery, blow up the Dives River crossings to the east to prevent German reinforcements getting up, and to hold the key high ground around Ranville. This was a very big ask, yet without these paratrooper successes the whole flank of the British landing grounds was wide open. Failure, as our US allies would say, was not an option. Grabbing the Orne River crossings was vital, these were the open road to Caen and 3rd Division was (improbably) tasked to take the city as a first-day objective. It's been said that while Monty had planned the assault on the beaches meticulously, he'd thought less about getting into Caen. This would have consequences.

Operation *Tonga* – midnight plus twenty; Major John Howard would lead the spearhead that was to seize the crossings by an audacious and highly risky attack. A company of 2nd Ox and Bucks Light Infantry, now part of 6th Airborne, would blitz the defences from three Horsa gliders towed in by Halifaxes and

Stirlings. Each fragile aircraft carried thirty men and the glider pilots had to land each of them on the narrow strip of ground between the two bridges. this was a very, very difficult assignment and getting it wrong could be fatal, not just to the mission but to the men inside.

C.J. Woodward flew with the RAF and witnessed the vast armada of gliders that would carry the airborne troops over to Normandy:

> I was the pilot of a Stirling aircraft of 161 Squadron on a mission dropping spies behind enemy lines on the night preceding D-Day. We had had a special briefing in which it was emphasised as being critically essential that there should be absolute radio silence, no matter what we saw. We were on our way out from Tempsford in Bedfordshire. At about 3,000 feet we were suddenly confronted by the most awesome sight. The whole night sky, high above and filling the whole area in front, was filled with myriads of red, blue and white lights rotating very slowly like one vast coloured whirlpool.
>
> It seemed impossible not to be engulfed, and we were without lights. Although we knew that something was going on, we were not aware that the French landings were imminent, so there was no explanation of the phenomenon. It was, of course, the gliders and their tug aircraft marshalling in the area, I believe, of Benson. All the members of my crew except the rear gunner crowded into the cockpit in amazed disbelief. Because of the dead quiet of the radio silence, the sight was unbelievably uncanny. Suddenly the spell was split wide open. 'What the f*****g hell are all those bleedin' lights?' It was a Canadian voice, and he must have accidentally pressed his transmitting button in his excitement. I have always wondered what any German listening watch who picked up that transmission must have thought.[56]

Howard instructed his glider pilot Staff Sergeant Wallwork, half in jest, to drop the Horsa with its nose actually inside the wire defences of the bridge. He wanted the other two bunched up behind and to the right. Gliders don't have engines; it was pitch dark and there would be no visible landmarks. Don't worry Wallwork reassured him, the [defensive] posts will take the wings off slowing us down, so we won't hit the bridge quite as hard (they'd be coming in at 100 mph). Behind Howard, the men were keyed up, their singing the balm for loose bowels and dry mouths. This was what they had trained and trained and trained for. The risks were so many, so varied and generally so lethal as to be hardly worth thinking about. At 12.26 hours, the tow was cast off, they were committed.

> The glider was too high as we came in and Jim Wallwork deployed the arrester parachute and adjusted the flaps as he tried to control the unwieldy aircraft and to steer it precisely where he wanted it to come in to land. He had cheekily asked me where I wanted the glider to finish up and, never imagining that he would take me seriously, I had told him 'Ideally Jim, right through the wire defences of the bridge!'[57]

No more reassuring sound of engines, just the steady hiss of the wind as they dove down through cloud cover. Lieutenant Den Brotheridge pushed the forward door open, a rush of cold air coming in: 'Hold tight.' Below the golden slash of the canal shone in the fitful moonlight. This wasn't a model, it wasn't training; this was it. The glider hit still doing 90 mph, tore along the ground, ripped itself to shreds like a cheap toy and then stopped.

Robert Ashby, Glider Pilot Regiment, describes his part in the operation:

> There below us, like a badly shaped question mark, was the mouth of the River Orne and neighbouring canal, exactly as I expected to find them. Further down into France, quite a way off ... was the blinking red light of the lead-in beacon positioned by the paratroops to mark our LZ [landing zone]... Our gliding approach required us to half-circle the beacon, and before we got there I began putting on flap as we were very much too high. I dived, with Jim, whose duty it was to keep an eye on the instruments, shouting that we were going too fast, at 100 mph, enough to tear the flaps off.
>
> I seemed to sense rather than see the ground coming up, and at the right moment levelled off and put her down. Once touched, she stayed on her wheels and came to a stop undamaged. We had expected that all hell would break loose when we landed, but it did not. In fact, when, after a very brief pause to catch our breath, we tumbled out, we were in mortal dread not of the enemy but of other gliders. They were landing all over the place, some with spotlights on, which was a dead give-away, others approaching from the wrong direction, and many crashing into posts or other gliders.
>
> All of this happened within the space of about a minute. When it was all over there was a sudden silence. All that could be heard was

> the gentle sighing of the wind through the grasses. Typically English, I suppose, but the first thing we did in Normandy was to get out the large Thermos flasks and have a mug of tea... But there were problems getting the 3-ton bulldozer out of the Horsa glider. It had been brought to clear any obstructions left by the Germans.
>
> Shouldering our rifles, bandoliers of ammunition and other kit, we gravitated towards the southern end of the sweep of meadowland which formed our LZ. There we found the church of Ranville at the top of a slight slope and the village beyond; this, though we did not know it at the time, was the first village in France to be liberated.

Perhaps the real value of training is you don't have to think, it's all instinctive. Howard and his boys were piling out of the wreckage almost before it shuddered to a broken halt. Howard struggled free tripping as he got out and as he looked up, there it was, the lattice tower of the bridge not twenty steps away. They were spot on target.

> The door opposite me seemed to collapse in on itself as we came to a halt and the tremendous impact caused me to pass out. As I came round, I found to my horror that I couldn't see anything. For a frightful second, I really believed that I might have been blinded, and then just as quickly realised that my helmet, the battle bowler, had rammed itself down over my eyes as I hit the roof or the side of the glider. Pushing it up I took in my surroundings – the smashed doorway, the air full of dust, the holes torn in the sides of the glider and the sound of the pilots groaning in the smashed cockpit. And then the glorious realization came to me that otherwise there was silence, complete silence, no gunfire at all – we had achieved our first objective of complete surprise.[58]

The adrenalin rush carried them forward, through the ripped wire, up ahead a pillbox erupted in MG fire but grenades took care of it, then up the embankment onto the echoing planks of the bridge and across. On the far side another MG opened up. Now the other two gliders came bursting in behind the shell of Howard's just as they were supposed to do. The paras hit the German trenches guarding the span like a tsunami, those who could bolted, it was over almost in seconds, but that other MG had killed Den Brotheridge who won the unenviable distinction of being the first British casualty.[59]

The canal bridge was secured in under three minutes and though the attack on the river crossing had run into problems, the drop wasn't as precise, Lieutenant Fox had got to the bridge in time to see its former custodians pelting off into the darkness. A British NCO capitalised on an abandoned MG42 and pointed the fleeing Germans on their way with a series of long swinging bursts.

> ... a message was received on Tappenden's [radio operator] wireless from Dennis Fox and Tod Swinney at the river bridge to say that they had captured it without any opposition but that there was no sign of the third glider containing Brian Friday and Hooper's platoon. They reported that the enemy had run off, leaving their guns in weapons pits, still warm from the sentries' occupancy. That the river bridge was secure was good news for me and I instructed Tappenden to send out the success signal for both bridges being captured intact. Jock Neilson and his sappers had searched the bridges, cut all wires, and found the chambers ready to receive their charges of explosive but in fact, no charges had been laid; they were later found in a magazine near the canal bridge.[60]

'Ham' was the code word for success at the Canal Bridge and 'Jam' for the river. Not very inspirational perhaps, but those two code words meant the first of the paras' D-Day objectives had been accomplished. Lieutenant D. J. Wood served with 'D' Company of 2nd Ox and Bucks:

> As luck would have it, I was commanding No. 24 Platoon... We went into action in Horsa gliders towed by Halifax bombers. Thanks to the superb timing and skill of the glider pilots, we landed within a very short distance of the objective. Admittedly, in my case, the landing was very rough, and the glider broke in half on touching down. Luckily it was made of plywood and we, its twenty-four passengers, more or less came out of the crash unscathed.
>
> What happened next is history; the leading platoon commanded by Lieutenant Den Brotheridge captured the bridge intact. My platoon took on the task of defending the Canal Bridge. Twenty-five minutes after landing I, my platoon sergeant 'Laces' Leather and my batman Private Chatfield were all wounded by a burst from a *Schmeisser* machine pistol – so as far as we were concerned, that was at least temporarily that...[61]

One veteran's family recorded:

> Bleaching the date of birth on his identity papers, and getting away with it, my father joined the Royal Ulster Rifles aged 17. He'd also heard

> they were recruiting for a new section that were going to be part of an airborne brigade and volunteered for it. Training endlessly, he told me years later that he found it hard, endless, but became incredibly fit. There were runs before breakfast, endless weapon training, assault courses and parachute jumps, often using Stonehenge as a reference point.
>
> He told me of having to jump out of a converted bomber where they had simply cut a circular hole in the flooring. The jumps were precarious as the wind would knock you forward as you went to jump out. You had to gauge it properly to allow for your parachute on your back to have clearance, but not jump too far forward as you would hit your chin on the rim. Many troops in training had broken and cut jaws. He became part of 6th Airborne Division, eventually landing during the night of D-Day right near Pegasus Bridge. He admitted he was petrified, but also excited. This was the first time he'd ever left Northern Ireland. Getting stories out of him was very hard. He met Madame Gondrée, owner of the now famous café at Pegasus Bridge, but the place was also crowded with soldiers.[62]

Robert Ashby, as a glider pilot vital to airborne operations, was on his way back to Britain the day after landing:

> The beach, all things considered, was remarkably tidy. Some disabled landing-craft were washing about in the surf and there was other wreckage. Round the bastions of the seawall were what looked like piled-up stores under canvas sheets. But they were not stores, but the bodies of men who had fallen in the first assault. I noticed a wax-like hand sticking out. It had on it a wedding-ring, token of some poor woman back home who did not know, and would not know for some time, that she was a widow.[63]

Located within the overall complex of IWM Duxford the Airborne Assault Museum brilliantly tells the very eventful history of the Paras, from beginnings to today's 16 Air Assault Brigade Combat Team. The collection dovetails very nicely into the Duxford collection. Overall, there's a lot to be said for blending regimental collections into a bigger whole. Visitors to IWM will generally include the Paras on their itinerary. On a more modest scale, the Fusiliers Museum of Northumberland, which occupies the Abbot's Tower of Alnwick Castle, benefits substantially from being where it is and drawing visitors from the many who come for the wider experience.

At IWM Duxford:

John: How did the Light Infantry get on with the Paras?
Graham: Not well.

Eden Camp, Malton

My German godfather, Kurt Block, was an ex-POW, captured in 1943 during the fighting in Sicily. A solid North German farmer's son, he always looked like you'd expect a *Fallschirmjager* to look like, stocky, square, physically powerful, hard as nails. He ended up in a temporary POW camp near where I now live in Mid-Northumberland and eventually married the local schoolteacher – my own children later attended the same primary. His was a success story, couldn't go back home as it was in the Soviet sector so made his life here, working in my old man's building business and doing very well from it. It's not a unique story. I've talked to farming families in the region who had Italian POWs assigned to them as agricultural labour. One, I recall, had three such sent, all were mechanics in their civvy jobs for Alfa-Romeo and our farmer's machines hummed for the duration like thoroughbreds!

Kurt never tried to escape. Few did. One event I find particularly resonant occurred on 1 April 1945, days before the end of the war in Europe, an attempted breakout from Camp 18, Featherstone Park, down on the flat alluvial plain before the wonderfully decrepit castle – home to the ancient line of Featherstanhaughs. The camp still survives as a few skeletal maintenance sheds surrounded by the footprints of huts and guardhouses. It opened first in 1944 to accommodate the anticipated flood of POWs harvested in the wake of the D-Day landings and spread to encompass over 200 barracks in four compound areas, able to house 4,000 officers and 600 ORs. this was an enormous spread and inmates were primarily officer class, including Luftwaffe, *Kriegsmarine* and desk wallahs.

Featherstone was for 're-education', 'denazification' of prisoners who were colour-coded according to their perceived adherence to

Nazism – 'white', sort of OK, 'grey', not too sure and 'black', serious fanatics.

> These denazification programmes at the camp were a huge success and towards the end of the 1940s prisoners were allowed to undertake work in the local community. Maureen Smith remembers German POWs being allowed to attend the village dance on a Saturday night... The POWs of Camp 18 also produced their own German newspaper *Die Zeit am Tyne* that was printed on the presses of the *Hexham Courant*.[64]

One who wasn't content to sit it out was Joseph Kirchdorfer, a 19-year-old Luftwaffe pilot who, with seven others, made their bid for freedom on that early spring evening. Joseph was already a colourful character. He'd been picked up behind Allied lines in Holland the previous year, unwise enough to have retained a letter from no less a Third Reich heroine than Hanna Reitsch.[65] The escape plan involved hi-jacking a plane and flying it back to Germany. Joseph later admitted the idea was born more out of a dangerous brew of boredom and bravado rather than any desire to serve the Fatherland, its days, at this point, strictly numbered.

A violent thunderstorm – suitably Wagnerian – was blasting as the eight, in their makeshift civvies and deploying home-made wire cutters, snipped through the perimeter fence. Once through they split up as agreed and each made his separate dash. Next obstacle was the foaming South Tyne, swollen by torrential rain. Karl Kropp, aged 24, attempted to wade the flood. He didn't make it and drowned.[66] The rest were soon recaptured. Not quite *The Great Escape* but dramatic enough.

This brings us to Eden Camp near Malton in North Yorkshire, the delightfully rolling spill from the Vale of York, as quintessentially English as it gets, before the ground northwards begins to swell up into wilder moorland. Early in 1942, an MoD detachment sent down from Castle Douglas in the Scottish West March set up a rather ad hoc stockade to house Italian POWs from the extensive haul netted in North Africa. As the camp was sited off Eden House Road, it was quickly dubbed 'Eden Camp'. Soon, the Yorkshire Lanes echoed to tramping boots as the first 250 residents arrived at their temporary home. Their initial job was to consolidate and extend what was already there, a total of 45 huts with ancillary canvas tents, were erected and around 1,200 prisoners were incarcerated, few if any showed any inclination to escape.

After the Normandy Landings, Italians were replaced with Germans. Like those at Featherstone and many other camps, most went to work in the fields. None absconded. By 1948 all had been released. Eden Camp had come to the end of its brief operational existence. A POW camp without POWs is pretty pointless and decay set in. There were brief flashes of light when the place was pressed back into service as a displaced persons centre and a holiday camp, not Butlins but a taste of agricultural life for urbanites. Stan Johnson bought up what was left in 1985, saving many of the huts from extinction. He was inspired by a visit from some former Italian inmates to turn Eden Camp into a museum.

It's fantastic. The refreshed huts house a series of exhibitions that take you through the war and the relative remoteness of the location adds to the effect. Once inside you could be back in 1943. The huts are themed: 'The U-boat Menace', The Home Front', Bomb Disposal' etc.

Most Axis POWs were just glad to be out of it and thousands of British captives would have empathised with their German/Italian equivalents. Moss Watertson, a DLI veteran and POW since the debacle of 1940 remembers getting home:

> I always used to tell my mother, 'When I come home, I'll open the door and throw in my cap, and you have that corned beef pie ready… I got on to the bus and the conductor came for my money and someone shouted, 'I'll pay his bus fare.' It was a lad I used to work with down the pit. I then bumped into a lad who was in my section at Warlus called Morgan Metcalfe. We got off at Houghton-le-Spring and just as I got off the bus, the church clock began to strike 10 o'clock.
>
> Oh, what a lovely sound. I sat on the curb and listened to it. Morgan said, 'Howay, let's have a pint.' I said, 'No I'm going home to my mum and dad.' We were just going past the pub, and he walked in and my dad was sitting there. He said, 'Your Moss has just walked past.' My dad got up and upset the beer and came out shouting along Newbottle Street. When we came to the house, the bunting was out, 'Welcome

> Home'. I opened the door, took off my cap and slung it in. My mother was standing there and said, 'I'm not going to throw it out.' We pulled the chairs up to the table and my mother went to the oven and said, 'Here it is' – a great big, corned beef pie.[67]

Clifford's Tower and York Army Museum

Quite often, I take history tour groups to Clifford's Tower in York. As good an illustration of motte and bailey as you'll find and wonderfully located in the heart of the North's ancient capital. During the medieval period, York housed a population of around 10,000, making it England's second city after London. Home during the Iron Age to the lordly Parisi, who felt themselves a tad superior to their rough and rustic Brigantian neighbours, it became a major Roman legionary base, Eboracum, then Saxon Eoforwic and Viking Jorvik, capital of the Kingdom of Northumbria and a 'northern powerhouse'.

William I built a timber motte there in 1068. His appointed governor was Robert Fitz Richard and the King dispatched Robert Comines to consolidate Norman control over Durham. This didn't end well. The newcomers behaved like brigands, so the locals immolated the lot of them. A good start for the fractious northerners who moved on to engulf York, meting out a similar fate to Fitz Richard. Despite the killing of their castellan, the garrison just managed to keep their walls. These insurgents became irritating enough to galvanise the Conqueror himself into action – if you want a job doing properly – he re-took the streets of York, bolstered defences, and appointed a senior knight William Fitz Osborn to controls things.

This new governor had learnt nothing, rapaciously plundering the burghers. The rebels, now on the backfoot, sent for help and Swein of Denmark responded. Warbands from Scandinavia and the Baltic Coast, arriving in 240 ships, quickly asserted a new local hegemony, burning out Fitz Osborn and his garrison. King William had by now had quite enough and this would be very bad news for northerners. He paid the Danegeld to get rid of his unwelcome guests, then set about devising his version of a Final Solution for the troublesome north; deserts are peaceful.

Clifford's Tower was to become notorious for another savage incident which occurred in March 1190, the destruction of the city's Jewish community. There had been Jews living in England during Roman and Anglo-Saxon times but not an organized community as such. After William successfully invaded England in 1066, he encouraged Jewish merchants and artisans from northern France to move to England. Jewish immigration consequently increased, most from France with some from Germany, Italy and Spain, seeking prosperity and a haven from embedded and pernicious anti-Semitism.

Acting as special representatives of the king, these Jews worked as moneylenders and coin dealers. Over the course of a generation, Jews established communities in London, York, Bristol, Canterbury, and other major cities. They generally lived in segregated areas (ghettoes) by themselves. However, until 1177, only one Jewish cemetery was allowed to be established in London. During the medieval era, usury, or lending money for interest, was considered a sin by the Catholic Church. Therefore, Christians were forbidden to work as moneylenders, so Jews filled the vacuum and having a monopoly were able to set high interest rates. They played a vital role in maintaining the British Treasury and for a time, the Crown watched over the Jewish financiers and their property, though they also taxed them onerously.

Disputes between Christian clerics and Jews in this period were supposedly encouraged by William Rufus (1087–1100). His younger brother and successor Henry I (1100–1135) granted Jews a charter of liberties. Generally, Jews could rely on the protection of the Church, which used their financial skills as an outlet for transactions otherwise prohibited – a commercial partnership in short. Yet Jews still faced persecution and were not fully protected by the Crown. In 1130, their community was fined £2,000 following an allegation that a Jew had killed a sick man. The first record of Jews in Oxford is from 1141 when they were caught in the political infighting of the Civil War, the 'Anarchy', between Stephen and Matilda. In 1144, the first blood libel charge of ritual murder was brought against Jews in Norwich.

During Passover, Jews were accused of torturing a Christian child named William, using his blood for the Passover Seder, eventually killing and burying him. Christians attacked Jewish settlements in retaliation. Despite Pope Innocent IV's protests about the ridiculousness of these allegations, the image of a murderous Jew out to hurt Christians developed in the public mind. These charges were repeated in Gloucester (in 1168), Bury St. Edmunds (1181), Bristol (before 1183) and Winchester (1192). In 1189, Richard I launched his Third Crusade and Jews were taxed at a much higher rate than the rest to finance his quest. Even though they comprised less than 0.25% of the English population, they provided 8% of the total income to the royal treasury. Despite this financial contribution, rabid pro-Christian ideology of the Crusade resulted in rioting in England and some Jewish businesses in London were burned.

One of the most notorious riots led to the massacre of the Jews of York. They'd lived in the city since 1170, seemingly secure among the city's resident elite, who were frequently also clients whilst feeling they could seek refuge in York Castle (Clifford's Tower). In July 1189 King Henry II, a protector of the Jewish community, died. His son Richard refused to grant Jewish representatives admission to Westminster Abbey when they came to offer him gifts. Riots began, mobs pelted Jews and burnt their property. Many Jews were murdered; some, to survive, allowed themselves to be baptized. Twenty-four hours later, King Richard found out about the riots and ordered that Jews once again be protected. It didn't last. As soon as the King left to join his beloved Crusade in 1190, attacks began again throughout England.

William of Newburgh, an Augustinian canon, active in Yorkshire, has left us an account of the terrible events in the City of York as they escalated towards a murderous denouement. Two Jews, Benedict and Joceus, had travelled down to London to attend King Richard's Coronation. They'd unwittingly stepped into a minefield as anti-Jewish disturbances flared. Of the two, Benedict didn't make it back. For his surviving family it soon got worse. Their property in Coney Street was raided and ransacked during a handy fire. This confusion was exacerbated by the Sheriff's absence – off on the King's grand jolly to Jerusalem. Most of Benedict's surviving family were murdered during the incident. No arrests followed.

What remained of the Jewish community, led by Joceus, fled to the sanctuary or hoped-for sanctuary of Clifford's Tower whilst their houses were stripped and torched. Any hope the tower would prove a safe refuge soon evaporated. There were maybe 150 survivors inside and relations with the keeper soon soured, to the extent Jewish survivors locked him out. He came back with an armed retinue and the stand-off slid into a siege as desperate Jews threw stones at his men-at-arms and the howling mob surging behind.

Friday 16 March was the Shabbat Hagadol, the 'Great Sabbath', which proceeds Passover. At the urging of their rabbi, Jews made ready for their own self-sacrifice. Clearly, they couldn't hold the Tower for much longer. No relief was coming. What passed for the forces of law and order were hand in hand with the mob. There was no happy ending in sight. Each head of family killed his dependents before turning the knife on himself. As they bled and died, they ritually set fire to whatever possessions they had salvaged. Flames consumed these, their bodies, and the very structure itself, a vast funeral pyre, leaving a dark pall of guilt hanging over the city.

An illusory offer of hope was thrown out by a local gentry member Richard Malebisse, also one of the prime ringleaders, who seduced some waverers out with offers of clemency. They got none and the rioters made sure they burnt all treasury accounts then held in the Minster. Repercussions for perpetrators, all considered, were rather muted. An enquiry was held and fines totalling £66 were imposed on nearly 60 leading families who were found to have participated. Despite the horror, a new Jewish community did take seed in York, only to be finally expelled during Edward Longshanks' pogroms at the end of the next century.[68]

This is not a slice of York's history that anyone locally really wants to commemorate, though a plaque was set up at the base of the mound in 1978. The annual planting of daffodils on the slope (their six-pointed form is said to echo the Star of David), is a poignant tribute. Recently, the Tower (probably not constructed till at least 60 years after the incident), has undergone a major and much needed £5 million makeover with a stunning new roof and the internal space superbly refreshed, with unmatched views over the great city. Nineteenth-century court and gaol, together with the rather magnificent Castle Museum, fill the ground formerly occupied by the bailey. Still, with the circuit of medieval walls intact, you still get a feel for the medieval layout.

Just don't forget what happened here.

Just by the northern flank of the Tower, in fact on Tower Street, stands the York Army Museum, a very neat late Georgian building sandwiched into the busy thoroughfare. This houses collections from the Prince of Wales Own Regiment of Yorkshire and the Royal Dragoon Guards. It's a lovely museum, cramming an awful lot of history and kit into a relatively modest space. It had a makeover too, part of the Great War Centenary commemoration. For me, one of the prime exhibits is 'Butcha' Hornby's 1908 pattern sword. I've mentioned Charles Hornby before and his laying waste of Uhlans in August 1914.

At York Army Museum:

John: I love that sword. Does that make me a bad person? One who glorifies war and fetishizes weapons?

Graham: Probably, but some weapons do become totemic and not just with armchair soldiers and military historians. When you graduate from RMA Sandhurst, the most successful cadet receives the Sword of Honour. Excalibur is still in fashion. Your daughter earned the Queen's Medal after all, that's pretty damn good.

John: We do like swords – the badge of an officer class from the gladius to the 1908 sabre. Firearms never quite attained the same allure, despite edged weapons being largely obsolete for centuries, though we still train our recruits to use the bayonet. Sam [my son-in-law], his Gurkhas still carry their kukris and they're not for decoration.

Graham: God no, if you ever been on the receiving end of a Gurkha charge and the ones we endured were only in training, we had to remind ourselves these blokes are on our side. 'Cold steel' still has this mystique, a badge of manhood and all that. Achilles or Hector would have got it if they'd seen our bayonet drill. If we didn't kind of worship our weapons, I'd not have bought a deactivated SLR, the constant of my army years.

Yorkshire Air Museum and Allied Forces Memorial, Elvington

During WWII, RAF Bomber Command sustained a total of 57,205 casualties, the highest rate of loss for any branch of the service. All were volunteers. One incident I've chronicled before was the air support for Operation *Agreement*, one of the big raids in the Western Desert against Tobruk, as an overture for a projected amphibious assault. In the first wave over the fortress harbour on that night were Wellingtons and Blenheims from 201 and 205 Groups. They were tasked firstly to provide cover for the assault troops, secondly to form a diversion against enemy radar, thirdly they were to detect the 'success' signals from the troops and transmit the messages back to HQ. Lastly, they were to undertake reconnaissance for naval forces both going to and returning from the raid.

This was the reverse of the earlier concept for Operation *Waylay*, which had reserved air support for the extraction phase. In hindsight, this was undoubtedly the preferable option. Air raids, whilst dramatic and terrifying, achieved very little in terms of tactical gains. Tobruk had been systematically bombed by both sides, yet, scarred and defaced, most buildings were still standing remarkably intact. An air raid would thoroughly alert the defenders and risked forfeiting the advantage of surprise. Something in the order of a hundred RAF planes would target Tobruk that night, including five Boeing B-17 Flying Fortresses, led by Major Max Fennell, contributed by the 7th Bombardment Group of the USAAF.

David Jefferson quotes from a memoir of the bombing raid compiled by S/Sgt. Wilbur (Bill) Mayhew who served aboard one of the B-17s, which had taken off from Lydda in Palestine. The US bombers, shining monsters, would fly in at varying altitudes so as to minimise the risk of mid-air collision. Mayhew's aircraft would be first in the bombing zone. He saw their role as keeping the AA and port gunners fully occupied so they'd not detect Force 'A' coming in. 'There was a 50 per cent overcast, and the reflections on these clouds of anti-aircraft fire, bombs bursting on the ground, and shells falling from naval vessels [in this he was mistaken, there was no naval bombardment at this stage] a few miles offshore made this an eerie scene for us.'[69]

Mayhew recalls a Wellington ('Wimpy') caught like a fly in the web of piercing searchlights, barking guns following the beams. The tornado of fire engulfed the unlucky bomber, 'in a few seconds the plane exploded.' Another Wellington was dispatched moments later as the Fortress began her bombing run: 'The line of tracers coming up ... looked like they were headed directly for my turret. It was very unnerving to watch the slow approach of that line of steel coming directly at me.'[70] The Fortresses were actually flying at far higher altitude than the older Wellingtons, effectively above the reach of the guns. This wasn't apparent to Bill Mayhew and his fellow fliers, who expected to be hit at any second.

Then it was bombs away, 'and a few moments later there was a terrific explosion on the ground directly beneath us. He [the bomb aimer] had hit an ammunition dump. I could read a magazine in my turret.'[71] After some forty minutes over their drop zone, the aircraft turned and headed for home. They could be forgiven for imagining they had dealt the defenders some very deep wounds. It was around midnight now and both *Tribal* class destroyers had swerved south to commence their run in towards the landing beach.

In terms of actual results, all that the air raid achieved was to bring the Axis defenders onto full alert. Italian *braggadocio* would later seek to claim credit for the successful outcome. In fact, as was usually the case, it was the German defenders who deserved any plaudits (which Rommel freely bestowed in the aftermath). Their after-action report noted that despite seven hours of continuous air attack, damage was minor and casualties few. DAK high command ensured its various detachments were all on the *qui vive* and primed units, encamped in the desert, to be ready to move. German reports suggested some six hundred bombs had fallen for the loss of seven 'kills'.

RAF Elvington was a Bomber Command Station in North Yorkshire. The early aerodrome formed an element of No. 4 Group, beginning its long operational life as a grass strip, the sort of thing Biggles would have been familiar with. During 1942, it received a major makeover with three hard runways laid. At 1.92 miles (3.08 kilometres), this runway was the longest in the north. The dispersal zone was one of the largest in the country.

By October 1942 the base was home to 77 Squadron, equipped with four-engined Halifax bombers, one of 30 such airfields in Yorkshire. Elvington was joined with RAF Pocklington and RAF Melbourne to form '42 Base'. During its time at Elvington, 77 Squadron flew regular sorties over the Ruhr, Germany's massively defended industrial underbelly; 80 planes and 500 crew of No. 77 Squadron were lost in this deadly battle. A month prior to D-Day the base was handed over to a Free French contingent, some 2,000 of them in two squadrons. The French pilots also flew Halifaxes against the Ruhr and Berlin, a final tally of nearly 2,500 sorties. After the end of the war, in October 1945 these gallant Frenchmen flew back to Bordeaux and their two squadrons merged into a new French Air Force. But they didn't forget, and survivors returned year upon year for an annual pilgrimage to pay homage to fallen comrades at their memorial located in the village.

This wasn't the end of RAF Elvington's story. Transferred back into UK service it became the HQ for 40 Group Maintenance Command before been handed over to USAF Strategic Air Command (SAC). Yet another makeover, lengthening the already long runways to take Cold War heavy bombers got underway, but B-52s never flew from Elvington, it was de-commissioned in 1958. Not completely, what is now BAE Systems used those extensive runways for test flying while RAF flying schools utilised them for training purposes. The flag was finally lowered just as the Falklands War erupted, but no planes to the

South Atlantic flew from Elvington. Buildings and infrastructure, like faded dowagers, slid into steady decline.[72]

All was not lost. A local resident, Rachel Semlyen, created a small, dedicated team who set to work. Their commitment and hard labour ultimately created the museum that formally came into being in June 1985. Since then, the charity has managed to buy up the whole of the original site and acquire a vast collection of artefacts. I managed to secure a short flight on a Dakota from there a decade or so ago. I loved the experience but once around the airfield was enough, the noise and discomfort reminded me of an airborne version of the rattling school-era buses we used to ride in.

At Yorkshire Air Museum:

John: I can't imagine what it was like to endure a tour of duty in bombers, I get nervous enough in the departure lounge but I'm none too keen on those who sneer at Bomber Command veterans as perpetrators of war crimes on account of demolishing Dresden et al. With industrial war an enemy city is a legitimate target. As you know, 'Bomber' Harris made no secret of his intention to level Germany through strategic bombing and civilians were fair game.

Graham: When fighting the nation which created Auschwitz, the Queensbury Rules don't apply. Didn't we reach Belsen just a couple of months after Dresden? The images, first anybody from here had seen of the camps, were handy for countering the critics, and there were more than a few even then of the bombing campaign.

John: We've sort of lost sight of just how shocking those pictures were, the piles of emaciated corpses, atrocities get less atrocious once you've seen the evidence so many times.

Graham: Strategic bombing, people like Harris and Spaatz thought, could win the war on its own. Initially the Germans thought so too. Thank God, they got it wrong, the Luftwaffe was a brilliant tactical but flawed strategic tool. Bad as it was here, that was lucky for us.

John: My Airfix Lancaster in 1/72nd scale was my all-time favourite, I know those who flew the rest of our heavy bombers, Halifaxes, Stirlings and the like, resented that 'iconic' status the Lanc acquired. I really, really, always wanted one, on me doorstep, like John Travolta has.

Graham: You'd need a bigger garden and a bloody good divorce lawyer first.

National Museum of the Royal Navy, Hartlepool

And we've come pretty much full circle, back on the north-east coast. At Hartlepool, south of Heugh Battery on its shrouded headland stands the Royal Naval Museum, one of the best certainly in the north. A whole Georgian dockland has been re-created with the magnificent HMS *Trincomalee* berthed there in splendour. A fully restored heavy frigate, just missed Boney and built in India, fashioned from teak, which endures. It's a fascinating insight into the world of Hornblower and Lucky Jack Aubrey. The displays inside the museum include a brilliant vignette of combat at sea. Spot on.

During the Great War the Admiralty was obliged to find an effective response to the U-boat menace. This form of economic warfare was proving highly effective. These sleek, silent predators were free to strike undetected and at will. It was impossible to provide sufficient warships to constantly patrol the coasts and even swift destroyers had a limited range of weapons and tactics to deploy. One possible remedy was to find a mongoose to take on the U-boat cobra and this was the 'Q' ship.[73]

Quite simply, this was a trawler or small merchantman, ostensibly too puny to merit the cost of a torpedo, best dealt with by gunfire from the U boats' deck-mounted weapons. But this target would not be as defenceless as it appeared. As the submarine surfaced and closed in for the kill, the victim would suddenly sprout an arsenal of its own and take the predator on in a surface gun-duel, hunter thus becoming hunted. By the spring and then summer of 1915 several such vessels were in operation but without scoring any significant hits. The armed collier *Prince Charles* was destined to do rather better.

Prince Charles was part of the vital collier fleet, which performed the vital if unglamorous role of maintaining coal supplies to the fleet at anchor in Scapa Flow. Lieutenant Mark Wardlaw RN commanded a ten-man naval detachment on the small vessel, armed with a brace of concealed small-calibre guns. On the evening of 24 July 1915, one of those endless days in the Orcadian summer, *U–36* had stopped the Danish vessel *Louise* but decided the collier offered a more tempting prize. As a chase developed, the U-boat opened fire at extreme range with her deck guns. Wardlaw then responded by ordering the civilian crew to heave to and abandon ship as though in panic. *U-36* slid in to a range of 600 feet, confident of the kill. She was simply not ready when Wardlaw unmasked his two guns to commence a steady and accurate fire. This vital element of surprise proved crucial. The Germans were badly shot up and as they attempted to dive Wardlaw's gunners finished the job, sending *U-36* to the bottom and taking 15 survivors prisoner.

This neat little action showed the worth of the Q-ship and northern waters became a frequent hunting ground. These were never proud men o' war. In fact, the more down-at-heel the Q-ship appeared, the more likely she was to successfully tempt a U-boat into a surface attack. Occasionally, such a craft might carry torpedoes but more typically a number of smaller guns that could be easily hidden along with machine guns and small arms.

Deception was key; naval crew had to blend perfectly into the less formal regime of a tramp merchantman. A U-boat would stalk her potential victim long before she surfaced, so sailors had to look scruffy and unmilitary. This façade had to be maintained in port lest enemy agents be vigilant. One could imagine Sir Andrew Wood or Cochrane fitting into such a role with gusto. The ship itself was an integral part of the ruse. She would be camouflaged and re-painted lest her former lines become too familiar and thus suspect.

When a U-boat surfaced and rode in for the kill, deception reached its theatrical denouement. A portion of the crew, equal in size to what might be expected of a merchant vessel of this class, would run around the decks, giving ample signs of incipient panic. After a decent interval, they would abandon ship with the captain or someone who looked like a captain, ostensibly clutching the tempting lure of her papers. This finely judged performance (as indeed it might be, as the sailors lives depended on the reactions of the predator), was intended to draw the U-boat on, until the range had closed and the Q-ship's guns were suddenly unmasked.

Memories of the slaughter of *Lusitania's* hapless passengers were still fresh when the Q-ship *Baralong* steamed to the aid of another liner, *Nicosian*, attacked in the Channel by *U-27*. The German decided to punish the shabby tramp steamer for her impudence but suddenly found herself under fire from the Q-ship's 12-pounders. Lt Commander Herbert, having crippled his opponent, showed no mercy to the desperate survivors. Numbers were shot in the water or as they sought to climb aboard the liner. Total war was by no means a one-sided business. As pressure on Germany mounted, the attraction of unrestricted submarine warfare re-emerged as strategic doctrine, the U-boat fleet could put over 130 craft into the water, with technical capability being continuously enhanced.

Typically, a hunter type U-boat now had four tubes forward with two aft and either a pair of 86 mm, or a single 105 mm, deck-mounted guns. These formidable predators had, by the end of 1916, accounted for a staggering 443,000 tonnes of Allied merchantmen. Jellicoe cautioned that with this rate of loss, Britain would be hard-pressed to maintain the war effort through the following year and on her knees by the summer. It was small wonder that Haig's plan for a summer offensive in Flanders included the notion of an amphibious assault on the U-boat pens located within the German-held Channel ports.[74] Sensing that the U-boats might yet achieve what German armies had not and break the will of the British Empire, Wilhelm II agreed to the resumption of totally unrestricted submarine warfare. Losses of allied merchantmen continued to spiral.

For all their swashbuckling, Q-ships proved the least effective of the Allies' responses. Depth charges and hydrophones, linked to more and more effective mines, and the shepherding of the convoy system, combined to defeat the U-boats. Of the 180 Q-ships deployed only ten managed to destroy German submarines, 14 in all. One of the most successful captains was Gordon Campbell who, commanding *Farnborough* in March 1916, sank *U-68*. Then, just less than a year later, accounted for *U-83*, losing his ship in the process but winning a Victoria Cross. On 8 August 1917 he was master of *Dunraven*, a converted merchantman, which refined the Q-ship concept by carrying some minor but visible armament, her additional firepower concealed. On that summer's morning she was engaged by *UC-71* and a rather desultory exchange of fire resulted in no damage or loss to either hunter or hunted.

When this opening exchange had spluttered on for half an hour or so, *UC-71* motored in for the kill. Her guns were now striking the superstructure, starting a fire that threatened to spread to a small magazine. It was now 12.10 hrs and the action had continued for nearly an hour. The 'panic crew' had already played their part, and one of the gun crews was horribly exposed if the magazine ignited. Nearly 50 minutes later the inevitable explosion occurred, wreaking carnage on the after decks and killing one of the gunners. Amazingly, the rest survived, though by no means unscathed.

An injured lieutenant apologised to Campbell for leaving his post without orders, he had in fact been blown overboard! The captain had already issued instructions for his remaining guns to open fire, but the U-boat dived unscathed. She was still the hunter and the badly damaged *Dunraven* still the prey. A torpedo struck at 1.30 pm, blowing out a section of hull. The master ordered all but the remaining gunners to abandon ship. Still, the German commander, *Leutnant* Saltzwedel, was not disposed to take chances. Over an hour later, as the Q-ship was already settling, *UC-71* surfaced astern to conclude matters with gunfire. Campbell described the ensuing bombardment as extremely unpleasant.

Dunraven was wallowing, without sufficient power to turn and return fire upon her tormentor, who now dived once more and circled, like a hungry shark. Campbell promptly loosed two torpedoes of his own but only one came close and that not close enough. *Dunraven* was clearly doomed. Campbell with only a single volunteer gun crew stayed aboard. But time was also running out for the Germans. *UC-71* could not afford to hang around indefinitely to finish the dying ship and most of her ammunition, shells, and torpedoes, was now expended. She withdrew and Campbell might at least claim the fight as a draw. A destroyer took the crippled *Dunraven* in tow. She eventually foundered, though without further loss of life. The lieutenant and petty officer who had stuck to their gun even as the magazine smouldered beneath their feet each won a VC. If the Q-ships were not a successful measure in terms of the strategy of the Battle for the Atlantic, this was no reflection on the extreme gallantry of their officers and crews.

6

The Future

The true soldier fights not because he hates what is in front of him, because he loves what is behind him.

G. K. Chesterton

Are you ready, paratrooper? Said the sergeant looking up,
The paratrooper feebly answered, the dispatcher hooked him up.
But when he hit the atmosphere, his chute became unstuck.
And he ain't gonna jump no more...
(Chorus)
Glory, glory, what a horrible way to die!
Glory, glory, paratrooper.
Glory, glory, what a horrible way to die!
And he ain't gonna jump no more.

Sung to the tune of *John Brown's Body*[1]

My generation lived, for all our somewhat distant youth, with the spectre of nuclear Armageddon so memorably lampooned in *Dr. Strangelove.*[2] When the Berlin Wall unexpectedly, came tumbling down and morphed instantaneously into a tourist attraction, smug pundits proclaimed, 'The end of history'. They were about as wrong as they could be.

During the same period, Britain experienced a stuttering retreat from Empire, sparking a slew of asymmetric conflicts across the globe, in Malaya, Kenya, Cyprus, Aden and Borneo. Simultaneously, there was scope for expeditionary warfare in Korea, Suez, the Falklands, two Gulf Wars and Afghanistan with thirty years of domestic counterinsurgency in Northern Ireland. No end to history in sight.

Post-war there has been an increasing emphasis on the use of Special Forces (however you define that). It is remarkable to think senior commanders immediately after 1945 envisaged a continuation of industrial warfare without the need for Special Forces, which had their genesis in the deserts of North Africa, as The Long Range Desert Group (LRDG) and Special Air Service (SAS). Today, we tend to view our Special Forces as the front line in asymmetric conflict:

> ... the period 1946–1991 could well be defined as one of an overarching confrontation ... maintained by industrial structures, containing non-industrial conflicts, the parallel wars. It is within these conflicts that we see the first signs of the new paradigm, especially in the nature and objectives of the opponents, and in the constant adaptation of the existing means – the industrial military machines – to non-industrial conflicts... It was the beginning of a trend we live with to this day.[3]

The late and great Brigadier Peter Young, DS MC and Two Bars, aside from a distinguished military and literary career was a founder of the Sealed Knot Re-enactment Society in 1968. Yes, I was, for quite a few years, a member. Once, at a public lecture, he was asked who the most formidable enemy was he'd ever faced in battle He responded: 'Well, if you've not fought the Germans, you've never been to war!' Having had a godfather who was ex *Fallschirmjager* I'm with him on that. When it comes to living history, if you've not been to the annual War and Peace show at the Hop Farm Country Park, you've seen nothing.

Graham and I were last there in, I think, 2012/2013, say a decade past and it was huge, some 4,000 vehicles of every type you could imagine, tanks, armoured personnel carriers, armoured cars, trucks, jeeps, motorbikes from 1914 upwards. It lasts over several days and can easily attract more than 100,000 visitors. That year the theme was mud, an awful lot of it, a proper slime-fest. As for re-enactors there were more groups than I could name, covering every war from the turn of the twentieth century, including s few that were still rumbling on.

At the Hop Farm Country Park:

Graham: I always knew you were bonkers but some of this lot make you seem quite ordinary.

John: Even I draw the line at current wars or anywhere you might encounter actual survivors who've been affected.

Graham: No shortage of choice here, we got Vietnam (brilliant display I have to say by Rolling Thunder), Arab-Israeli wars, might be a tad contentious and even the Civil War in Bosnia. I know loads of blokes who still have the nightmares.

John: Yonder fellow striding about in full SS gear, long leather coat and all, seems to have translated himself from aberrant fantasy into some pastiche of reality. He's found his spiritual home, which could be seen as worrying.

Graham: If you don't support Man U, you don't wear the T-shirt. At least he's not a Sunderland supporter.

John: I did once get a salutary lesson in that. We were doing a reiver day in the borders and I was being the officious, arrogant warden, no method acting needed and I sentenced someone to drowning and we had fun, shoving this poor sod's head into a bucket of water. Someone in the audience freaked out, ex Polish resistance and he'd suffered water torture from the Germans, or maybe the Russians, or both. Even made me think and we never did it again.

An ex-Russian BTR charged through the sodden lanes at speed, deluging all with glutinous spray.

Graham: Bloody typical, you re-enactors take on the characteristics of the people you portray, like an act of homage, arrogant Soviet bastards (they were from Essex in fact but anyway).

John: You're not wrong, I fear. We used to do stuff along Hadrian's Wall, Chesters, Corbridge and the like. I was with a Celtic group, and we appeared alongside some Romans. And they were proper Romans an' all, sneering arrogant, kit-obsessed and contemptuous of all us *Brittunculi*, they never drank at an event for God's sake. They wouldn't fight with us in case their lorica got scratched and besides, we weren't worth the effort being such a low form of life. No matter how much we tried to interact they ignored us, even the insults.

Graham: Bit boring.

John: Totally. So, we Celts ended up fighting amongst ourselves, fully period authentic.

End of Empire

The sharp end of SAS operations was arduous and dangerous work. An operation would be for up to three months with re-supply by parachute

> every fourteen days, security permitting. Fresh meat and bread would be dropped and consumed the same day before the beasties got at it. It was not unusual for supply 'chutes to 'whistle in' – plummet to earth without fully opening or become hung up in the trees or be otherwise inaccessible. These would invariably be the ones with the mail or rum ration onboard.[4]

The Malayan Emergency, a twelve-year counter insurgency campaign against Communist Terrorists (CTs). At the outset, CTs were able to strike almost at will from superbly camouflaged jungle bases virtually impossible to detect. Part of the riposte was to re-form the SAS, who took the fight to the enemy, a major factor in their final defeat. Obviously, the Chindit experience was invaluable and expert practitioners in the art of jungle warfare such as 'Mad' Mike Calvert were brought in to train recruits. Malaya established the enduring value of a Special Forces component, one which despite the nightmare terrain could penetrate the heart of enemy-held areas and direct telling air strikes.

> From Changi the squadron moved to the jungle warfare school at Kota Tingi in Malaya. British and Australian instructors taught basic jungle tactics with assistance from a company of Gurkhas. This course ended with the squadron's first operation off the east coast of Southern Malaya. Landing from Navy MTBs, the squadron stormed ashore into flat, dense, sea-swamped jungle. Unfortunately, nobody had considered the water supply (never usually a problem in jungle). Here the water was salty, so we lived on coconut milk, a miserable experience.[5]

During the long and vicious years of the Troubles in Northern Ireland, the SAS learned to function covertly in an urban/rural environment. In temperament this sectarian conflict was the very reverse of *Krieg ohne Hass*.[6] When an SAS detachment very publicly stormed the Iranian Embassy in May 1980, ending a violent siege and hostage scenario, theirs became a household name. In the thirty odd years since, this status has been elevated to the iconic. Writers like Chris Ryan and Andy MacNab who took part in the epic failure of Bravo Two Zero during the first Gulf War have together sold millions of books since.

The killing of three Provisional Irish Republican Army (PIRA) terrorists on the rock of Gibraltar during Operation *Flavius*[7] led to some disquiet. LRDG had only ever been called upon to fight in an honourable war and took prisoners wherever possible. Shooting to kill is part of war, but the lines are less clear cut when facing terrorists or insurgents – 'war among the people' as General Smith would define it.[8]

This slow death of *Imperium*, protracted, complex and not infrequently bloody, offered the perfect training ground for Special Forces. Each conflict threw up different scenarios, Kenya, Aden, the Radfan, and Borneo. All demanded that UK Special Forces learn to operate across a range of terrain, that they should work closely with local forces and pursue the Holy Grail of 'hearts and minds'.

With the 'War on Terror', the rise of asymmetric and now hybrid warfare, the perceived role of Special Forces has changed. Interstate wars are an increasingly rare phenomenon (or were until President Putin decided to invade Ukraine). International rivalries, as illustrated by long decades of the Cold War, are often fought out vicariously. Conflicting ideologies, control of diminishing resources and the rise of nationalism fuel modern conflicts which continue to proliferate. Special Forces are now deployed as the front line rather than purely for intelligence gathering and raiding. The clear and high moral plateau offered by the fight against fascism no longer exists. Civilians are very often first in the firing line, the 'enemy' a shifting, faceless chiaroscuro of changing factions and alliances.

Britons, already alerted by a spate of terrorist attacks, including those of 7 July 2005, woke up to the increasing tempo of this Islamist threat at 5.50 am on 16 January 2013 when terrorists attacked the giant Tigantourine natural gas installation near In Amenas, Algeria. The sudden eruption of the crisis and the costly ruthlessness of the government's response highlighted the potential danger to British workers in the region. Such facilities are isolated, sprawling and virtually impossible to defend, apart from by massive military deployment. The terrorist will always have the advantage that only he knows where, when and exactly how he will strike. The incident left many hostages dead and injured, including a number from the UK.

In the past, guerrilla armies, from David seeking sanctuary in The Cave of Adullam to Patrick Leigh Fermor and Stanley Moss spiriting Axis commander General Kreipe across the Cretan mountains, have sought refuge on the high ground. Geronimo led an entire US army a merry dance through the sharp peaks and ravines of Arizona for decades with a mere handful of followers. Now, modern surveillance and satellite tracking systems have largely destroyed that advantage. It is estimated that by 2050 some 75 per cent of the world's population, which will by then have increased by another three billion, will be in urban centres. In the developing world, sprawling cities, an irresistible magnet to the rural poor, will grow exponentially in a riot of slums and shanties, perfect conditions for the rise of extremism. Kilcullen cites four main drivers priming the powder keg: huge growth in populations; concentration of people in cities; movement towards the world's coastlines; a spiralling revolution in 'technology driven connectivity'.

He foresees in some instances (Somalia being a case in point perhaps), doomsday outcomes where states and their governments fail, and cities become battlegrounds. The major powers, rather than deploying troops for counter insurgency across barren plains and rock-strewn deserts, may find themselves fighting in the tangled rookeries of urban hives where all order has gone. The US experience in Mogadishu vividly recalled in *Black Hawk Down* [9] illustrates the perils such conflicts hold.

Street-fighting is inevitably harder, far costlier and demands infinitely more boots on the ground. Firepower, aerial observation, and heavy weapons are far less effective. Ask General von Paulus. Worse, there is unlikely to be a defined 'enemy' as such, no Saddam or bin Laden to focus on. The current, awful civil war in Syria is being fought on the rebel side by a multiplicity of diverse groups, at best uneasy allies, very often mutually antipathetic, floundering in a quicksand of changing allegiances.

Conventional armies like conventional enemies, but western forces will find themselves up against Al-Qaeda/ISIS lookalikes or franchises, drug cartels, or sectarian mercenaries such as Hezbollah, fighting under the instructions and in the pay of a hostile or rogue state, war by proxy. The Nairobi attack, Dennis Kilcullen argues, like Mumbai, demonstrates how a small gang of well prepared, murderously ruthless terrorists can bring a whole city to a standstill.[10]

No-one anticipated the Arab Spring, fuelled largely by internet connectivity; nobody really understands how it may finally play out. Libya has descended into anarchy and Egypt remains balanced on the edge of chaos, the void of failure which provides an open door for extremists. Kilcullen takes a view that whilst huge areas of urban slum housing are nothing new, the internet changes everything, alters the dynamic in a way we have never seen before. Tinder for these fires is provided by the deep wells of inequality, lack of opportunity, 'permanent exclusion and marginalization'. Awareness without hope is a dangerous pairing.

None of this is glorious, not really stuff you can commemorate in museums; complex and convoluted with players swapping roles and sides faster than *Game of Thrones*. Computer games and online commentary seem to fit the scenario better, what do ancient, faded red coats in glass cases have to say? When I was last working in IWM during 2019, they had a special exhibition on the current conflict in Yemen. A worthy try and an important subject, yet it didn't really seem to connect with visitors, too complex, too 'foreign' and too far away.

Hybrid Warfare

'Hybrid Warfare' is a relatively new buzz expression. It involves warfare at every level, not the type of all-out industrial war we saw in the twentieth century but a new breed of conflict where media manipulation and the use of the internet is as important as bullets and boots on the ground. The expression came into vogue when President Putin was manoeuvring to wrest Crimea back from Ukraine in February 2014. Everyone at the time thought he'd stop there. We heard of 'little green men', not from Mars but from Moscow. The expression may have gained something in translation, but it derives from a colloquial term describing masked, military looking types with no insignia but plenty of state-of-the-art weapons. President Putin naturally denied that these had anything to do with him, asserting they were purely locally raised militia units. This bland denial rather lost credibility when a Finnish expert (and the Finns have many reasons to be watching Russia), confirmed their kit was exclusive to Russian Special Forces.

None of this was particularly subtle, but the deployment represents an aspect of hybrid conflict. War isn't any longer a business of armies facing each other, relying on armour, airpower and battlefield tactics. Such confrontations will become much rarer. For one thing, industrial war is just too expensive. The UK campaign in Afghanistan, from the intervention in Helmand in 2006, cost £14m a day just to keep going.

We can't afford that. Britain has now formed a bespoke unit specialising in the doctrine. This new 77th Brigade (of Chindit fame, see below), will have a complement of around 2,000. These will be military personnel, but they'll be using words and media more than ammunition. Some 'civvies' may also be recruited independently to bring additional expertise in related fields such as psychology and marketing. This represents a significant shift in strategic and tactical doctrine. Readers may recall the ill-fated attempts by David Cameron and President Obama to get the west involved in Syria's fratricidal war. This would have been a deeply flawed policy had representatives on both sides of the Atlantic not vetoed the idea. In such conflicts – and Syria is very much the shape of things to come – it will be very difficult to identify clear factions. No more good guys and bad guys but the shifting sands of fractious temporary allegiances, a mix of ideologies and theocratic doctrines, seasoned by ruthless barbarity. Nuclear submarines may have little or no role to play.

This nebulous battlefield will not replace military action but may well dictate the timing and use of conventional forces. Those little green men might be there to quietly put some stick about or to harness and focus local militias. An ability to move and shape economic factors has always influenced warfare. Germany blamed its defeat in 1918 not on the superiority of the BEF but on a shadowy fifth column, a Jewish banking cartel which stabbed the generals in the back. Hitler knew all about hybrid warfare – using concentration camp victims, murdered, and then uniformed, to create a myth of enemy action.

This designation, 77th Brigade, evokes the potent image of the original formation, far better known as the Chindits; a name to conjure with. Appropriate, too, as Wingate's brainchild launched a new and unexpected type of jungle warfare behind Japanese lines. In such circumstances, the burden of the west's military response must fall on Special Forces. In the Western desert, LRDG were a tiny proportion of the whole Allied war effort, but today their successors might well *be* the war effort. The pioneers of LRDG created British Special Forces as we know them, and that resonance has a global implication. To that extent, the story continues: 'Certainly, as we prepare to tackle the wars of the future, it is clear we need more investment in our special forces, not ill-judged budget cuts. For, as George Orwell memorably remarked, 'Those who "abjure" violence can only do so because others are committing violence on their behalf.'[11]

The full terrible story of the Chindits in Burma is provided in *War in the Wilderness* by Tony Redding. He quotes Chindit Corporal Ted Treadwell:

> I will never talk about this, because I am a murderer. That's how I look at it. We were all murderers... Before we went in Mountbatten told us, in effect, not to take prisoners... In any case, we couldn't take prisoners. We had no means of getting them out.

> ...Our badly wounded were a serious problem. We couldn't get them out, so our MO gave them morphine and then they were shot. There was nothing else we could do.

That's Special Forces.

In 2016, Peter Nesser's *Islamist Terrorism in Europe* was published.[12] This work follows in subjectLorenzo Vidino's *Al Qaeda in Europe* (2006)[13] and *Europe's Angry Muslims* by Robert Leiken (2102).[14] Nesser, a Scandinavian academic, states that 'London was at the heart of the jihadist sub-culture in Europe.' He refers more than once to the baleful influence of such rebarbative fanatics as Abu Qatada, Omar Bakri Mohammed and Abu Hamza – 'all were ultimately deported only after protracted efforts by British human rights lobbyists and activist lawyers to keep them in our midst.'

While teaching in prisons some years back, I was made acutely aware by staff and resident clergy of the spread of Radical Islam amongst the prison population, fertile ground and a problem which has escalated sharply in recent years. Nesser reminds us that the tag 'Londonistan' was coined by French security services and 'quickly became slang for the tendency of the British authorities to turn a blind eye to extremism and to offer sanctuary to terrorists'. This is pretty strong stuff, but Nesser, who is no mere ranter, agrees. I can't see us having a museum to record any of this, there's no honour or glory here.

Nesser puts the blame, and while we're very quick to apportion blame there is clear culpability, squarely on the shoulders of 'patrician liberal complacency'. His case is compelling: 'Rarely can so much damage have been done by so few.' He condemns so-called 'covenants of security' where fanatics apparently undertake not to bite the hand currently feeding them. Abu Hamza, much cherished by well-meaning if naïve human rights activists, 'privately described such bargains as "crap."' On that I think we could agree. Nesser ponders how many of those bureaucratic heads, which entertained such fiction, have rolled. His suggestion, impossible to refute, is very likely none.

Meantime, there is no universally accepted definition of hybrid warfare, which leads to some debate whether the term is useful at all. Some argue that the term is too abstract and only the latest catch-all to refer to irregular methods to counter a conventionally superior force. The abstractness of the term means that it is often used as a term for all non-linear threats.

Hybrid warfare is warfare with the following aspects:

- **A non-standard, complex, and fluid adversary.** A hybrid adversary can be state or non-state. For example, in the Israel–Hezbollah War and the Syrian Civil War the main adversaries are non-state entities within the state system. These non-state actors can act as proxies for countries but have independent agendas as well. For example, Iran is a sponsor of Hezbollah but it was Hezbollah's, not Iran's, agenda that resulted in the kidnapping of Israeli troops that led to the Israel–Hezbollah war. On the other hand, Russian involvement in Ukraine can be described as a traditional state actor waging a hybrid war (in addition to using a local hybrid proxy).
- **A hybrid adversary uses a combination of conventional and irregular methods.** Methods and tactics include conventional capabilities, irregular tactics, irregular formations, terrorist acts, indiscriminate violence, and criminal activity. A hybrid adversary also uses clandestine actions to avoid attribution or retribution. These methods are used simultaneously across the spectrum of conflict with a unified strategy. A current example is the Islamic State's transnational aspirations, blended tactics, structured formations, and cruel use of terror as part of their arsenal.
- **A hybrid adversary is flexible and adapts quickly.** For example, the Islamic State's response to the US aerial bombing campaign was to quickly reduce the use of checkpoints, large convoys, and cell phones. IS militants also dispersed among the civilian population. Civilian collateral damage from airstrikes can be used as an effective recruiting tool.
- **A hybrid adversary uses advanced weapons systems and other disruptive technologies.** These weapons can be now bought at bargain prices. Moreover, other novel technologies are being adapted to the battlefield such as cellular networks. In 2006, Hezbollah was armed with high-tech weaponry, such as precision guided missiles, that nation-states typically use. Hezbollah forces shot down Israeli helicopters, severely damaged a patrol boat with a cruise missile and destroyed heavily armoured tanks by firing guided missiles from hidden bunkers. The organization also used aerial drones to gather intelligence, communicated with encrypted cell phones and watched Israeli troop movements with thermal night-vision equipment.
- **Use of mass communication for propaganda.** The growth of mass communication networks offers powerful propaganda and recruiting tools. The use of fake news websites to spread false stories is an element of hybrid warfare.
- **A hybrid war takes place on three distinct battlefields:** the conventional battlefield, the indigenous population of the conflict zone, and the international community.

How to commemorate such a murky story. It's very difficult to do so by object association, we museum lovers all know what a Bren gun looks like and we grew up on a diet of classic war movies. The Good Guys (that was us), against the Bad Guys (usually Germans or Japanese). It's no wonder we, by whom I mean the post-war generation, remain obsessed with the Second World War – it was very much the 'just' war. Just, because the Swastika and Rising Sun flags represented the embodiment of evil. I know we can't really say that any more, but they did and we, the Allies, fought a good fight to rid the world of fearful tyranny. In 1982, we went to war in the Falklands after (admittedly facilitated by intelligence failures on our part), a vicious, far right dictatorship which was oppressing and murdering its own people sought to enslave our citizens. That was another 'Just' war. We knew who the enemy was, we had a worthy cause, and we won a clear-cut victory (which like all such victories, came with a high price tag).

What is a 'Just' War?

'Just War' (*jus bellum iustum*) is a function of military/political ethics, which has been debated over centuries. The core of the doctrine seeks to create a mechanism of ensuring war, relying on a given set of criteria, is justifiable; all of which must be met. These stipulations are split into two distinct elements: 'the right to go to war' (*jus ad bellum*) and 'right conduct in war' (*jus in bello*). The first details the morality of going to war and the second focuses on moral conduct within war as it is being fought. Recently, there have been calls for the inclusion of a third facet of just war theory – *jus post bellum* – the morality of post-war settlement and reconstruction.

This third arm of the doctrine would have an impact on, say, the Anglo-American invasion of Iraq in 2003 where clearly there had been no viable attempt made to plan a post-war settlement and this deficiency created the anarchic void and a descent into sectarian violence. Rowan Williams, former Archbishop of Canterbury, now Lord Williams of Oystermouth, recently grappled with a modern interpretation of Just war theory in a lecture he gave to the Royal College of Defence Studies. In doing so, he follows in the footsteps of such theologians as St Augustine and St Thomas Aquinas. His conclusion was that, in terms of modern wars, the concept could no longer be defined 'because not only had warfare changed but so had the possibility of controlling conflict.'[15]

The Battle of Waterloo in 1815 was fought in a day and decided the war, one side won, the other lost and the decisive outcome brought hostilities to an end. The impact on civilian populations was not particularly great and the fighting was spread over an area equal to a certain number of football pitches. Today, the civil war in Syria has left the nation devastated, hundreds of thousands have died, millions have been turned into traumatised refugees, the infrastructure will take a generation to restore, and the country's heritage has been desecrated by Daesh.[16] Most of us have no real idea who is fighting who. So many groups and factions are involved and these, like amoeba, constantly split, re-form, and realign themselves. The conflict has been extended and deepened not just by Daesh imperialism but by clumsy Russian and faltering western meddling.

How do you define a 'just' war in such a maelstrom of madness? You can't. Most conflicts in the 21st century will follow the Syrian precedent and will spill out over regions and even continents. Recent horrors enacted in Paris and Brussels are part of this and there will be more. There are no 'civilians' anymore – contemporary asymmetric and civil war is total. Find the 'just' equation for any of this if you can! Would anyone wish to visit a museum which focused on an unjust war?

At the same time, we're getting exercised over the possible renewal of the Trident nuclear missile system. Is it needed? Can we afford it? All relevant questions and virtually impossible to find anything resembling a definitive answer unless, following one recent suggestion we renew the nuclear missile subs but send them out without missiles, presumably just torpedoes, *We Dive at Dawn* style – the kind of compromise that's similar to buying a guard dog and feeding him valium.

Our most recent overseas military commitments, despite the magnificent efforts of our armed forces, were, from a political viewpoint, ill-judged and flawed. In Iraq there was no post war strategy and in Afghanistan the coalition, initially successful, took its collective eye off the ball and gave the enemy time to regroup and switch tactics. These were serious and foreseeable errors and they cast a long shadow. The Coalition's final retreat or rout from Afghanistan, prompted by the US, was a disgrace. The defeat was total and the Taliban are now back in control of a starving nation, apparently destined to be dragged back into the Middle Ages.

Over a quarter of a century ago and Britain was involved in a major inter-state war – 'The 1st Armoured Division advanced into Iraq 'to smash Saddam Hussein's Republican Guard and to help restore sovereignty of

Iraq-occupied Kuwait.' The operations were completely successful, and Saddam's army was utterly defeated – 'those who chose to flee along the Basra Road were chopped down in ways that made it abundantly clear that any violation of international law would be dealt with ruthlessly by the free world.' At some point we will have to deal similarly with Daesh, a ruthless bunch of savage ideologues. Every war has its 'back door' conversations and channels, but not this one. It really is them or us and by 'us' I mean the civilized world regardless of race or creed.

The 1st Iraq War was a conventional conflict won by 'shock and awe', a modern version of *blitzkrieg* using tactical air power, massive armoured punch and prior deployment of Special Forces. The political support provided by governments, all backed by a clear UN mandate, was also crucial. It wasn't perfect; far from it, there were serious mistakes but as one senior commander aptly put it, 'battles are won by the side that can best absorb the cock-ups.' In the field it comes down to qualities of leadership and the grit, training and courage of the servicemen and women involved. In this we are second to none.

Things changed; the earth shifted on 24 February 2022 when Russia attacked Ukraine with massive conventional forces. The low intensity conflict that had ground on without conclusion for eight years exploded into a major European interstate war. At the time of writing this is continuing and the final consequences for Europe and indeed the wider world have yet to be felt. But it won't end well, already energy prices have rocketed and will continue their trajectory. Interstate war is back and whilst the UK and indeed NATO are not directly involved (yet), the potential for escalation is unlimited.

Back at the Beresford Arm, in Whalton, no enemy in sight:

John: Where do we see the perceived threat to us, i.e., UK coming from?

Graham: Hard to say, always has been. Who'd have thought Old Vlad the Invader was so far up himself to invade Ukraine?

John: There are plenty here, despite everything, who still cleave to Mother Russia and think he's justified with his Special Military Operation, resisting intolerable provocation from NATO.

Graham: They're not the poor sods living there and being bombed, shelled, tortured raped and murdered.

John: They never are, deep irony in that it's our Armed Forces that protect us and allow the sneerers to stay safe in their ivory towers and battle bravely on social media.

Graham: It's a cliché that armies are always ready for the last war, not the next one. I spent my military career, Northern Ireland aside, getting ready, as did the whole of NATO, for the Third World War against the Soviet Empire which, thank Christ, never happened. At least not yet and I don't see Ukraine as triggering Armageddon. Hopefully someone on Putin's side will have the decency to shoot him before too long.

John: There were plenty who said that when the Berlin Wall came down it was 'The end of history' how wrong can you get? We've had two Gulf Wars, Bosnia, Kosovo, Afghanistan and Ukraine which if, from our perspective isn't yet a shooting war, it's war by proxy, we do their supply, and they do our dying.

Graham: If you read Kilcullen, he's right, we face an unprecedented rise in populations, most in impoverished, often failing states, various forms of terrorist activity will flourish, and I don't just mean the loons gluing their noses to lamp posts here. ISIS may be beat but there will be ever emerging and morphing groups with associated franchises, mainly within the Muslim world but not exclusively, and we've got fanatical far right nutters trying to keep up with them.

John: And there's resources?

Graham: For sure. The Six-Day war was prompted by a threat to water, as populations multiply, as poverty deepens and climate change intensifies, these are the factors developing future threats.

John: How will we commemorate any of this? In the past we fought interstate wars with regular forces in uniforms, 'us and them', and we generally won. In both world wars we could say, rightly, we were striving to destroy vicious authoritarian regimes. We did the same with the Falklands, a totalitarian junta had invaded our sovereign territory, aiming to enslave our citizens. It was a good clean fight.

Graham: Damn right, 1982 was a good year to be in uniform even if you never got near nearer to the South Atlantic than North Shields, women were crazy for us boys, not much chance of that now. No specific museums to the Gulf Wars or to Afghanistan. The moral dilemmas are too complex and there's no 'victory' in the sense we used to understand the term. No stiff parades and proud salutes etc.

John: I can't think there's been a totally new military museum built for several decades, plenty of makeovers, some damn good ones. We have to feel guilty about past wars and be careful of our statues to our military heroes, they're no longer in fashion.

Graham: A sad observation. It's still your round.

The Way Ahead?

For a long time, defence pundits had been proclaiming that the age of major interstate wars is over and that asymmetric conflicts, often at sub-state level are the new paradigm. As we have said, this assertion took a significant knock on 24 February 2022 when Russian forces invaded Ukraine. This attack was a real throwback in strategical terms, huge tank columns (soon mired in winter slush) lumbered across the landscapes, and a full-on interstate war ensued. At the time of writing this continues locked, as it would seem, in deadly stalemate. Russian forces dug in behind their gains, a new Siegfried Line, unable to advance but too entrenched to be dislodged. No ending is currently in sight and few of us would dare to speculate what that conclusion might look like. For the first time in over a generation, the spectre of nuclear war looms.

Despite this, Ukraine pretty much disappeared from the media after 7 October 2023 when Hamas sub-state forces, a devil's porridge of freelance Jihadi terrorists, carried out a slew of atrocities in Israel, a killing spree unequalled since the days of Hitler's Third Reich. Inevitably, these atrocities sparked massive retaliation, something presumably Hamas was hoping for, and inflicted untold suffering on the ordinary people of Gaza. Hamas, like ISIS, is an ideology, it cannot be defeated by conventional military tactics, a terrorist hydra, cut off one head and another appears. Special Forces, as Israel has amply demonstrated in the past, can carry out surgical strikes against identified targets, to focus on infrastructure and, above all, gather intelligence.

Ideologies are best defeated from within, spread cancer throughout the organisation's cells, foster in-fighting and generate paranoia. Ultimately, such tactics will be far more successful and far less costly than conventional hammer blows. Hamas, despite the brutish savagery of their murders, still has many supporters in the west, plenty of evidence exists that antisemitism is rife on both fringes of far left and right. The 7 October *Einsatzgruppen* were, horrible as it is, a PR success and the nature of Israel's response plays very much to the terrorists' agenda. Propaganda has always been an element of war, ever since Joshua learnt to sound the trumpet, but media manipulation is now a major tool. Surgical strikes deny the ideologues these PR opportunities and are often lauded. But how do we commemorate such struggles? There is no glory, the forces involved divide and multiply like amoeba. By and large, the bulk of casualties are inflicted on the civilian populations, innocents suffer terribly, voyeuristic manna for the world's media.

Why would we want a museum to chronicle such horrors? Perhaps because they are so horrible. Germany has numerous 'Documentation Centres' attached to former concentration sites. These are sensitively handled, underplaying displays of weaponry and uniform, rightly so as these could send out the wrong messages. But SS gear was designed by Hugo Boss and yes, there is an appeal, however we may recoil from the fact. We have a duty to teach history, all of it. A revulsion for conflicts like the Israeli/Hamas war, which cannot offer a clear and satisfying outcome and in which civilians suffer horribly, combined with the desire to re-write history to reflect a particular narrative, combine to make this difficult. But we should not walk away from the task. If we do, military museums become irrelevant.

The heritage sector took a pounding during the Covid pandemic, and a cost-of-living squeeze has an inevitable knock-on effect. Museums also face cultural pressures from the 'Woke' movement. Now 'Woke' can be defined as 'being aware of and actively attentive to important facts and issues (especially issues of racial and social justice).' No one could argue with that. But regarding military heritage, we are encouraged to shy away from weapons, warfare and death. I may be cynical and instinctively hostile to such a dogma, but it has reinforced hostility to our military heritage, which is now something we should be ashamed of and ready to apologize for to anyone we, or our ancestors may have offended in the distant past. Finding oneself offended is very popular. History, I would suggest, is not there to be censored or politicized, it's there for us to learn from, not preach against.

In 2017, I gave a lecture at the National Army Museum in Chelsea, hitherto an enthusiast's paradise. British wars unfolded in a chronological sequence, from Crecy to Helmand. At that point, the place had just had a very extensive and expensive makeover. The new look, on which I was asked to comment, was dire. Themes had replaced timelines, vast amounts of empty space and a feebly apologetic approach to history. War *is* hell, we all knew that Empire involved brutality, exploitation and, not infrequently oppression. But history isn't about apologies, it's about realities.

The editor of this book told me a story he had heard about the IWM many years back. A new boss (no names, no pack drill) suggested during a boardroom meeting that some of the heavy stuff, artillery, tanks etc, cluttering up the place be removed to make room for different displays. Her first suggestion was 'Warfare and Children's Art.' A silence descended for a moment, and then one old IWM hand said: 'I've got a suggestion for a new exhibition. Let's do something on young men killing other young men.' He retired soon after.

Bad things happen in the past, just as they do in the present but I for one, don't feel the need to say sorry for something my great-great-grandfather might or might not have done, it was all a couple of centuries and more before I was born. Heavy-handed attempts to make military museums 'relevant' to the perceived needs of a modern generation are bound to fail. Dragging people out of their comfort zones and opening debates is not necessarily a bad thing.

There is a deeper problem in that we generally think the appeal of a military/regimental museum declines as those who have served die off and their family interest becomes increasingly remote, and this is clearly a factor. At the same time, MoD and other funding is drying up; museums cannot rely on central support or a certain stream of people coming in through the door. I have mentioned the contentious case of the DLI Museum in Durham. This must and hopefully will reopen, but the experience cannot simply be the same as before, it has to be restructured and reinvigorated.

I'd like to bang the drum for living history here. Firstly, I have to declare an interest as a historical interpreter who makes a living from appearing in museums and galleries. Nonetheless, I'm a firm believer in the notion that living history invigorates and illustrates a museum's offer. For instance, for a good number of years running we had a regular vehicle and reenactment *fest* at DLI Museum. Some academic commentators recognise the value of living history as both an enhancement to a museum and as a readily accessible learning tool. Others display a grand disdain, missing the point entirely. But I really believe this is how military museums can both survive and retain relevance in an age generally without heroes.

In his magisterial autobiographical account of Chindit Operations in 1944, John Masters recounts the story of a desperate withdrawal in the face of overwhelming odds, his exhausted men hopelessly outnumbered and outgunned: 'A Cameronian lay near the ridge top, near death from many wounds. "Gi' me a Bren," he whispered to his lieutenant. "Leave me. I'll take a dozen wi' me."'[17]

And that's why we need never to forget.

Endnotes

Chapter 1

1. *Daily Mail*, August 1923, trustees of the Green Howards, MA 364.
2. Hewitson, T.L., *Weekend Warriors from Tyne to Tweed* (Gloucs., Tempus 2006), p. 60.
3. Ibid., p. 64
4. Wilfred Owen: *Disabled.*
5. Hewitson, p. 66.
6. Ibid. p. 63.
7. Notorious as the 'Hexham Massacre'.
8. Hewitson, p. 63.
9. Ibid. p. 65.
10. Pease, H., *the History of the Northumberland (Hussars) Yeomanry 1819 – 1923* (Constable, London, 1924) p. 60.
11. Sadler, J., *The Field Gun Run* (Royal Military Tournament, London, 2010) p. 5.
12. An artillery innovation from the 1890s; 'Quick-firers' were distinguished from earlier ordnance in that they were fitted with buffers to limit recoil, the breech was adapted to allow rapid re-loading and shell and propellant were combined in cartridge form. Consequently, in the early C20, the killing power of artillery increased exponentially.
13. The Short Magazine Lee Enfield (SMLE) was the standard British service rifle of the era, called 'short' because of the barrel length being shorter than the earlier 'Long' Lee-Enfield which did service in the Boer War and with which many territorial units were still

armed in 1914. The SMLE was intended both for cavalry and infantry, firing 10 rounds of .303 ball cartridge. The weapon reached its final pre-war variant, the Mark III in 1907 and confounded its critics, doing good service throughout the conflict and, depending on circumstance for decades afterwards.

14. Pease, op. cit., pp 73 – 74.
15. Ibid., p. 74.
16. Ibid.
17. Ibid., p. 74.
18. Ibid., p. 76.
19. Source: Tyne and Wear Archives and Museums NEWNH: 0688.
20. Ibid.
21. Ibid.
22. Ibid.
23. Ibid.
24. Ibid.
25. Ibid.
26. Ibid.
27. Ibid.
28. Ibid.
29. Ibid.
30. Ibid.
31. Ibid.
32. The Pennyman family lived at Ormesby Hall on Teesside for nearly 400 years, the magnificent 18th-century house is now maintained by the National Trust.
33. Trustees of the KOSB Museum
34. Ibid.
35. Ibid.
36. Ibid.
37. Ibid.
38. Ibid.
39. Ibid.
40. Holmes, R., *The Western Front* (London, BBC 1999), pp 48–50.
41. Trustees of the KOSB Museum.
42. Ibid.
43. Holmes, op. cit., pp 49–50.
44. Trustees of the KOSB Museum.
45. Ibid.
46. Ibid.

47. Ibid.
48. Ibid.
49. Ibid.
50. Holmes, op. cit., p. 51.
51. Trustees of the KOSB Museum.
52. Ibid.
53. Ibid.
54. Ibid.
55. Ibid.
56. Ibid.
57. Ibid.
58. Ibid.
59. Ibid.
60. Ibid.
61. Heugh Gun Battery Trust Limited.
62. Ibid.
63. Ibid.
64. D/DLI, 7/478/S (17).
65. Heugh Gun Battery Trust Limited.
66. D/DLI, 7/478/ S (17).
67. Heugh Gun Battery Trust Limited.
68. D/DLI, 7/478/ S (17).
69. Ibid.
70. Ibid.
71. Heugh Gun Battery Trust Limited.
72. Ibid.
73. Ibid.
74. Ibid.
75. D/DLI, 7/478/ S (17).
76. Heugh Gun Battery Trust Limited.
77. D/DLI, 7/478/ S (17).
78. Ibid.
79. Morgan, M., *D-Day Hero* (Stroud, Spellmount 2004), p. 59.
80. Ibid., p. 60.
81. Ibid.
82. Ibid., p. 61.
83. Ibid., p. 62.
84. Ibid., p. 63.
85. Projector Infantry Anti-tank – 'PIAT' a weapon manufactured by ICI of all people, a development of the Spigot Mortar and which entered service in 1943, an absolute swine to operate.

86. Morgan, op. cit., p. 67.
87. Ibid.
88. Ibid, quoted on p. 68.
89. D/DLI.7115/17.
90. Ibid.
91. Ibid.
92. Ibid.
93. Ibid.
94. Ibid.
95. Ibid.
96. Rissik, D., *The DLI at War – History of the DLI 1939–1945* (Durham, 1952), p.94.
97. Gerrard, C., with P. Greaves, A. Millard, R. Annis and A. Caffell *Lost Lives, New Voices – Unlocking the Stories of the Scottish Soldiers from the Battle of Dunbar 1650* (Oxbow, Oxford 2018).

Chapter 2

1. https://www.edinburghcastle.scot/see-and-do/highlights/scottish-national-war-memorial accessed 26 October 2022.
2. Paterson, J., *Autobiographical Reminiscences* (Glasgow, Maurice Ogle and Co 1876) pp 205–213.
3. The Eagle is still on show in the Regimental Museum.
4. Paterson, op. cit., p. 213.
5. Quoted in Sadler, D. J., *Scottish Battles* (Edinburgh, Canongate 2010), pp 274–275.
6. Ibid., p. 275.
7. Ibid., p. 276.
8. Directed by Peter Watkins, released on 15 December.
9. Sadler, op. cit., p. 263.
10. Ibid.
11. Ibid., p. 264.
12. Ibid.
13. Ibid., p. 265.
14. Ibid., p. 266.
15. Ibid., p. 267.
16. https://tunearch.org/wiki/Annotation:Merrily_Danced_the_Quaker, accessed 26 October 2022.
17. https://www.historicenvironment.scot/visit-a-place/places/fort-george/, accessed 28 October 2022

18. https://www.historicenvironment.scot/visit-a-place/places/clava-cairns/, accessed 28 October 2022.
19. Directed by Valerio Zulini, released 1976.
20. https://www.scallowaymuseum.org/, accessed 28 October 2022.
21. Ibid.
22. Ibid.
23. https://www.warmemorialsonline.org.uk/memorial/116837/, accessed 28 October 2022.
24. Peter Sidney Ernest Lawford (1923–1984), a British born actor, member of the infamous Hollywood 'Rat Pack'.
25. Lovat, S., *March Past* (London Weidenfeld and Nicolson 1978) pp 309–310.
26. Ibid., p. 311.
27. Ibid., p. 313.
28. Ibid.
29. https://theddaystory.com/, accessed 29 October 2022.
30. Lovat, op. cit., pp 313–314.
31. Ibid., p. 316.
32. Ibid.
33. Ibid.
34. Miller, M., R. Laycock, J. Sadler and R. Serdiville, *As Good as Any Man; Scotland's Black Tommy* (Stroud, The History Press 2014), p. 10.
35. Ibid., p. 11.
36. Ibid., p. 84.
37. http://www.stobscamp.org/, accessed 29 October 2022.
38. Ibid.
39. Ibid.
40. Ibid.
41. https://www.devilsporridge.org.uk/, accessed 29 October 2022.
42. Ibid.
43. Ibid.
44. Ibid.
45. https://www.thecelticgoldsmith.com/the-borderlands-museum, accessed 27.10.2022. The website promised that the museum may reopen soon.
46. Sadler, op. cit., pp 117 -119.
47. Ibid.
48. Ibid.
49. Ibid.

50. Ibid., p. 122.
51. BBC 1968, two seasons.
52. http://www.hawickreivers.com.

Chapter 3

1. 'The Shawfield Manuscript', as quoted in *Scott's Minstrelsy of the Scottish Border* (London, Ward Lock 1892), p. 259.
2. In Cumberland's Museum of Military Life, just left of the entrance stands a brilliantly realised mannequin display of a mounted trooper from 1914, complete with every item of kit.
3. https://collection.nam.ac.uk/detail.php?acc=1971-02-33-106-1, accessed 31 October 2022.
4. A redan is a form of field fortification, angled towards an enemy but with the rear left open to facilitate rapid reinforcement, certainly an effective outwork before Sebastopol.
5. Baring-Pemberton, W., *Battles of the Crimean War* (London, B.T. Batsford 1962), p 195.
6. Ibid., p. 196.
7. Ibid., p. 199.
8. Ibid., p. 203.
9. Macdonald Fraser, G., *Quartered Safe Out Here* (London, Harper Collins 2000), p. 61.
10. https://www.weareexplore.org.uk/.
11. Grid ref. NY 6339 7556.
12. These were the Vickers Valiant, Avro Vulcan and Handley Page Victor.
13. See Boyes, J. (2013) 'The Blue Streak Underground Launcher' in *Airfield Research Group Review*, no. 140.
14. Directed by Stanley Kubrick, 1964.
15. Bristol Bloodhound; a ram jet surface to air missile, developed in the 1950s.
16. With the former Royal Northumberland Fusiliers and Royal Warwickshire Fusiliers.
17. Haythornthwaite, P. J., *Gallipoli*, No. 8 in Osprey Campaign Series (London Osprey 1991) p. 41.
18. Carlyon L. A., *Gallipoli* (London, Doubleday 2002), p. 206.
19. He died in 1966, aged 89. He'd had to sell his medal.
20. Losing the limb didn't stop him serving in the Home Guard during the next war. He died in 1953.

21. From Stearck, C., *Gallipoli Association*, 30 October 2002, accessed 16 September 2022.
22. Keneally's award was announced in the London Gazette on 23 August to the immense pride of his family, a parade was to be held in his honour, the toast of Wigan. By then he was already dead, killed on 28 June, though his family didn't get to hear until October (Carlyon, op. cit., p. 207). Grimshaw was like two of the others (Bromley and Stubbs), originally recommended for a Distinguished Conduct Medal, but after some manoeuvring the higher awards were confirmed by the London Gazette on 13 March 1917.
23. 'I serve.'
24. 2nd Lieutenant David Cuthbert Thomas (1895–1916) influenced Sassoon, who, with Graves, featured him in several works after his death.
25. Graves, R., *Goodbye to All That* (London, Cassell 1957), p. 236.
26. Von Falkenhayn, the German C in C, when launching his attack against the French at Verdun in February 1916, had already calculated that Britain was now becoming the greater foe.
27. https://www.nam.ac.uk/explore/battle-somme, accessed 31 October 2022.
28. https://oldfrontline.co.uk/2022/01/15/forgotten-front-cuinchy-brickstacks/ with Paul Reed, accessed 31 October 2022.
29. Graves, op. cit. p. 214.
30. Ibid.
31. Ibid., p. 215.
32. Ibid., p. 219.
33. Ibid., p. 220.
34. Ibid., p. 221.
35. Ibid., p. 227.
36. Ibid., p. 228.
37. Ibid.
38. See, Walter, G. (editor) *In Flanders Fields, Poetry of the Great War* (London Penguin 2004), p. 127.
39. https://www.historyireland.com/the-capture-and-death-of-major-geoffrey-lee-compton-smith-1921/ accessed 30 September 2022.
40. https://www.broadwayworld.com/article/VIDEO-FLASHBACK-1965-Peter-Sellers-Mimics-Laurence-Oliviers-Richard-III-Reciting-The-Beatles-A-Hard-Days-Night-20160405, accessed 07 September 2022.
41. Ibid.

42. This wasn't complimentary. Richard's famous badge was the white boar. William Catesby 'The Cat', Richard Ratcliffe 'the Rat' and Viscount Lovell 'the Dog' – his badge was the wolf – were the King's henchmen.
43. https://www.battlefieldstrust.com/resource-centre/warsoftheroses/battleview.asp?BattleFieldId=8, accessed 07 September 2022.
44. https://www.thefreelibrary.com/Uncovering+the+true+Battle+of+Bosworth.-a0246001784, accessed 07 September 2022.
45. Not a good date for kings to be undertaking risky new ventures; James IV of Scotland crossed the Tweed on that day in 1513, heading for disaster at Flodden on 9 September, and Charles I raised his standard at Nottingham in 1642, thus igniting the Civil War in England and Wales that ultimately cost him his head.
46. Farrar-Hockley, A., *The Edge of the Sword* (Watford, Odhams 1954), p. 21.
47. Ibid., p. 26.
48. Ibid., p. 31.
49. Ibid., p. 33.
50. Ibid., p. 35.
51. Ibid., p. 70.
52. Walter, G., op. cit., p. 99.
53. See Reed, P., *Walking the Somme; A Walker's Guide to the Somme Battlefields* (Barnsley, Pen and Sword 1997), pp. 123–124.
54. https://www.keepmilitarymuseum.org/history-of-the-regiments/devonshire-regiment/accessed 13.10.2022. In terms of losses, these included the CO, major, adjutant, 7 captains, 5 lieutenants and 14 2nd lieutenants. See Rogerson, S., *The Last of the Ebb* (London, Arthur Barker 1937), p. 116.
55. Walter, G., op. cit., p. 99. This poem 'Before Action' by Lt Noel Hodgson MC was first published on 29 June 1916 in the weekly paper *The New Witness*, two days before he went into action with the 9th Devons on 1 July 1916.
56. https://www.bridgwatermuseum.org.uk/battle-of-sedgemoor.htm.

Chapter 4

1. The Worthington quote is kindly supplied by Major S. T. Meadows of 2 RGR.
2. 'Dacoit' is a general term used for any bandit in Bengali, Odia, Hindi, Kannada and Urdu.

3. Saharanpur is both a city and Municipal Corporation in the state of Uttar Pradesh in Northern India, the administrative headquarters of Saharanpur District and Division.
4. Forbes (1799–1858) born in India, became a distinguished botanist, latterly professor of Materia Medica at King's College London.
5. Shore was a Daniel Dravot-like figure, something of a mystic with a tendency to 'go native' at times.
6. Shakespear, Colonel L.W. and Lieutenant-Colonel G.R. Stevens, *The History of the 2nd King Edward VII's Own Goorka Rifles (The Sirmoor Rifles)* (London Gale and Polden 1952), p.6.
7. Ibid.
8. Ibid.
9. Ibid.
10. Farwell, B., *The Gurkhas* (London, Allen Lane 1984), p. 49.
11. Ibid., p. 51.
12. General Sir John Moore (1761–1809), 'Moore of Corunna' was an inspirational leader whose innovative training methods for light infantry at Shorncliffe barracks after 1803 had a lasting effect.
13. The Baker Rifle was designed by Ezekiel Baker in 1800 and became the first standard issue rifle in service with the British army, calibre .615 inch (15.9 mm).
14. Young, Brigadier P., *The British Army* (London, William Kimber 1967), p. 124.
15. Glover, David, 'Thomas Plunkett: A Pattern for the Battalion'. 2nd Bn. 95th Rifles, Battle Re-enactment and Living History Society, accessed 14 November 2022.
16. Warner, P., *Alamein – Recollections of the Heroes* (London, William Kimber 1979), p. 175.
16. Ibid., p. 175.
17. Carver, M., *El Alamein* (London, Batsford 1962), p. 146.
18. Ibid., p. 146.
19. Warner, op. cit., p. 175.
20. Carver, op. cit., p. 147.
21. Ibid. p. 147.
22. Lucas, J., *War in the Desert – the Eighth Army at El Alamein* (London, Arms and Armour Press)., p. 221.
23. Carver, op. cit., p. 146.
24. For their gallantry both men were decorated, Colonel Turner received the VC and Sergeant Callistan the DCM.

25. Lucas, op. cit., p. 222.
26. The Churchill Tank ('I' Tanks Mark IV (A22)) was intended to replace both the Matilda and Valentine, it was heavily armoured and equipped with the 6-pdr gun. Later, some were fitted with 75 mm guns salvaged from disabled Shermans.
27. The Deacon was a development of the earlier portee arrangement which mounted a 6-pdr on an AEC Matador truck chassis.
28. Warner, op. cit., p. 175.
29. Lewis, P.J., and I.R. English, *Into Battle with the Durhams: 8 DLI in World War II* (London Stamp Exchange, 1990), p. 42.
30. Ibid., p. 175.
31. Published Stroud, Amberley 2018.
32. Published London, Penguin 1994.
33. Published London, Collins 1959.
34. Sadler, D. J., *D-Day – The British Beaches* (Stroud, Amberley 2019), pp 33–34.
35. https://www.loc.gov/exhibits/churchill/interactive/_html/wc0020.html, accessed 7 November 2022.
36. Directed by Joe Wright and released in January 2018.
37. Famously commemorated by Picasso.
38. The actual total was 43,000 for the entirety of the bombing.
39. 9 June 1983, Conservative majority of 144.
40. Designed by Clash Architects in partnership with consulting engineers Price and Myers.
41. http://britains-smallwars.com/campaigns/oman/page.php?art_url=battle-of-mirbat, accessed 14 November 2022.
42. https://royalartillerymuseum.com/, accessed 14 November 2022
43. A Monitor is neither that big or that fast, nor is its lines pleasing. They were a form of floating battery with a heavy armament intended for coastal bombardment, beginning in the 1860s and continuing throughout the Great War.
44. https://www.britannica.com/biography/Bertram-Home-Ramsay, accessed 14 November 2022.
45. The Hurricane was the workhorse of the RAF in 1940. It had been introduced three years earlier, but its box girder primary fuselage structure was already out of date. It was slower than the ME109 but handier and more agile. Despite being outclassed, it would still account for some 60% of the RAF's total kills during the battle. It served throughout the war and over 14,500 were built.

46. The Messerschmitt Bf 109 was the backbone of German fighter command at the start of the war. a superb all-metal monocoque design, it was a world-class plane. Synchronized machine guns were fitted to the fuselage to keep the wing structure light and latterly a 20 mm cannon was fitted to fire through the propeller shaft.
47. The Bf 110 (or *Jabo* as it was nicknamed) was a heavy twin-engined fighter introduced in 1937. It was an impressive gun bus sporting two 20 mm cannon and four 7.92 mm machine guns. Its abiding weakness was a lack of agility.
48. Sadler, D. J., and R. B. Serdiville, *Fighter Aces, Knights of the Sky* (Oxford, Casemate 2017), pp 87–88.
49. Sadler, D. J., and R. B. Serdiville, *Tommy at War 1914–1918* (London, Biteback 2014), p. 237.
50. Sir John Norton-Griffiths, first Baronet (1870–1930), 'Empire' or 'Hellfire Jack', an immensely colourful character, served in colonial conflicts, the South African War and during 1914–1918. Based on his extensive engineering experience, he designed and built field fortifications and supervised mining activity on the Western Front.
51. Royal Engineers Archive A.231.2.
52. Ibid.
53. Ibid.
54. Ibid.
55. Ibid.
56. Ibid.
57. Ibid.
58. Ibid.
59. Ibid.
60. Ibid.
61. Ibid.
62. Ibid.
63. Ibid.
64. Ibid.
65. Ibid.
66. That sum would probably equate to over £400,000 today.
67. See *Bletchley Park*, the Bletchley Park Trust (London, Pitkin 2018).
68. Asa Briggs speaking on UK Channel 4 TV *Britain's Greatest Codebreakers*, 21.11.2011.

69. Thomas Harold Flowers (1905–1998) – the modest colossus who built Colossus, the first programmable electronic computer, ended the war in debt as he'd had to pay for many of the components himself. His design was still active within GCHQ certainly until 1960.
70. Directed by Morten Tyldum and released in the UK on 8 October 2014.
71. The Ferret Armoured (or Scout) Car was chosen in 1949 to provide the kind of armoured reconnaissance function undertaken by the Daimler Dingo during WWII. Daimler also produced the Ferret, which was manufactured in significant numbers (c. 4,500) between 1952 and 1971. These vehicles saw worldwide service over a long period with many of the world's armies.

Chapter 5

1. 'We who are about to die salute you.'
2. 12 July 1943.
3. 30 August 1945.
4. ROTOR was a post-war UK response to a perceived threat from Russian bombers and was generally located around wartime radar stations.
5. Scotland's Secret Bunker – Visitor attraction of Scotland's best kept secret, accessed 30 November 2022.
6. 'Vulcan', 'Victor' and 'Valiant'.
7. https://www.bcwm.org.uk/ accessed 22 November 2022.
8. 1965, director, Ken Annakin.
9. Yeomanry were the mounted arm of militia and volunteer units raised for home defence, comprised usually of the rural gentry and middle classes, 'yeomen'. The Yeomanry squadrons now form the TAVR element of the RAC.
10. 'Hobo' was Montgomery's brother-in-law.
11. The emblem of 7th Armoured Division, the jerboa or desert rat, was to become a potent symbol.
12. Hobart was at one stage reduced to serving as a corporal in the Home Guard, but his career was resurrected before D-Day when his eccentric brand of ingenuity produced the legendary 'funnies' – adapted AFVs which did good service in the Normandy campaign and after.
13. The difference between the Lee and Grant tanks was that on the former a cupola housing a machine gun was mounted onto the turret.

14. White, B.T., *German Tanks and Armoured Vehicles 1914–1945* (London, Ian Allan 1966), pp 43–45.
15. Ibid. pp 45–47.
16. Playfair op. cit. p. 500.
17. Lucas, op. cit., p. 241; the tank pennants were of course significant in that they detailed seniority.
18. A 'stonk' means a bombardment usually by shell or mortar fire upon an enemy position.
19. Lucas, op. cit., p. 253.
20. This was the UK's primary main battle tank of the post-war period and one of the most successful designs of the era, it saw service around the globe.
21. Armoured Vehicle Royal Engineers, an armoured vehicle designed to offer protection to engineering personnel during frontline operations.
22. Catchphrase from BBC TV's show *Dad's Army*.
23. Played by Clive Jones in the series.
24. We merely have to consider the fate of the Polish AK or Home Army in their desperate struggle in Warsaw during the later summer of 1944.
25. A small, squat concrete bunker constructed, usually at key tactical points, after 1940.
26. See http://blythbattery.org.uk.
27. See http://www.heughbattery.com.
28. DLI Sound Recording Project.
29. Ibid.
30. Amazon Defence Corps: a martial organisation where experienced female shots taught others marksmanship.
31. A rare example of the badge can be viewed at: http://www.iwm.org.uk/server/show/Con MediaFile.5690.
32. DLI Sound Recording Project.
33. The P17 was .300 cal. whilst the Lee-Enfield was .303.
34. The Thompson used a .45 cal. pistol round and served Allied forces through WWII and after.
35. DLI Sound Recording Project.
36. BBC Radio Broadcast 14 May 1940.
37. The German airborne arm, unlike the British, came under the aegis of the Luftwaffe.
38. The HG was formally stood down in December 1944.
39. Peter Fleming writing in the *Spectator*, 8 July 1966.

40. *Home Guard Manual 1941*, introduction by M. McCutcheon.
41. Warwicker, J., *Churchill's Underground Army* (Barnsley, Pen and Sword 2008), p.25.
42. There is an excellent website on this subject built by Brian Pears: http://www.bpears.org.uk/Misc/War_NE/w_section_04.html, accessed 30 November 2022.
43. Ibid.
44. SOE: Special Operations Executive – their mission from the Prime Minister was to 'set Europe ablaze'.
45. Brian Pears' website (see above).
46. Warwicker, op. cit., p.208.
47. Ibid. p.223.
48. Interview in Prudhoe Library, March 2010.
49. Directed by John Guillermin.
50. De Havilland DH4, designed by De Havilland for Airco (hence 'DH'), the first British two-seater day bomber with effective defensive armament. Designed in 1916, it came into service the next year.
51. 1964, directed by Walter Grauman.
52. Morane Parasol, this was the 1914 Morane-Saulnier, single-winged, one- or two-seater scout. Austrian Albatross, presumably the B1 which entered service in 1913. The Voisin-Farman Biplane, an early model designed in 1907. The Aviatik B1 Biplane was a two-seater reconnaissance plane designed by Automobil and Aviatik AG.
53. *Daily Telegraph*, reprinted in *Flight* magazine 16 October 1914.
54. Cecil Lewis.
55. Belgium's answer to the Maginot Line and even less successful!
56. The D-Day Story, Portsmouth – How do you get an army across the English Channel and into France when your enemy is well armed, well trained and expecting you? Find out how at the D-Day Story, Portsmouth (theddaystory.com), accessed 01 December 2022.
57. From John Howard's *Pegasus Diaries* originally published by Pen and Sword and featured in *D-Day and Normandy, Sixty-Five Years On*, Wharncliffe History Magazines, Barnsley 2009), p. 61.
58. The D-Day Story, Portsmouth (theddaystory.com), accessed 1 December 2022.
59. *Sixty-five Years On*, p. 62.
60. Ibid., p. 64.

61. D-Day 50th Anniversary *Soldier Magazine* Supplement.
62. The D-Day Story, Portsmouth (theddaystory.com), accessed 1 December 2022.
63. Ibid., Ashby went on to take part in the Battle of Arnhem in September 1944 and the crossing of the Rhine in March 1945. Married in 1944, he retired as Surrey's county librarian in 1978.
64. The Northumberland Archives hold original copies along with some transcriptions of issues one to six of the newspaper; prisoners in the Camp also staged plays, radio broadcasts and football matches. communityhttps://www.northumberlandarchives.com/2019/02/15/escape-from-camp-18/ work, accessed 28 November 2022.
65. Hanna Reitsch (1912–1979), a celebrated Third Reich test pilot.
66. His body was found three days later by John Walton, a local shooter. Kropp was buried with full honours in the CWGC plot within Hexham Cemetery, lying there until 1958 when his remains were repatriated.
67. DLI Sound Recording Project.
68. https://www.english-heritage.org.uk/visit/places/cliffords-tower-york/history-and-stories/massacre-of-the-jews/, accessed 29 November 2022.
69. Jefferson, D., *Tobruk: A Raid too Far* (London, Robert Hale 2013), pp 188–189.
70. Ibid.
71. Ibid.
72. https://yorkshireairmuseum.org/, accessed 30 November 2022.
73. So called because their home port was Queenstown in Ireland, though they were also known as decoy vessels, special service ships and, quite alluringly, as mystery ships. The Germans referred to them as *U-Boot-Fallen* (traps).
74. The notion was part of the package Haig offered to secure cabinet support for his planned 3rd Battle of Ypres in 1917.

Chapter 6

1. Army doggerel verse/song
2. From 1964, directed by Stanley Kubrick with, most memorably, Peter Sellers and re-released in 1994.
3. See Smith, General Sir Rupert, *The Utility of Force* (London, Allen Lane 2005), p. 147.

4. Bill Sculthorpe, quoted in Neillands, R., *In the Combat Zone* (London, Weidenfeld and Nicolson 1997), p. 109. He served with 2 SAS in Malaya, 1955–1957.
5. Ibid.
6. Literally, 'war without hate' – possibly also wishful thinking.
7. March 1988 and these terrorists were Sean Savage, Daniel McCann and Maireed Farrell.
8. See Rupert Smith op. cit.
9. See Bowden, M., *Black Hawk Down* (London, Corgi 2021 edition).
10. See Kilcullen, D., *Out of the Mountains, the Coming Age of the Urban Guerrilla* (London, Hurst Books 2015).
11. Orwell 1945 article, 'Notes on Nationalism'.
12. (London, Hurst Books 2018).
13. (Guilford, Prometheus 2006)
14. (Oxford University Press 2012)
15. As reported in the *Saturday Times* on 16 April 2016.
16. ISIS/ISIL The Islamic State of Iraq and the Levant, also known as the Islamic State of Iraq and Syria, the Islamic State of Iraq and ash-Sham, or simply Islamic State. 'Daesh' is an adapted acronym of the Arabic *Dawlat al-Islamiyah f'al-Iraq w Belaad al-Sham*, similar to another Arabic word, *das*, which means 'to trample down' or 'crush'; *Saturday Times* 05.12.2015.
17. Masters, J., *The Road Past Mandalay* (London, Michael Joseph 1961), p. 258.

Gazetteer

As Julian Barnes observes in *Metroland*, 'objects contain absent people.' Museums bring to life the stories of cultures, objects and peoples past and present. The recent and lengthy controversy in Durham, following the county council's decision to shut the DLI Museum became a *cause célèbre* and cost them their majority. When regiments are disbanded or amalgamated, the immediate connection is lost, but it is the museums that keep the memory of service, sacrifice and honour alive.

Those collections featured in this book range from nationally important major exhibitions run by English Heritage, say, or Historic Environment Scotland, to those set up and managed by local volunteers and enthusiasts, both and all in between are important. So much of our heritage is curated by volunteers, those who give of their time and commitment to ensure our connection to the past survives. As veterans die, their words, their kit and their medals remain the tangible reminder of their achievement. Below are locations and contact details for those museums currently open to the public.

Northern England and Border

Charge – the Museum of the Northern Cavalry: located in Discovery Museum, Blandford Square, Newcastle upon Tyne NE1 4JA, open weekdays and weekends. Tel. (0191) 2326789: https://discoverymuseum.org.uk/. The gallery is situated within the Discovery Museum which also houses one of Armstrong's early breechloading, rifled cannon. Little on-site car parking but spaces nearby, easily accessible from the Central Station.

King's Own Scottish Borderers Museum: located in Berwick Barracks, The Parade, Berwick on Tweed TD15 1DF (managed by English Heritage). Check website for opening times. (Tel. 01289 33181). https://kosb.co.uk/museum/. A visit here combines nicely with EH's 'Soldier's Life' exhibition within the barrack piles. Can be combined with a more strenuous tour of the wholly intact Elizabethan/medieval walls. No on-site parking but a large public car park opposite.

Heugh Battery Museum: located on Moor Terrace, Headland Hartlepool TS24 0PS, open Friday–Monday inclusive. (Tel. 01429 270746). https://www.heughbattery.co.uk/. This is an evocative setting with well-structured commemoration for the battle and bombardment of 1914. Very accessible and activity level easy. No on-site car parking but plenty of spaces adjacent.

The Green Howards Museum: located in Trinity Church Square, Richmond, DL10 4QN, open Monday–Saturday. (Tel 01748 826561); https://greenhowards.org.uk/visit/. Superbly located within the older church building, very accessible. Twins nicely with a visit to the castle (below). Nearby car parking available.

Richmond Castle: located on Tower Street, Richmond DL10 4QW, open daily. (Tel. (0370 331181); https://www.english-heritage.org.uk/visit/places/richmond-castle. This is the ancient Norman keep and bailey, curated by English Heritage and a visit here combines well with one to the Green Howards Museum.

Scotland

The Royal Scots Dragoons Museum, Edinburgh: 'Scotland's Cavalry – second to none.' Ensign Ewart is the star! Open daily. (Tel. 0131 3105138). https://www.scotsdg.org.uk/. Situated within the austere grandeur of Edinburgh Castle (EH1 2YR) this should form part of a visit to the great fortress itself, combined with the Scottish IWM and **Royal Scots Museum** (Tel. 0131 3105040), https://www.theroyalregimentofscotland.org/. All three are well worth a visit and very different, though that involves a good deal of walking (a lot of it uphill on unyielding granite cobbles). Car parking is extremely difficult, and the castle is a brisk uphill walk from Waverley Station.

The Black Watch Museum, Perth: located in the Castle, Hay Street, Perth, PH1 5HR. Having benefitted from a major makeover just prior to the Great War Centenary (2013), this is an extensive and very well mounted collection, open daily but with slightly restricted winter hours (01 November–31 March). (Tel. 01738 638152); https://theblackwatch.co.uk/visit/. There is car parking around.

Culloden Visitor Centre and Museum: location Culloden Battlefield Centre, Inverness, IV2 5EU. In my opinion the best interpreted battlefield site in the UK and under the aegis of National Trust for Scotland, a breathtaking gallery of superb Jacobite artefacts and weaponry, open daily. (Tel. 01463 796090). https://www.nts.org.uk/visit/places/culloden. The ground has been cleared and marked with troop dispositions for both sides. Good on-site car parking.

Fort George, Ardersier: Inverness IV2 7TD. spectacularly located on the south shore of the wide Moray Firth, undoubtedly the finest C18 artillery fort in the UK, a splendidly presented, still military establishment, but heritage is superbly curated by Historic Environment Scotland. There is much to see, the mighty walls and gun terraces, military museum and powder magazine, open daily. (Tel. 01667 460232); https://www.historicenvironment.scot/visit-a-place/places/fort-george/. Good on-site car parking.

Shetland Bus Museum, Scalloway: located on Castle Street Scalloway, Shetland ZE1 0TP, it is closed over the winter season though the memorial situated on Main Street (ZE1 0XJ) is accessible all year round. (Tel. 01595 880734). https://www.scallowaymuseum.org/get-in-touch.html. Plenty of available car parking.

West Highland Museum Fort William and Spean Bridge Commando Memorial: the museum is located in the centre of Fort William on Cameron Square PH33 6AJ, accessible by train and with car parks nearby. Check website for seasonal opening times, https://www.westhighlandmuseum.org.uk/visit.jsp (Tel. 01397 702169). The WWII Commando gallery is especially commended. The monument at Spean Bridge (PH34 4EG) is accessible by car with good parking.

Stobs Camp, Hawick: the site is outdoors and located close to Hawick (TD9 9SG) but requires a degree of detective work, and visitors need to be properly clothed and able to walk at least a mile unaided.. Contact is Archaeology Scotland, Suite 1a, Stuart House, Eskmills, Musselburgh EH21 7PB. (Tel 0300 0129878); http://www.stobscamp.org/contact-us/. Car parking is available in lay-by at the base of the climb.

The Devil's Porridge Museum: located on Annan Road Eastriggs (west of Gretna), DG12 6TF, open daily. (Tel. 01461 700021). https://www.devilsporridge.org.uk/. A recent, purpose-built and interactive museum, a fitting commemoration and interpretation of a remarkable story, activity level easy. Opens visitor access to a wider hinterland and an opportunity to track down buildings in the area dating from the Great War complex. Good on-site car parking available.

Northwest and West of England

Cumbria's Museum of Military Life, Carlisle: located within Alma Block, the Castle, Carlisle, CA3 8UR (Tel. 01228 532774). This excellent museum had a major makeover and relocated within the last few years. It is open daily during summer but closed on Fridays during the winter, the wider setting within the great medieval castle which is under the care of English Heritage offers a combined visit, activity level moderate. https://www.cumbriasmuseumofmilitarylife.org/. There's a large public car park nearby.

Blue Streak Spadeadam RAF Ranges: no public access, can only be visited by arrangement with the MoD, generally group access only, activity level strenuous. https://www.raf.mod.uk/our-organisation/stations/raf-spadeadam/.

Lancashire Fusiliers Museum: located on Moss Street, Bury BL9 0DF (Tel. 0161 7638950) is a purpose-built museum but with a long history, the current exhibition opened in 2009 and features a range of interactive exhibits, it's open six days a week, Monday to Saturday, activity level easy: https://www.fusiliermuseum.com/about-the-fusiliers/about-the-museum. Car parking available.

Royal Welch Fusiliers Museum: located within the Castle, Caernarfon, LL55 2AY (Tel. 01286 673363), opened in 1960 and with admission coming free with castle entry, it makes sense to combine the tour. The castle remains one of Longshanks' great feats of military architecture. Open daily. https://www.rwfmuseum.org.uk/index.php. There is some car parking nearby.

Bosworth Battlefield Centre, Market Bosworth: located on Ambion Lane, Sutton Cheyney, Nuneaton, CV13 0AD (Tel. 01455 290427) and open daily (with restricted hours November–March). Now in the care of Leicestershire County Council, the site has much to offer and deals well with the controversy over where the fight actually took place. The museum and exhibition space is extensive and very well laid out, while the café is a delight, activity level moderate. https://www.bosworthbattlefield.org.uk/. Excellent on-site car parking.

Soldiers of Gloucestershire Museum: located in Custom House the Docks, 31 Commercial Road, Gloucester (Tel. 01452 522682). Open daily throughout the year, the collections had a significant makeover prior to the Great War centenary and re-opened in April 2014. https://www.soldiersofglos.com/visit-us/. There is some car parking nearby.

The Keep Military Museum (Devonshire Regimental Museum): located in the Keep, Bridport Road, Dorchester DT1 1RN (Tel. 07586 161872). One of our older military museums in the wholly gothic keep, once the barracks gatehouse from 1877 and now accessible daily, activity level easy. https://www.keepmilitarymuseum.org/. Car parking nearby.

South of England

Gurkha Museum: located in Peninsula Barracks, Romsey Road, Winchester SO23 8TH (Tel. 01962 842832). Open Monday–Saturday, all year round. I recently wrote a history of the Gurkhas so spent some time in the excellent archives, teasing out personal stories and like so many others this museum has a treasure trove of primary source material in its archives (opened 1987). Activity level easy. https://thegurkhamuseum.co.uk/. Car parking on site.

Royal Green Jackets (Rifles) Museum: also located in Peninsula Barracks, Romsey Road, Winchester SO23 8TH (Tel. 01962 825549). Open Monday–Saturday, all year round. The regimental museum was, for many years, peripatetic and didn't find a permanent home until Peninsula Barracks benefitted from a major makeover in 1964. Activity level easy; https://rgjmuseum.co.uk/. Car parking on site.

D-Day Museum, Portsmouth (the D-Day Story): located on Clarence Esplanade, Southsea PO5 3NT (Tel. 023 9288 2555) and open daily throughout the year. This really is the UK portal for a visit to the battlefields and museums of Normandy. Activity level easy. https://theddaystory.com/. Car parking nearby.

Churchill's War Rooms: located at Clive Steps, Charles Street, London, SW1A 2AQ (Tel. 020 7416 5000), open daily. This is a unique insight into the wartime governance of Britain and how the war effort was led and coordinated, it feels as though the people had just stepped out and there's a marvellous exhibition dedicated to the life of Winston Churchill. Activity level easy. https://www.iwm.org.uk/visits/churchill-war-rooms. Limited car parking, best accessed via the Tube (Lambeth North).

Firepower, Royal Artillery Museum: currently not open.

Dover Castle (Operation *Dynamo*): located atop Hill Road, Dover CT16 1HU, this is one of the key British fortresses, towering above the port. The Great Keep is a mighty Angevin edifice, raised on the site of William the Conqueror's motte and bailey, itself thrown up on the site of an earlier Anglo-Saxon fort which surrendered rapidly to Duke William after his great victory at Hastings. Tel. 0370 333 1181 and open daily, activity level strenuous. https://www.english-heritage.org.uk/visit/places/dover-castle/?utm_source=Google%20Businessandutm_campaign=Local%20Listingsandutm_medium=Google%20Business%20Profilesandutm_content=dover%20castle. There is a car and coach park within 500 metres, steep climb back!

Spitfire and Hurricane Memorial Museum, Ramsgate: located at RAF Manston, Manston Road, Manston, Ramsgate CT12 5DF (Tel. 01843 821940), open Tuesdays to Sunday. Home to those magnificent Battle of Britain aircraft, activity level easy. https://www.spitfiremuseum.org.uk./. Car parking available.

Royal Engineers Museum, Chatham: located on Prince Arthur Road, Gillingham, ME7 1UR and first opened in 1987 (Tel. 01634 822839).

This is one of my personal favourites and displays Lieutenant John Rouse Chard's Webley-Pryse revolver, something for all fans of the movie *Zulu*. Activity level easy. https://www.re-museum.co.uk/visit-us/. Some car parking.

Bletchley Park: located in the Mansion, Sherwood Drive, Bletchley, Milton Keynes, MK3 6EB (Tel. 01908 640404). Open daily from 1 March–31 October, slightly restricted hours during winter, activity level moderate. This is a fascinating museum and there is a great deal to see and experience, hub of Britain's wartime code-breaking. https://bletchleypark.org.uk. Plenty of car parking spaces on site.

Bovington Tank Museum: located on Linsay Road, Bovington, Wareham, Dorset, BH20 6JG (Tel. 01929 405096) Set up in 1947 and open daily, spiritual home for all enthusiasts of armoured warfare, much interactive stuff and live demos. The annual Tankfest is worth watching out for, activity level moderate. https://tankmuseum.org/. Plenty of car parking on site.

East of England

RAF Bentwaters Cold War Museum: located at Building 134, Bentwaters Parks, Rendlesham, Woodbridge IP12 2TW. Contact via email: info@bcwm.org.uk and check website for opening times, a very interesting volunteer-led museum, activity level easy. Good car parking nearby.

Norfolk Tank Museum: located at Station Road, Forncett St. Peter, Norwich NR16 1HZ (Tel. 01508 532650), opening times from Easter to November are on Tuesdays, Wednesdays, Thursdays and Sundays. On Wednesdays, visitors can see volunteers working on the vehicles, activity level easy. http://norfolktankmuseum.co.uk/opening-times-admission-costs/. Good car parking.

Dad's Army Museum: located on Cage Lane, Thetford, IP24 ZDS (Tel. 07470 165795). Check website for opening times. Inspired by the totemic TV series, this is an engaging volunteer-run museum, activity level is easy to moderate. https://dadsarmythetford.org.uk/index.php?option=com_contentandview=articleandid=1andItemid=181. No on-site parking.

IWM Duxford and The Parachute Regiment and Airborne Forces Museum: located in what was Duxford Airfield (Tel. 020 7091 3067) and opened in 1976, this is an epic museum, it will take at least a full day to get around all the galleries and the place still has a real WWII feel. There are usually period aircraft flying and the site is open daily, activity level moderate. https://www.iwm.org.uk/visits/iwm-duxford. Plenty of on-site car parking.

Clifford's Tower and York Army Museum: located at respectively Tower Street York, Y01 9SA (Tel. 0370 333 1181) and 3 Tower Street YO 9SB (Tel. 01904 461010). Clifford's Tower has benefitted from a massive revamp which has greatly enhanced the visitor offer, activity level, with a very steep hill (motte) to climb, strenuous. https://www.english-heritage.org.uk/visit/places/cliffords-tower-york/. But the museum is far easier: https://yorkarmymuseum.co.uk/. Very limited car parking in adjacent car park – usually very busy.

Eden Camp: located at Malton, North Yorks YO17 6RT (Tel. 01653 697777), open daily. This large site is unique and extensive, activity level moderate. Visitors should allow up to four hours for a thorough visit; https://edencamp.co.uk/contact/. Plenty of car parking available.

Yorkshire Air Museum and Allied Forces Memorial: Halifax Way, Elvington, YO41 4AH (Tel. 01904 608595). A brilliant museum and easily accessed from York, open daily during summer hours bu, in winter, shorter hours and closed on Wednesdays. Activity level easy/moderate. https://yorkshireairmuseum.org/plan-your-visit/. Plenty of on-site car parking.

National Museum of the Royal Navy: located in the re-created Georgian dock, Jackson Dock, Marine Avenue, Hartlepool, TS24 0X2 (Tel. 01429 860077). Open daily, this great exhibition houses HMS *Trincomalee*, a post-Napoleonic heavy frigate and provides a wholly immersive experience, activity level moderate. https://www.nmrn.org.uk/visit-us/hartlepool. Plenty of on-site car parking.

Select Bibliography

Allen, K., *The Wars of the Roses* (London, Jonathan Cape 1973)
Ashley, M., *The English Civil War* (London, Thames and Hudson 1978)
Banks, F.R., *Scottish Border Country* (London, Batsford 1951)
Barbour, R., *The Knight and Chivalry* (London, Sphere Books 1974)
Baring-Pemberton, W., *Battles of the Crimean War* (London, B.T. Batsford 1962)
Barr, J., *Border Papers* 2 vols. Eds: Revd J. Stevenson and A. J. Crosbie (Edinburgh 1894)
Bletchley Park, the Bletchley Park Trust (London, Pitkin 2018)
Boardman, A. V., *The Battle of Towton 1461* (Gloucs., Alan Sutton 1996)
Bowden, M., *Black Hawk Down* (London, Corgi 2021 edition)
Boyes, J. (2013) 'The Blue Streak Underground Launcher' in *Airfield Research Group Review*, no. 140
Burne, Colonel A.H., *Battlefields of England* (London, Methuen 1950)
- *More Battlefields of England* (London, Methuen 1952)
- and P. Young, *The Great Civil War 1642–1646* (London, Eyre and Spottiswoode 1959)

Carlyon L. A., *Gallipoli* (London, Doubleday 2002)
Carpenter C., *The Wars of the Roses: Politics and the Constitution in England c.1437–1509* (Cambridge, University Press 2002)
Carver, M., *El Alamein* (London, Batsford 1962)
Chesterton G. K., *A Short History of England* (London, Biblio-Bazaar, 2008)
Corfe, T., *Riot: The Hexham Militia Riot 1761* (Northumberland, Hexham Community Partnership 2004)
Deary, T., *Dirty Little Imps – Stories from the DLI* (Durham, County Record Office 2004)
Ducklin, K., and J. Waller, *Sword Fighting* (London, Robert Hale 2001)
Ellis, J., *Eye Deep in Hell* (London, Crook Helm, 1976)
Farrar-Hockley, A., *The Edge of the Sword* (Watford, Odhams 1954)

Farwell, B., *The Gurkhas* (London, Allen Lane 1984)
Fiorato, V., A Boylston and C. Knussel (ed.) *Blood and Roses: The Archaeology of a Mass Grave from the Battle of Towton AD 1461* (Oxford, OUP 2000)
Foster, J., *Guns of the North-East Coast* (Barnsley, Pen and Sword 2004)
Gerrard, C., with P. Greaves, A. Millard, R. Annis and A. Caffell *Lost Lives, New Voices – Unlocking the Stories of the Scottish Soldiers from the Battle of Dunbar 1650* (Oxbow, Oxford 2018)
Graves, R., *Goodbye to all That* (London, Cassell 1957)
Haigh, P.A., *The Military Campaigns of the Wars of the Roses* (Boston, Da Capo Press 1995)
Haythornthwaite, P. J., *Gallipoli* No. 8 in Osprey Campaigns Series (London, Osprey 1991)
Hewitson, T.L., *Weekend Warriors from Tyne to Tweed* (Gloucs., Tempus 2006)
Holmes, R., *The Western Front* (London, BBC 1999)
Jackson, D., *The Northumbrians* (London, Hurst, 2019)
Jefferson, D., *Tobruk: A Raid too Far* (London, Robert Hale 2013)
Keegan, Sir J., *The Face of Battle* (London, Penguin 1976)
Kilcullen, D., *Out of the Mountains, the Coming Age of the Urban Guerrilla* (London, Hurst Books 2015)
Leadman, A.D., 'The Battle of Towton' *Yorkshire Archaeological Journal* vol. 10 (1889)
Lewis, P.J., and I.R. English, *Into Battle with the Durhams: 8 DLI in World War II* (London, Stamp Exchange, 1990)
Lovat, S., *March Past* (London, Weidenfeld and Nicolson 1978)
Lucas, J., *War in the Desert – the Eighth Army at El Alamein* (London, Arms and Armour Press 1982)
Lynch, M., *Scotland, A New History* (London, Pimlico 1992)
McCutcheon, C., *Home Guard Manual 1940* (Gloucs., History Press 2007)
Macdonald-Fraser, G., *the Steel Bonnets* (London, Barrie and Jenkins 1971)
- *Quartered Safe Out Here* (London, Harper Collins 2000)
Masters, J., *The Road Past Mandalay* (London, Michael Joseph 1961)
Miller, M., R. Laycock, J. Sadler and R. Serdiville, *As Good as Any Man, Scotland's Black Tommy* (Stroud, The History Press 2014)
Morgan, M., *D-Day Hero* (Stroud, Spellmount 2004)
Moses, H., *The Fighting Bradfords* (Durham, County Durham Books, 2003)
Neillands, R., *In the Combat Zone* (London, Weidenfeld and Nicolson 1997)
Paterson, J., *Autobiographical Reminiscences* (Glasgow, Maurice Ogle and Co 1876)
Pearson, H., *Achtung Schweinehund!* (London, Abacus 2008)

Pease, H., *The History of the Northumberland (Hussars) Yeomanry 1819–1923* (Constable, London, 1924)
Ransome, C., 'The Battle of Towton' *English Historical Review* vol. 4 (1889)
Redding, Tony, *War in the Wilderness: The Chindits in Burma 1943–1944* (Gloucs., Spellmount 2011)
Reid, S., *All the King's Armies – A Military History of the English Civil War 1642–1651* (Gloucs., History Press 2007)
Reed, P., *Walking the Somme; A Walker's Guide to the Somme Battlefields* (Barnsley, Pen and Sword 1997)
Richardson, D., *The Sword-makers of Shotley Bridge* (Newcastle upon Tyne, Frank Graham 1973)
Rissik, D., *The DLI at War – History of the DLI 1939–1945* (Durham, DLI 1952)
Rogers, Col. H.C.B., *Artillery through the Ages* (London, Military Book Society 1971)
Sadler, D. J., *The Field Gun Run* (Royal Military Tournament, London, 2010)
- and R. B. Serdiville, *Tommy at War 1914 – 1918* (London, Biteback 2014)
- and R. B. Serdiville, *Fighter Aces, Knights of the Sky* (Oxford, Casemate 2017)
- *D-Day – The British Beaches* (Stroud, Amberley 2019
- *Scottish Battles* (Edinburgh, Canongate 2010)

Shakespear Colonel L.W., and Lieutenant-Colonel G.R. Stevens, *The History of the 2nd King Edward VII's Own Goorka Rifles (The Sirmoor Rifles)* (London Gale and Polden 1952)
Smith, General Sir Rupert, *The Utility of Force* (London, Allen Lane 2005)
Smurthwaite, D., *The Ordnance Survey Guide to the Battlefields of Britain* (London, Michael Joseph 1984)
Tough, D.L.W., *The Last Years of a Frontier* (Oxford, OUP 1928)
Walter, G. (ed.) *In Flanders Fields, Poetry of the Great War* (London Penguin 2004)
Warner, P., *Alamein – Recollections of the Heroes* (London, William Kimber 1979)
Warwicker, J., *Churchill's Underground Army* (Barnsley, Pen and Sword 2008)
White, B.T., *German Tanks and Armoured Vehicles 1914–1945* (London, Ian Allan 1966)
Young, Brigadier P., *The British Army* (London, William Kimber 1967)

Index